Applications of Heterogeneous Catalysts in Green Chemistry

Applications of Heterogeneous Catalysts in Green Chemistry

Guest Editors

Anand Kumar
Siham Y. Al-Qaradawi
Mohamed Ali S. Saad

MDPI

Basel • Beijing • Wuhan • Barcelona • Belgrade • Novi Sad • Cluj • Manchester

Guest Editors
Anand Kumar
Department of Chemical Engineering
Qatar University
Doha
Qatar

Siham Y. Al-Qaradawi
Department of Chemistry and Earth Sciences
Qatar University
Doha
Qatar

Mohamed Ali S. Saad
Department of Chemical Engineering
Qatar University
Doha
Qatar

Editorial Office
MDPI AG
Grosspeteranlage 5
4052 Basel, Switzerland

This is a reprint of the Special Issue, published open access by the journal *Catalysts* (ISSN 2073-4344), freely accessible at: https://www.mdpi.com/journal/catalysts/special_issues/Heterogeneous_Catalysts_Green_Chemistry.

For citation purposes, cite each article independently as indicated on the article page online and as indicated below:

Lastname, A.A.; Lastname, B.B. Article Title. *Journal Name* **Year**, *Volume Number*, Page Range.

ISBN 978-3-7258-6578-9 (Hbk)
ISBN 978-3-7258-6579-6 (PDF)
https://doi.org/10.3390/books978-3-7258-6579-6

Contents

About the Editors

Anand Kumar

Anand Kumar is a Professor in the Department of Chemical Engineering at Qatar University. His primary research focus is on the synthesis of catalysts and understanding how heterogeneous catalysts function. Over the past years, he has extensively used combustion-based techniques for synthesizing porous nanomaterials, used in different applications such as catalysts for hydrocarbon reforming, hydrogen production, carbon dioxide conversion, and electrodes for fuel cell reactions. In addition, his recent work includes reactor design and optimization for methane reforming processes, particularly tri-reforming and cracking of methane. His research emphasizes integrating advanced catalysis with innovative reactor configurations to enhance efficiency, stability, and improve hydrogen yield.

Siham Y. Al-Qaradawi

Siham Y. Al-Qaradawi is currently a professor of organic chemistry at Qatar University. She is presently an elected member of the QU Faculty Senate. She received her BSc in 1981 from Qatar University and her Ph.D. in 1992 from the University of Reading, UK. She served as Dean of the College of Arts & Sciences for five years. During her deanship, she established the College of Pharmacy in 2006. In addition, she worked hard to earn accreditation for the chemistry program from the Canadian Society of Chemistry (CSC) in 2009 and for the Biomedical program from NAACLS in 2008. Prof. S. Al-Qaradawi received the Fulbright Award as a Visiting Scholar at Temple University, USA, in 2011, and the DAAD in Germany in 2003. She has authored over 100 papers, holds three patents, has an H-index of 33, and has more than 3,700 citations. She presented her research findings at numerous international and regional conferences. She is the lead PI on several research grants totaling more than USD 8 million. Prof. Al-Qaradawi has successfully supervised the work of one PhD student and four MSc students, as well as ten final-year research project students. She held several fellowships at the University of Strathclyde, Glasgow, UK (summer 2000), the University of Hamburg, Germany (Summer 2002), the University of Hannover, Germany (Summer 2003), and the California Institute of Technology, Caltech, CA, USA (Summer 2013). Her research interests are carbon dioxide conversion, Lithium-ion batteries, and wastewater treatment. She is a reviewer for several prestigious journals, including the *Journal of Molecular Structure*, the *Journal of Molecular Liquids* and the *Journal of Spectrochimica Acta*. She is currently a guest speaker for the *Catalyst* journal.

Mohamed Ali S. Saad

Mohamed Ali S. Saad is the Director of the Gas Processing Center at Qatar University and an Associate Professor in Chemical Engineering. He leads research focused on CO_2 conversion, catalysis, and developing cleaner and more efficient solutions for the oil and gas industry. His work includes creating new catalysts, green chemicals, and advanced materials that help reduce pollution, improve oilfield operations, and support sustainable energy technologies. He has many funded research projects and has worked on several patents related to corrosion protection and scale removal. He has also published widely in well-known scientific journals. Through his leadership, he maintained the high standards of the Gas Processing Center, expanding its research activities and building strong collaborations.

Preface

Green chemistry, often referred to as sustainable chemistry, has emerged in response to growing concerns over the environmental impact of chemical manufacturing, particularly the release of harmful emissions. Its central aim is to design products and processes that minimize or eliminate the use and generation of hazardous substances.

In chemical plants, where raw materials are transformed into valuable products, green chemistry plays a crucial role in selecting reaction environments and optimizing reaction conditions. Among the key facilitators of these transformations are heterogeneous catalysts, which are considered the backbone of the chemical industry. These catalysts not only facilitate reactions, but also significantly influence feed conversion, product purity, and the continuity of production—factors that generally determine the feasibility, economic viability, and environmental footprint of a process.

The discovery and implementation of suitable catalysts can make chemical processes greener by enhancing conversion rates and product selectivity, thereby reducing the need for extensive downstream separation and purification. This special issue focuses on recent advancements in the application of heterogeneous catalysts within the framework of green chemistry. We have assembled ten invited articles that explore various dimensions of this topic, including catalyst synthesis, characterization, performance evaluation, and their roles in improving chemical conversion, selectivity, and stability. These contributions collectively explore a diverse list of chemical manufacturing.

We extend our sincere thanks to all the authors for their valuable contributions and commitment to this Special Issue. We are also deeply grateful to the peer reviewers whose rigorous evaluations helped uphold the scholarly standards of this collection. Special appreciation goes to the editorial team and support staff for their dedication and timely coordination throughout the publication process.

We hope that this Special Issue serves as a meaningful platform for researchers to appreciate the role of heterogeneous catalysts in advancing green chemistry.

Anand Kumar, Siham Y. Al-Qaradawi, and Mohamed Ali S. Saad
Guest Editors

catalysts

MDPI

Editorial

Applications of Heterogeneous Catalysts in Green Chemistry

Anand Kumar [1,*], Mohammed Ali H. S. Saad [1,2,*] and Siham Y. Al-Qaradawi [3,*]

1 Department of Chemical Engineering, College of Engineering, Qatar University, Doha P.O. Box 2713, Qatar
2 Gas Processing Center, College of Engineering, Qatar University, Doha P.O. Box 2713, Qatar
3 Department of Chemistry and Earth Sciences, College of Arts and Sciences, Qatar University, Doha P.O. Box 2713, Qatar
* Correspondence: akumar@qu.edu.qa (A.K.); m.saleh@qu.edu.qa (M.A.H.S.S.); siham@qu.edu.qa (S.Y.A.-Q.)

In our day-to-day lives, the utilization of industrial products, including chemicals and fuels, is essential. The vast majority of these products are manufactured in chemical plants that transform raw materials into products that are suitable for the market based on the requirements of local and international regulatory authorities [1–4]. A chemical plant, in general, consists of many reactors and separator units, which are essential for the purification of the feed, its chemical conversion into the desired products and the subsequent purification of the products [5–7]. Each unit of a chemical plant is carefully designed and optimized to achieve a certain level of conversion and separation efficiency that eventually determines the volume and purity of the final products and plays a crucial role in bringing them to the market [5,8,9]. The efficiency of the entire chemical plant helps with minimizing the production cost and maintaining a competitive advantage in the market [5,7,10–12]. However, based on the process selected, the units of chemical plants may generate harmful emissions, such as liquid discharge, gases and particulate matter, that need to be controlled or treated before being released into the environment to avoid severe consequences [13–16]. Green chemistry [17–21] has emerged as an alternative strategy for producing chemicals in a more environmentally friendly and sustainable manner, as the environmentally harmful chemicals are replaced by more benign ones [22–27].

Heterogeneous catalysts play a central role in chemical industries, driving the molecular transformations required to generate the desired products [28,29]. Their role is even more significant in green processes [12,30–34]. A highly active, selective and stable catalyst can help with achieving sustainability goals and minimizing harmful emissions [35–40]. An effective catalyst must be efficient at converting the feedstock into the targeted products and will require minimum process interruptions caused by catalyst deactivation [41–46]. High activity ensures energy efficiency and the ability to process large quantities of feedstock in reactors with smaller footprints, while high product selectivity ensures fewer undesired byproducts that need to be separated after production. A stable catalyst with a steady performance is ideal for any industry, as it eliminates frequent plant shutdowns and restarts, with the overall impact of lowering operational costs and boosting plant productivity [47–50]. Additionally, efficient catalysts reduce the separation requirement by generating a lower number of byproducts. In short, the role of heterogeneous catalysts is vital for making sustainable and environmentally friendly processes a reality.

In this Special Issue, our objective was to gather original articles and review papers highlighting the critical roles of heterogeneous catalysts in green processes. Various topics were considered to cover the various aspects of chemistry and chemical industries, such as green synthesis techniques, pollution control, energy efficiency, environmental catalysis, clean energy, CO_2 conversion, hydrogen production, automobile emission catalysts etc. As we finalize the Special Issue, we can see that it provides a balanced list of review and original articles covering various applications of catalysis in green chemistry [36,51–59].

This collection includes three review articles [36,51,58] and seven original research works [52,54–58,60]. Hydrocarbon reforming, energy conversion and carbon dioxide emission control via capture, utilization and conversion are the primary goals in industries for address pressing environmental issues [60–64]. The review articles provide an overview of the development of biomass-based catalysts for biodiesel production [36], catalyst development for the conversion of CO_2 into syngas [51] and other products including olefins and aromatics using structured catalysts [58]. Biomass-derived catalysts show their potential as green ways of achieving for catalyst and chemical product synthesis. CO_2's conversion into syngas via the reverse water gas shift (RWGS) reaction provides a pathway for converting industrially generated CO_2 back into value-added chemicals and fuels. There has been considerable growth in the field of catalyst synthesis to ensure a precise control of nanostructures up to the atomic level [65,66], as well as in our understanding of the roles of defects [67], acidic/basic supports [68,69] and oxygen storage materials [70], which can be exploited for improving catalytic performances of various reactions. The reviews included in this Special Issue cover a broad spectrum of heterogeneous catalysts supported by robust structured zeolites and oxides with mobile surface oxygen.

The compilation of original articles is based on their application of catalysis for addressing challenging reactions related to various aspects of green chemistry. Bioderived catalysts show their potential to use the transesterification reaction to produce alkyl esters [60], as well as the synthesis of furfural from corncob [57]. Studies on the applications of solid acid catalysts in the esterification process [55], the promotional effects of bimetallic catalysts for methane decomposition [56], RuO_x/MoS_2 catalyst use for sugar's conversion into lactic acid [54] and the degradation and mineralization of pesticide wastewaters in laboratory- and pilot-scale reactors [53] are included to demonstrate the breadth of the applications of heterogeneous catalysts in addressing challenging issues. Additionally, some fundamental aspects of heterogeneous catalysts such as the impact of site geometry and local composition on surface carbon's conversion into methane over transition metal and bimetal (Ni, Co and NiCo) catalysts were also included [52] for readers interested in understanding the core issues of catalysis related to the structure–activity relationship. In conclusion, the present Special Issue provides a well-rounded collection of reviews and original research articles addressing various aspects of heterogeneous catalysis applied in green chemistry. We hope that this collection satisfies the curiosity of many researchers interested in catalysis and green chemistry.

Funding: This research received no external funding.

Conflicts of Interest: The authors declare no conflict of interest.

References

1. Cavani, F.; Centi, G.; Perathoner, S.; Trifiró, F. *Sustainable Industrial Processes*; Wiley-VCH: Weinheim, Germany, 2009; 599p.
2. Wittcoff, H.A.; Reuben, B.G.; Plotkin, J.S. *Industrial Organic Chemicals*, 3rd ed.; John Wiley & Sons, Inc.: Hoboken, NJ, USA, 2012. [CrossRef]
3. Chenier, P.J. *Survey of Industrial Chemistry*, 2nd ed.; Vch Pub: Weinheim, Germany, 2012.
4. Soetaert, W.; Vandamme, E. The impact of industrial biotechnology. *Biotechnol. J.* **2006**, *1*, 756–769. [CrossRef] [PubMed]
5. Grossmann, I.E.; Sargent, R.W.H. Optimum Design of Multipurpose Chemical Plants. *Ind. Eng. Chem. Process Des. Dev.* **1979**, *18*, 343–348. [CrossRef]
6. Subrahmanyan, S.; Pekny, J.F.; Reklaitis, G.V. Design of Batch Chemical Plants under Market Uncertainty. *Ind. Eng. Chem. Res.* **1994**, *33*, 2688–2701. [CrossRef]
7. Ramos, J.L.; Duque, E. Twenty-first-century chemical odyssey: Fuels versus commodities and cell factories versus chemical plants. *Microb. Biotechnol.* **2019**, *12*, 200–209. [CrossRef] [PubMed]
8. Singh, R. Facts, growth, and opportunities in industrial biotechnology. *Org. Process Res. Dev.* **2011**, *15*, 175–179. [CrossRef]
9. Siirola, J.J. Industrial Applications of Chemical Process Synthesis. *Adv. Chem. Eng.* **1996**, *23*, 1–62. [CrossRef]
10. Agrawal, R.; Offutt, M.; Ramage, M.P. Hydrogen economy-an opportunity for chemical engineers? *AIChE J.* **2005**, *51*, 1582–1589. [CrossRef]
11. Dunn, R.F.; El-Halwagi, M.M. Process integration technology review: Background and applications in the chemical process industry. *J. Chem. Technol. Biotechnol.* **2003**, *78*, 1011–1021. [CrossRef]
12. Chenier, P.J. Introduction to the Chemical Industry: An Overview. In *Survey of Industrial Chemistry*; Topics in Applied Chemistry; Springer: Boston, MA, USA, 2002; pp. 1–22. [CrossRef]

13. Gabrielli, P.; Rosa, L.; Gazzani, M.; Meys, R.; Bardow, A.; Mazzotti, M.; Sansavini, G. Net-zero emissions chemical industry in a world of limited resources. *One Earth* **2023**, *6*, 682–704. [CrossRef]
14. Ioannou, I.; Galán-Martín, Á.; Pérez-Ramírez, J.; Guillén-Gosálbez, G. Trade-offs between sustainable development goals in carbon capture and utilisation. *Energy Environ. Sci.* **2022**, *16*, 113–124. [CrossRef]
15. D'Angelo, S.C.; Cobo, S.; Tulus, V.; Nabera, A.; Martín, A.J.; Pérez-Ramírez, J.; Guillen-Gosalbez, G. Planetary boundaries analysis of low-carbon ammonia production routes. *ACS Sustain. Chem. Eng.* **2021**, *9*, 9740–9749. [CrossRef]
16. Galán-Martín, Á.; Tulus, V.; Díaz, I.; Pozo, C.; Pérez-Ramírez, J.; Guillén-Gosálbez, G. Sustainability footprints of a renewable carbon transition for the petrochemical sector within planetary boundaries. *One Earth* **2021**, *4*, 565–583. [CrossRef]
17. Cathles, L.M.; Brown, L.; Taam, M.; Hunter, A. A commentary on "The greenhouse-gas footprint of natural gas in shale formations" by RW Howarth, R. Santoro, and Anthony Ingraffea. *Clim. Chang.* **2012**, *113*, 525–535. [CrossRef]
18. Andraos, J.; Matlack, A.S. *Introduction to Green Chemistry*; CRC Press: Boca Raton, FL, USA, 2022; pp. 1–624. [CrossRef]
19. Clark, J.H.; Luque, R.; Matharu, A.S. Green chemistry, biofuels, and biorefinery. *Annu. Rev. Chem. Biomol. Eng.* **2012**, *3*, 183–207. [CrossRef] [PubMed]
20. Sheldon, R.A. Fundamentals of green chemistry: Efficiency in reaction design. *Chem. Soc. Rev.* **2012**, *41*, 1437–1451. [CrossRef]
21. Sankar, M.; Dimitratos, N.; Miedziak, P.J.; Wells, P.P.; Kiely, C.J.; Hutchings, G.J. Designing bimetallic catalysts for a green and sustainable future. *Chem. Soc. Rev.* **2012**, *41*, 8099–8139. [CrossRef]
22. Barbaro, P.; Bianchini, C. *Catalysis for Sustainable Energy Production*; Wiley-VCH: Weinheim, Germany; Chichester, UK; John Wiley Distributor: Hoboken, NJ, USA, 2009.
23. Dunn, S. Hydrogen futures: Toward a sustainable energy system. *Int. J. Hydrogen Energy* **2002**, *27*, 235–264. [CrossRef]
24. Anwar, M.N.; Fayyaz, A.; Sohail, N.F.; Khokhar, M.F.; Baqar, M.; Khan, W.D.; Rasool, K.; Rehan, M.; Nizami, A.S. CO_2 capture and storage: A way forward for sustainable environment. *J. Environ. Manag.* **2018**, *226*, 131–144. [CrossRef]
25. Manzano-Agugliaro, F.; Alcayde, A.; Montoya, F.G.; Zapata-Sierra, A.; Gil, C. Scientific production of renewable energies worldwide: An overview. *Renew. Sustain. Energy Rev.* **2013**, *18*, 134–143. [CrossRef]
26. van Haveren, J.; Scott, E.L.; Sanders, J. Bulk chemicals from biomass. *Biofuels Bioprod. Biorefin. Innov. A Sustain. Econ.* **2008**, *2*, 41–57. [CrossRef]
27. Zhao, H.; Yu, C.; You, H.; Yang, S.; Guo, Y.; Ding, B.; Song, X. A green chemical approach for preparation of Pt_xCu_y nanoparticles with a concave surface in molten salt for methanol and formic acid oxidation reactions. *J. Mater. Chem.* **2012**, *22*, 4780–4789. [CrossRef]
28. Davis, J.; Ertl, B.H.; Knozinger, G.; Weitkamp, H. *Handbook of Heterogeneous Catalysis*; Wiley-VCH: Weinheim, Germany, 1997; Volume 4.
29. Thomas, J.M.; Thomas, W.J.; Anderson, J.R.; Boudart, M. *Principles and Practice of Heterogeneous Catalysis*; VCH: Weinheim, Germany, 1997; Volume 638.
30. Dale, B.E. 'Greening' the chemical industry: Research and development priorities for biobased industrial products. *J. Chem. Technol. Biotechnol.* **2003**, *78*, 1093–1103. [CrossRef]
31. Panke, S.; Held, M.; Wubbolts, M. Trends and innovations in industrial biocatalysis for the production of fine chemicals. *Curr. Opin. Biotechnol.* **2004**, *15*, 272–279. [CrossRef] [PubMed]
32. Erickson, B.; Nelson, J.E.; Winters, P. Perspective on opportunities in industrial biotechnology in renewable chemicals. *Biotechnol. J.* **2012**, *7*, 176–185. [CrossRef]
33. Tang, W.L.; Zhao, H. Industrial biotechnology: Tools and applications. *Biotechnol. J.* **2009**, *4*, 1725–1739. [CrossRef]
34. Clarke, C.J.; Tu, W.C.; Levers, O.; Bröhl, A.; Hallett, J.P. Green and Sustainable Solvents in Chemical Processes. *Chem. Rev.* **2018**, *118*, 747–800. [CrossRef]
35. Whipple, D.T.; Kenis, P.J.A. Prospects of CO_2 Utilization via direct heterogeneous electrochemical reduction. *J. Phys. Chem. Lett.* **2010**, *1*, 3451–3458. [CrossRef]
36. Chandra Kishore, S.; Perumal, S.; Atchudan, R.; Sundramoorthy, A.K.; Alagan, M.; Sangaraju, S.; Lee, Y.R. A Review of Biomass-Derived Heterogeneous Catalysts for Biodiesel Production. *Catalysts* **2022**, *12*, 1501. [CrossRef]
37. Royer, S.; Duprez, D.; Can, F.; Courtois, X.; Batiot-Dupeyrat, C.; Laassiri, S.; Alamdari, H. Perovskites as substitutes of noble metals for heterogeneous catalysis: Dream or reality. *Chem. Rev.* **2014**, *114*, 10292–10368. [CrossRef]
38. De, S.; Dokania, A.; Ramirez, A.; Gascon, J. Advances in the Design of Heterogeneous Catalysts and Thermocatalytic Processes for CO_2 Utilization. *ACS Catal.* **2020**, *10*, 14147–14185. [CrossRef]
39. Razali, N.A.M.; Lee, K.T.; Bhatia, S.; Mohamed, A.R. Heterogeneous catalysts for production of chemicals using carbon dioxide as raw material: A review. *Renew. Sustain. Energy Rev.* **2012**, *16*, 4951–4964. [CrossRef]
40. Domen, K. Photocatalysts: Nanostructured Photocatalytic Materials for Solar Energy Conversion. *Sel. Nanocatal. Nanosci. Concepts Heterog. Homog. Catal.* **2011**, 169–191. [CrossRef]
41. Farrauto, R.J. Industrial catalysis: A practical guide. In *Handbook of Industrial Chemistry and Biotechnology*, 12th ed.; Springer: Boston, MA, USA, 2012; Volume 1–2, pp. 201–230. [CrossRef]
42. Abdo, S.F.; Wilson, S.T. Chapter 9: Zeolites in Industrial Catalysis. In *Zeolites in Catalysis: Properties and Applications*; Catalysis Series; The Royal Society of Chemistry: Cambridge, UK, 2017; pp. 310–350. [CrossRef]
43. Lloyd, L. *Handbook of Industrial Catalysts*; Fundamental and Applied Catalysis (FACA); Springer: New York, NY, USA, 2011. [CrossRef]

44. Armor, J.N. A history of industrial catalysis. *Catal. Today* **2011**, *163*, 3–9. [CrossRef]
45. Bartholomew, C.H.; Farrauto, R.J. *Fundamentals of Industrial Catalytic Processes*, 2nd ed.; John Wiley & Sons: Hoboken, NJ, USA, 2010; pp. 1–966. [CrossRef]
46. LeRoy, R.L. Industrial water electrolysis: Present and future. *Int. J. Hydrogen Energy* **1983**, *8*, 401–417. [CrossRef]
47. Leach, B.E. *Applied Industrial Catalysis*; Academic Press: Cambridge, MA, USA, 1983; Volume 1.
48. Hagen, J. *Industrial Catalysis: A Practical Approach*; John Wiley & Sons: Hoboken, NJ, USA, 2015; pp. 1–522. [CrossRef]
49. Mortensen, P.M.; Dybkjær, I. Industrial scale experience on steam reforming of CO_2-rich gas. *Appl. Catal. A Gen.* **2015**, *495*, 141–151. [CrossRef]
50. Mears, D.E. Tests for transport limitations in experimental catalytic reactors. *Ind. Eng. Chem. Process Des. Dev.* **1971**, *10*, 541–547. [CrossRef]
51. Ebrahimi, P.; Kumar, A.; Khraisheh, M. A Review of CeO_2 Supported Catalysts for CO_2 Reduction to CO through the Reverse Water Gas Shift Reaction. *Catalysts* **2022**, *12*, 1101. [CrossRef]
52. Godoy, S.; Deshlahra, P.; Villagra-Soza, F.; Karelovic, A.; Jimenez, R. Effects of Site Geometry and Local Composition on Hydrogenation of Surface Carbon to Methane on Ni, Co, and NiCo Catalysts. *Catalysts* **2022**, *12*, 1380. [CrossRef]
53. Berberidou, C.; Kokkinos, P.; Poulios, I.; Mantzavinos, D. Homogeneous Photo-Fenton Degradation and Mineralization of Model and Simulated Pesticide Wastewaters in Lab- and Pilot-Scale Reactors. *Catalysts* **2022**, *12*, 1512. [CrossRef]
54. Li, Z.; Wu, P.; Pang, J.; Li, X.; Zhai, S.; Zheng, M. Catalytic Conversion of Sugars into Lactic Acid via a RuO_x/MoS_2 Catalyst. *Catalysts* **2023**, *13*, 545. [CrossRef]
55. Yu, X.; Shi, C.; Cheng, Y.; Zhu, Y.; Song, R.; Hu, S. Synthesis of a Series of Methyl Benzoates through Esterification with a Zr/Ti Solid Acid Catalyst. *Catalysts* **2023**, *13*, 915. [CrossRef]
56. Hasnan, N.S.N.; Pudukudy, M.; Yaakob, Z.; Kamarudin, N.H.N.; Lim, K.L.; Timmiati, S.N. Promoting Effects of Copper and Iron on Ni/MSN Catalysts for Methane Decomposition. *Catalysts* **2023**, *13*, 1067. [CrossRef]
57. Yang, L.; Li, Y.; Wu, Y.; He, Y.; Ma, C. Efficient Synthesis of Furfural from Corncob by a Novel Biochar-Based Heterogeneous Chemocatalyst in Choline Chloride: Maleic Acid–Water. *Catalysts* **2023**, *13*, 1277. [CrossRef]
58. Cutad, M.B.; Al-Marri, M.J.; Kumar, A. Recent Developments on CO_2 Hydrogenation Performance over Structured Zeolites: A Review on Properties, Synthesis, and Characterization. *Catalysts* **2024**, *14*, 328. [CrossRef]
59. Gogoaşă, C.I.; Răducanu, C.E.; Petraş, L.E.; Cioroiu Tîrpan, D.R.; Vasilievici, G.; Mîrţ, A.L.; Dobre, T.; Pârvulescu, O.C. Bacterial Cellulose and Biodegradable Superbase for Heterogeneous Transesterification to Alkyl Esters. *Catalysts* **2023**, *13*, 1431. [CrossRef]
60. Botte, G.G. Electrochemical manufacturing in the chemical industry. *Electrochem. Soc. Interface* **2014**, *23*, 49–55. [CrossRef]
61. Farhat Ali, M.; El Ali, B.M.; Speight, J.G. *Handbook of Industrial Chemistry: Organic Chemicals*; McGraw-Hill Education: New York, NY, USA, 2005.
62. Centi, G.; Quadrelli, E.A.; Perathoner, S. Catalysis for CO_2 conversion: A key technology for rapid introduction of renewable energy in the value chain of chemical industries. *Energy Environ. Sci.* **2013**, *6*, 1711–1731. [CrossRef]
63. Gangadharan, P.; Kanchi, K.C.; Lou, H.H. Evaluation of the economic and environmental impact of combining dry reforming with steam reforming of methane. *Chem. Eng. Res. Des.* **2012**, *90*, 1956–1968. [CrossRef]
64. Hansen, J.; Johnson, D.; Lacis, A.; Lebedeff, S.; Lee, P.; Rind, D.; Russell, G. Climate impact of increasing atmospheric carbon dioxide. *Science* **1981**, *213*, 957–966. [CrossRef]
65. Wang, A.; Li, J.; Zhang, T. Heterogeneous single-atom catalysis. *Nat. Rev. Chem.* **2018**, *2*, 65–81. [CrossRef]
66. Mitchell, S.; Thomas, J.M.; Pérez-Ramírez, J. Single atom catalysis. *Catal. Sci. Technol.* **2017**, *7*, 4248–4249. [CrossRef]
67. Huang, H.; Wang, X. Pd nanoparticles supported on low-defect graphene sheets: For use as high-performance electrocatalysts for formic acid and methanol oxidation. *J. Mater. Chem.* **2012**, *22*, 22533–22541. [CrossRef]
68. Ahmad, Y.H.; Mohamed, A.T.; Kumar, A. Al-Qaradawi SY. Solution combustion synthesis of Ni/La_2O_3 for dry reforming of methane: Tuning the basicity via alkali and alkaline earth metal oxide promoters. *RSC Adv.* **2021**, *11*, 33734–33743. [CrossRef]
69. Ahmad, Y.H.; Mohamed, A.T.; El-Sayed, H.A.; Kumar, A.; Al-Qaradawi, S.Y. Design of Ni/La_2O_3 catalysts for dry reforming of methane: Understanding the impact of synthesis methods. *Int. J. Hydrogen Energy* **2022**, *47*, 41294–41309. [CrossRef]
70. Ebrahimi, P.; Kumar, A.; Khraisheh, M. A review of recent advances in water-gas shift catalysis for hydrogen production. *Emergent Mater.* **2020**, *3*, 881–917. [CrossRef]

catalysts

MDPI

Review

A Review of CeO_2 Supported Catalysts for CO_2 Reduction to CO through the Reverse Water Gas Shift Reaction

Parisa Ebrahimi, Anand Kumar * and Majeda Khraisheh

Department of Chemical Engineering, Qatar University, Doha P.O. Box 2713, Qatar
* Correspondence: akumar@qu.edu.qa

Abstract: The catalytic conversion of CO_2 to CO by the reverse water gas shift (RWGS) reaction followed by well-established synthesis gas conversion technologies could be a practical technique to convert CO_2 to valuable chemicals and fuels in industrial settings. For catalyst developers, prevention of side reactions like methanation, low-temperature activity, and selectivity enhancements for the RWGS reaction are crucial concerns. Cerium oxide (ceria, CeO_2) has received considerable attention in recent years due to its exceptional physical and chemical properties. This study reviews the use of ceria-supported active metal catalysts in RWGS reaction along with discussing some basic and fundamental features of ceria. The RWGS reaction mechanism, reaction kinetics on supported catalysts, as well as the importance of oxygen vacancies are also explored. Besides, recent advances in CeO_2 supported metal catalyst design strategies for increasing CO_2 conversion activity and selectivity towards CO are systematically identified, summarized, and assessed to understand the impacts of physicochemical parameters on catalytic performance such as morphologies, nanosize effects, compositions, promotional abilities, metal-support interactions (MSI) and the role of selected synthesis procedures for forming distinct structural morphologies. This brief review may help with future RWGS catalyst design and optimization.

Keywords: reverse water gas shift reaction; mechanism and kinetics; CeO_2 support; CO_2 conversion; affecting parameters

1. Introduction

Carbon dioxide has been identified as the primary anthropogenic greenhouse gas that has resulted in catastrophic climate change and ocean acidification [1,2]. Various approaches have been employed to reduce the amount of CO_2 in the atmosphere. For example, power-to-liquid (PtL) sustainable aviation fuel (SAF) was recently proposed as a long-term and scalable solution to minimize aircraft CO_2 emissions. The procedure turns CO_2 into a synthetic fuel with less sulfur and fewer aromatics, which enhances local air quality and minimizes the effect of aviation at high altitudes [3]. On the other hand, since enormous amounts of low-cost, relatively pure carbon dioxide are available from carbon sequestration and storage facilities, more efforts have been made to utilize CO_2 as an alternative C1 source rather than merely considering it as waste [4]. A unique and appealing alternative to storing CO_2 through sequestration would be recycling the gas into energy-rich compounds via carbon capture, storage and utilization (CCSU) [5,6]. E-fuels, also known as electrofuels or powerfuels, are hydrocarbon fuels produced from hydrogen and CO_2 in which hydrogen is generated from water and electricity through electrolysis and CO_2 is either captured from fossil sources (such as industrial sectors) or the atmosphere [7–9]. E-fuels aim to directly electrify a system without the demand-side adjustments necessary for a direct electrification by substituting fossil fuels with renewable power [7]. However, the CO_2 molecule is a relatively inert and unreactive molecule with a high level of thermodynamic and chemical stability due to its linear chemical structure with double bonds connecting the carbon and oxygen atoms, so converting it to the more reactive CO is energy-intensive [10]. Among the systems currently available for CO_2 conversion, catalytic conversion to CO, commonly known as the reverse water-gas shift (RWGS) reaction (Equation (1)), is one of the most promising reversible hydrogenation methods that offer a high potential efficiency [11].

$$CO_2 + H_2 \leftrightarrow CO + H_2O \; \Delta H^0 = +41.3 \text{ kJ/mol} \quad (1)$$

RWGS reaction is recognized as an important intermediate stage in a number of key CO_2 hydrogenation reactions such as the Sabatier process [12] and methanol synthesis [13], and is hence referred to as the "building block stage" [14]. Synthesis gas (CO + H_2), a crucial precursor in the field of C1 chemistry, can be produced using the RWGS reaction in the presence of an appropriate catalyst. The syngas can further be used as a feedstock for the Fischer–Tropsch synthesis reaction (FTS) to produce organic compounds, such as methanol (a crucial component of synthetic fuels and polymers), hydrocarbons, or oxygenated hydrocarbons [15,16]. ExxonMobil recently revealed that its "methanol-to-jet" technology can provide SAF from methanol derived through waste, biomass, captured carbon dioxide, and low-carbon hydrogen [17]. However, further side reactions, such as CO methanation (Equation (2)) [18], could emerge under the same reaction conditions, consuming a large amount of hydrogen. The CO_2 methanation reaction is an exothermic catalytic process that normally takes place at temperatures from 150 °C to 550 °C in the presence of a catalyst [19]. The CO_2 conversion and CH_4 selectivity can almost approach 100%; however, as the temperature rises, the reaction rate increases [20], with preference for RWGS at higher temperature. Therefore, at low reaction temperatures, the highly exothermic methanation reaction is thermodynamically more preferred to the slightly endothermic RWGS reaction [21,22]; hence, reducing methanation throughout RWGS has been a challenging issue. A remaining concern seems to be either the RWGS reaction should be operated at high temperature (over 900 °C), which is thermodynamically favorable, but carbon and undesirable byproducts may also be present; or it should be performed at low temperatures (below 500 °C), in which case it is not kinetically favored but may be made up for by extensive catalyst use [23]. Over the temperature range of 100 to 1000 °C, Kaiser et al. investigated the equilibrium composition of the gaseous products in RWGS reaction for a three-to-one molar H_2/CO_2 input ratio [24]. Based on the results, methanation was thermodynamically preferred at low temperatures (below 600 °C), while, the only product that could form at temperatures beyond 700 °C was CO and very little to no methane. However, to cut down on the energy losses and investment expenses, the temperature must be kept as low as feasible [24]. They proposed that using RWGS at greater pressures in conjunction with high temperature and high-pressure steam electrolysis might be an alternative [24]. Additionally, the FTS normally operates at 2.5 MPa, and the produced syngas or the RWGS supply gas must be compressed [25]. Kaiser et al. came to the conclusion that at this pressure (2.5 MPa), the methane curve was pushed to higher temperatures; for instance, at 900 °C, the equilibrium methane level was 4 mol% as opposed to 660 °C at 1 atm pressure [24]. When Unde et al. tested the Al_2O_3 catalyst through RWGS reaction, they discovered that the CO_2 to CO conversion equilibrium was reached at a high temperature of 900 °C. Reaction was controlled kinetically between 300 and 700 °C, and thermodynamically above this temperature range [26]. As a result, production of active RWGS catalysts operated at low-temperature with higher CO selectivity and limited CH_4 production was required. Insight into the mechanisms of CO production is also vital for rational catalyst design in such processes. Various reaction routes could lead to various kinetic parameters and selectivity variations for CO_2 hydrogenation [27].

$$CO_2 + 3H_2 \leftrightarrow CH_4 + H_2O \; \Delta H^0 = -206.5 \text{ kJ/mol} \quad (2)$$

According to the concept of microscopic reversibility and the fact that the RWGS reaction is typically carried out at equilibrium, the active catalysts in the water gas shift (WGS) process are also effective in the RWGS reaction, but may be under different reaction conditions, suggesting that similar catalysts should enhance both reactions [28]. Some typical features of WGS catalysts include the presence of oxygen vacancies, the strength at which CO can be adsorbed, and activity for dissociation of water [29]. In our earlier study, a thorough overview of the most recent advancements of catalysts utilized in low-temperature WGS reactions is presented [30]. In various CO_2 conversion processes, many types of catalysts have been used, including oxide-supported metal catalysts and oxide catalysts in which cerium oxide (CeO_2) has had a key role [30,31]. CeO_2 is a typical rare earth metal oxide with a face-centered cubic (FCC) fluorite structural pattern, and has oxygen storage capacity (OSC) and a number of intriguing features that can be exploited to improve catalytic efficiency [28,32]. In comparison to other reducible oxides, oxygen vacancies on the surface of CeO_2 are more easily formed during the reduction process owing to its unique electron arrangement [33]. Besides, the reversible redox pair Ce^{3+}/C^{4+} and the acid basic surface properties of CeO_2 are effectively leading to its broader catalytic application [34]. It was found that the reducibility of ceria had an inverse connection with the bimetallic cluster promoted local electronic band, which caused the stability of germinal OH groups and was assumed to be the reason for higher WGS activity [35]. Besides, the RWGS reaction has been reported to work well with noble metal–loaded CeO_2 catalysts [36]. In a comparative study, Castao et al. looked at the efficiency of platinum and gold catalysts on ceria supports [37]. Transition metals supported on CeO_2 also have greater RWGS activity than metals supported on non-reducible supports. Moreira et al. investigated the sorption-accelerated WGS process at low temperature (125–295 °C) over Cu-CeO_2/HTlc catalysts; Cu supported on polyhedral nanoparticle-sized ceria displayed a high conversion of 87.6% [38]. Comparing the performances of 1.7% Pt-CeO_2 and Pt-Al_2O_3 at 573 K in WGS reaction, Porosof and Chen examined the amount of CO uptake as an indicator of the dispersion of Pt metal. They found out that the amount of CO uptake using Pt-CeO_2 is ~5.7 times higher than that on Pt-Al_2O_3 [39].

As it is shown in Figure 1, the CeO_2 nanocrystal surface consists of three low-index lattice planes: (100), (110), and (111) [40,41]. The three planes have distinct activity and follow the sequence (100) > (110) > (111), while showing an opposite trend for stability [42–45]. The basis of the face impact of ceria on catalytic function is the variation in electronic properties and surface atomic configuration of ceria, both of which affect the ability of oxygen vacancy formation and the structure of surface intermediates [46–50]. Based on catalytic activity evaluation, CeO_2 alone as a catalyst does not perform well and can only generate a limited number of oxygen vacancies; therefore, adding metal components to CeO_2 improves its reducibility and capacity to create oxygen vacancies [51,52]. The role of ceria in distributing the active phase appears to be particularly important in defining CO selectivity. According to recent findings, large metal particles are generally selective for methane, but well-dispersed smaller nanoparticles are more in favor of the RWGS reaction than methanation [53–55]. The potential of ceria surface to initiate a variety of CO_2 reaction pathways leading to various products is also important in this regard [56,57]. Different synthesis methods [58] and treatment procedures can be used to design and control the shape and size of ceria nanocrystals in order to increase redox characteristics and catalytic activity [59,60].

The use of different catalysts in the RWGS reaction has been considered in some review articles related to the catalytic conversion of CO_2 to chemicals and fuels [61,62]. Porosof et al., for example, looked at several catalysts for converting CO_2 to CO, methanol, and liquid hydrocarbons [63]. Kattel et al. investigated the conversion of CO_2 to the CO products containing C1 only, such as CO, methanol, and methane, with an emphasis on the effect of metal/oxide catalyst interfaces in these processes [64]. Moreover, attempts have been made to review the RWGS reaction mechanism on supported metal and oxide catalysts. For example, Su et al. looked into the RWGS reaction pathways over different types of catalysts (supported metal catalysts (including ceria), mixed oxide catalysts, and transition metal carbide catalysts) [65]. After going through many papers utilizing various catalysts for each type, they found that large volumes of water are generated in all CO_2

hydrogenation processes, and the creation of hydroxyls can poison the catalyst. Therefore, due to the fact that RWGS reaction is typically carried out at a moderate to high temperature, catalysts with great water tolerance are desirable for practical applications [65]. As they suggested, this shortcoming might be resolved by converting CO_2 to CO using the RWGS method and then treating the CO_2 with a moisture separator unit before it enters the subsequent reactor systems [65]. In any case, stable, modern water-tolerant catalysts with affordable facilities are still appealing for direct CO_2 conversion. On the other side, there are several studies describing the principles of CeO_2, and a number of recent reviews detail their catalytic uses in different CO_2 hydrogenation reactions [66–69].

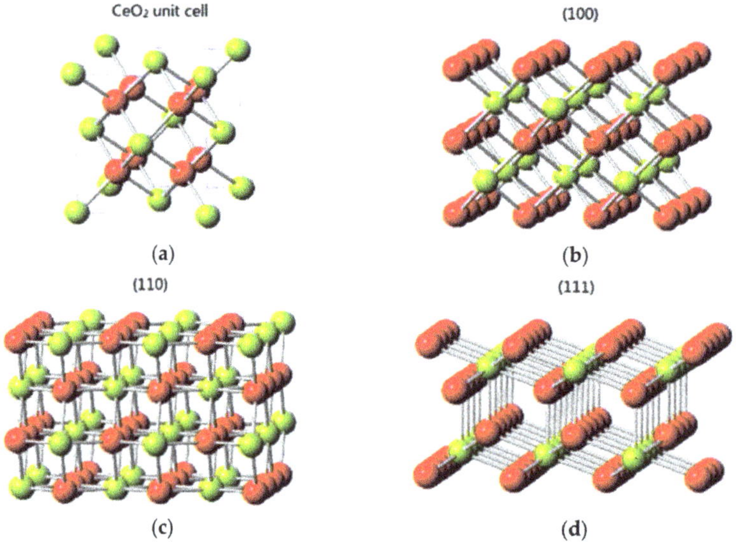

Figure 1. (**a**) CeO_2 FCC Unit cell, (**b–d**) the (100) [or (200)], (110), and (111) planes of the CeO_2 structure, reprinted with permission from [41]. Copyright Royal Society of Chemistry, 2014.

Unlike prior studies that provided an overview of several types of catalysts for the RWGS reaction, this review focuses on the function of ceria-supported catalysts in the RWGS reaction for CO_2 valorization, to highlight the benefits and drawbacks of using selected catalytic systems. Besides, the physicochemical properties of supported catalysts, which affect the catalytic activity/stability and CO selectivity in the RWGS reaction, such as the morphologies, metal loading, and metal size, are discussed to explain the structure–activity correlations. The unique significance of ceria in RWGS reaction mechanisms is also discussed with the prospect of developing cost-effective formulations based on ceria oxides to help relevant CO_2 conversion technologies gain market acceptance. In order to comprehend how CeO_2 performs in the RWGS reaction system, a detailed review of reaction mechanism is performed in addition to considering the thermodynamic factors. With all these aspects, the objective of this review is to provide a comprehensive framework for understanding the development of heterogeneous catalysts based on CeO_2 support that successfully catalyze the RWGS reaction. The review also examines the associated challenges and presents the future prospects of this field.

2. Mechanism

According to the literature, the nature of the support has a substantial influence on the reaction mechanism [70,71]. Two RWGS mechanisms have been suggested based on reaction kinetics, spectroscopy, and density functional theory (DFT) simulations; Surface Redox Mechanism and Associative Mechanism [72,73]. Nevertheless, redox and associative mechanisms have been highly controversial since the mechanism is so susceptible to the catalyst types and reaction conditions. Hence, no clear consensus on the RWGS reaction mechanism has been established yet [74]. The primary step that is different between the two pathways is whether the dissociated H atoms are implicated in the production of carbon-containing intermediates, e.g., formates and carboxyls [75]; the potential chemical pathways are depicted schematically in Figure 2 [76].

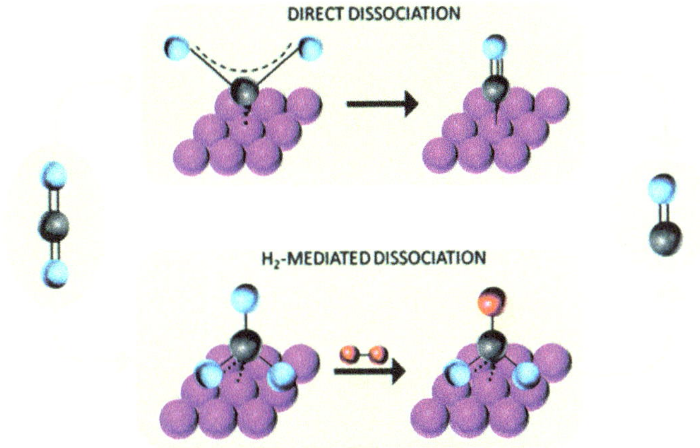

Figure 2. Simple schematic of RWGS mechanism for direct and H-assisted CO_2 dissociation processes, reprinted with permission from [76]. Copyright Royal Society of Chemistry, 2021.

In the redox mechanism, CO_2 is reduced to CO while the catalyst is directly oxidized, and subsequently H_2 reduces the catalyst to fulfill the catalytic turnover (Equations (3) and (4)) [77]. This mechanism is only viable when two essential criteria are met: (i) H_2 can reduce the catalyst surface and (ii) the partially reduced catalyst surface can be reoxidized effectively by CO_2 under RWGS experimental parameters [78]. Reducible oxide catalysts, which have the advantage of being easily reduced and oxidized during the reaction settings due to the fact that oxygen is provided by the support itself, have been found to largely follow this mechanism [79]. In the case of CeO_2, CO adsorbs on a metal surface and gets oxidized by oxygen atoms from the ceria, before being re-oxidized by water [80]. This mechanism is most commonly reported on Cu-based catalysts. For example, Lin et al. investigated Cu-CeO_2 nancatalysts in the RWGS reaction and discovered that it follows the redox mechanism to effectively improve CO_2 dissociative activation (Figure 3a) [81].

In another study by Lee et al. a series of catalysts with different CeO_2 and TiO_2 support proportions were synthesized via impregnation [82]. They discovered that in the RWGS reaction over Pt/CeO_2-TiO_2 catalyst, the redox and dissociation mechanisms coexisted, and more importantly, adding a proper amount of CeO_2 improved the redox properties and catalytic activity (Figure 3b) [82]. According to Kim et al., the sequencing of the injections affects how much CO is produced [79]. Therefore, Lee et al. measured the CO produced by feeding CO_2 after H_2 for the first run of their step reaction tests and the CO produced by feeding H_2 after CO_2 for the second run in order to investigate the impact of the injection sequence [82]. The results showed that CO production increases when H_2 is added first [82]. The catalytic mechanisms of RWGS on ceria nanocrystal were explored by Liu et al., with a function of oxygen vacancies in ceria being particularly highlighted [83]. CO can be produced by two alternative CO_2 dissociation routes, one of which takes place over ceria oxygen vacancy. As it is concluded by them, in case of ceria, the oxygen vacancy could travel through bulk active oxygen. The oxygen-ion conductivity and density of surface oxygen vacancies were boosted by the apt nanostructure of ceria, which included more exposed active planes and bulk mobile oxygens [83]. Furthermore, adsorbed or dissociated hydrogen can interact with the surface active oxygen of ceria to form water and generate a new oxygen vacancy at the initial location (Figure 3c) [83].

They investigated the mechanism using Ni as an active site supported on CeO_2 and found that CO is directly formed by the dissociation adsorption of CO_2 on the nickel surface (Figure 3d) [83]. Absorbed hydrogen, whether dissociated or not, combines with oxygen species provided by CO_2 dissociation to form water, and then with CO to make CH_x species, e.g., CH_4 [83].

$$CO_2 + * \rightleftharpoons CO + O - * \tag{3}$$

$$H_2 + O - * \rightleftharpoons H_2O + * \tag{4}$$

where $*$ stands for a surface site, such as a metal surface site.

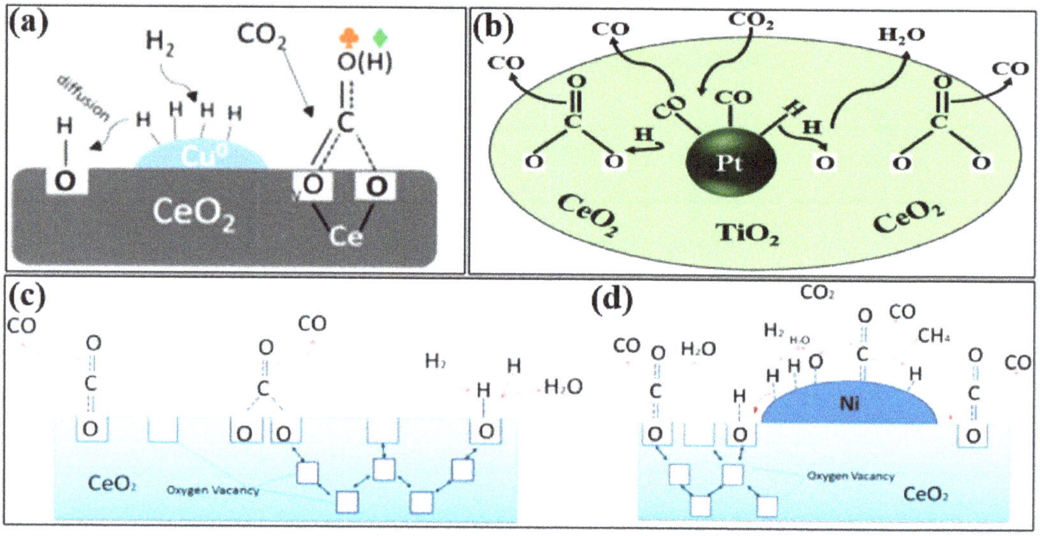

Figure 3. The Redox reaction mechanism of the RWGS reaction over (**a**) Cu-CeO_2, reprinted with permission from [81]. Copyright ACS, 2018, (**b**) Pt-CeO_2-TiO_2, (**c**) ceria nanocube, reprinted with permission from [83]. Elsevier, 2016, and (**d**) Ni-CeO_2 nanocube, reprinted with permission from [83]. Elsevier, 2016.

The adsorption and dissociation of CO_2 on defects of ceria surface followed by the generation of surface carbonate and formate species are also often postulated as reaction mechanism for the RWGS reaction over ceria-supported catalysts [84]. Cu/CeO_2 catalysts have been found to have significant catalytic activity, probably due to the emergence of intermediates like formates and bidentate carbonates [81]. In this mechanism, known as the "associative mechanism", CO_2 adsorbs onto the catalyst surface and interacts with dissociated H in order to create transitional molecules such as formate (*HCOO), carboxyl (*COOH), carbonate (CO_3^{-2}), and bicarbonate (HCO_3^{-}), that will then be degraded to the final RWGS reaction products, CO and H_2O [65]. For example, in a study led by Zhu et al. the reaction over the Au/CeO_2 catalyst was found to follow an associative pathway through a surface formate intermediate, which was proven by DRIFTS and mass spectrometry analysis [85]. In another study conducted by Lu et al., the RWGS reaction mechanism over liquid nitrogen (LN)-CeO_2 catalyst was evaluated resulting in formation of intermediate formates as shown in Figure 4a [86]. They examined the reaction intermediates in the RWGS reaction using in situ FTIR to identify the types of reactants adsorbed and carbonate species generated on LN-CeO_2 catalysts and understand the pathway of the CO_2 hydrogenation process (Figure 5a) [86]. The spectra peaks that they described are presented as follows [86]. The notable peaks between 2200 and 2500 cm^{-1} could be explained by different types of CO_2 or gas molecules that have been adsorbed on the LN-CeO_2 surface. Two sharp peaks were visible at 3730 and 3628 cm^{-1}, and many small peaks between 3800 and 3000 cm^{-1} can be attributed to the stretching vibration of the OH group on the LN-CeO_2-100 catalysts. The large peaks that appear at frequencies from 2500 and 2800 cm^{-1} centered near 2560 cm^{-1} were caused by the stretching vibration of the COO group of the carboxylate on the LN-CeO_2-100 catalysts. The broad peak, which occurred at 2078 cm^{-1}, was related to CO-Ce^{3+} and was caused by CO adsorption on LN-CeO_2 catalysts. The sharp peak at 1036 cm^{-1} and many smaller peaks around 1020–1100 cm^{-1} may be linked to the vibration modes of carbonate species produced on the LN–CeO_2 catalyst surface. It is possible to attribute the sharp peak at 1278 cm^{-1} to the bidentate carbonates O-C-O stretching vibration. The peaks, which were the typical peak of formate species, were found approximately at 1600 cm^{-1}, and can be linked to the asymmetric stretching vibration of the O-C-O group. These results indicate that the formate ($HCOO^-$) species was formed by LN-CeO_2 catalysts. In this scenario, the intermediate formate ($HCOO^-$) species in the CO_2 hydrogenation process could be configured on the LN-CeO_2-100 catalysts, and the formate ($HCOO^-$) species passed through deconstruction to release CO at high temperature [87].

The decomposition of formate has long been regarded as a significant pathway for the RWGS reaction. Nonetheless, one of the most essential paths, particularly for reducible oxide-supported metal catalysts, is the carbonate route [88]. Surface carbonates are designated as reaction intermediates, while formates are characterized as "minor intermediates" owing to a stronger bonding and lower exchange rate found in transient isotopic studies [89]. Using in situ DRIFTS experiments, Zhang et al. revealed that CO_2 may be activated efficiently over the Cu-CeO_2 catalyst via the formate and carbonate pathways in the presence of oxygen vacancies [88]. To discover reaction intermediates and speculate the probable reaction routes, Dai et al. performed in situ DRIFTS analysis on a series of $Ga_2O_3CeO_2$ composite oxide catalysts with varied Ga_2O_3 and CeO_2 ratios [90]. They concluded that, in case of pure Ga_2O_3 with no CeO_2 the system creates bicarbonate intermediates, which are more likely to hydrogenate to CH_4 at high H_2 levels. The reaction mechanism for composite oxides and CeO_2 may involve the formation of formate, which reacts with H* to generate HCOOH*, dehydrates to produce CO*, and then eventually transforms to CO (Figure 4b) [90]. Li. et al. used Cu and Cu-In active metals to get insight into the RWGS reaction mechanisms over the bimetallic catalysts on CeO_2 support [91]. In case of Cu-CeO_2, the rich contact between a highly dispersed Cu, for H_2 dissociation (Equation (5)), and defective CeO_2, for CO_2 activation (Equation (6)), results in a high RWGS activity on CeO_2 (Figure 4c). On the other hand, Cu5In5-CeO_2 has lower RWGS activity than Cu10-CeO_2 due to the presence of In, which enhances the level of Cu aggregation and fills oxygen vacancies on CeO_2 (Figure 4d). The catalysts were followed the routes shown in Equations (5)–(8) [91].

$$H_2 \rightleftharpoons 2H^* \tag{5}$$

$$CO_2 + O^* \rightleftharpoons CO_3^* \tag{6}$$

$$2H^* + CO_3^* \rightleftharpoons HCOO^* \tag{7}$$

$$H^* + HCOO^* \rightleftharpoons CO + H_2O \tag{8}$$

Although noble metal–loaded CeO_2 catalysts have been proven to be active for the RWGS reaction, the reaction mechanism remains a point of contention. Earlier research on Pt-CeO_2 catalysts, for instance, has revealed that intermediates like carbonates and formates have an essential part in the reaction, implying an associative mechanism [89,92]. At 300 °C, however, WGS process is predominantly controlled by a redox mechanism over Pt-CeO_2 catalyst [93]. Chen et al. in their research work on Pt-CeO_2 catalyst identified that despite the fact the redox and dissociation mechanisms coexisted in the RWGS reaction over Pt/CeO_2 catalyst, the dissociation mechanism, which involves the creation of intermediates, has been identified as the main pathway for CO generation due to difficulty of the CO_2 molecules adsorbing on Ce^{3+} active sites to directly generating CO (Figure 6a) [94]. In situ reflectance FTIR studies were carried out to gain more understanding of the reaction process and to analyze the generated intermediates during the RWGS reaction (Figure 5b) [94]. The peaks at 2045 and 1772 cm^{-1} were thought to be responsible for the linear- and bridge-bonded CO on the reduced Pt particles' surface. The peak at 1940 cm^{-1} was attributed to the linearly adsorbed CO at the Pt and CeO_2 contact areas. The spectrum at 2130 cm^{-1} was compatible with the Ce^{3+}−CO assignment, where the CO is adsorbed on Ce^{3+} sites. The peaks of 2830 and 1600 cm^{-1}, respectively, emerged by formation of the formate and carbonate species. Purged by He stream, the evidence of the formates and CO adsorbed on Ce^{3+} sites diminished immediately, whereas the strength of the peaks due to the linearly adsorbed CO at the contact sites and the Pt-CO species faded gradually. This indicates that even while inert gas was flowing, the produced CO was highly adsorbed on certain metals or interfacial sites. In the meantime, there is a weak connection between the produced formate species and the CO adsorbed on Ce^{3+} sites. The findings reveal that it is challenging for the CO_2 molecules adsorbed on Ce^{3+} active sites to produce CO directly (surface redox mechanism). The alternate dissociative pathway is more likely, where the synthesis of formate species as major intermediates is possible via the interaction of adsorbed CO_2 with H_2 that ultimately decomposes into CO [94].

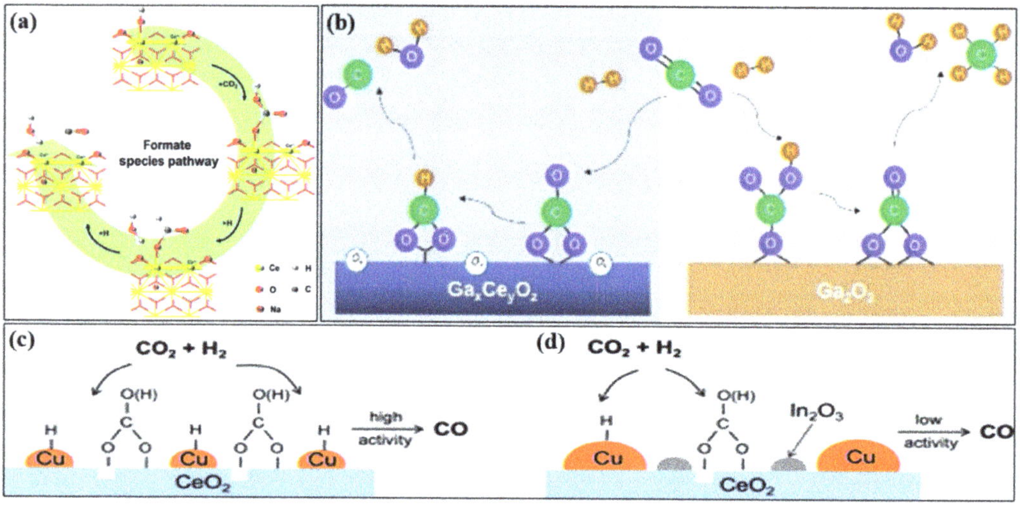

Figure 4. The associative mechanism of the RWGS reaction over (**a**) LN-CeO_2, reprinted with permission from [86]. Copyright Elsevier, 2022, (**b**) Ga_2O_3−CeO_2, reprinted with permission from [90]. Copyright Wiley, 2022, (**c**) Cu-CeO_2, reprinted with permission from [91]. Copyright ACS, 2022, (**d**) Cu5In5-CeO_2, reprinted with permission from [91]. Copyright ACS, 2022.

Goguet et al. investigated the reaction mechanism and intermediates in RWGS using a Pt/CeO_2 catalyst and reported that it followed the redox mechanism [84]. They stated that the RWGS reaction took place through surface carbonate intermediates, with carbonate species forming at the reduced CeO_2 surface (Figure 6b) [84]. On Au-CeO_2 catalysts, heterolytic dissociation is feasible at 150 °C, leading to a hydroxide and gold hydride groups [95]. In order to have a better understanding of the mechanism, Wang et al. conducted research on Au-CeO_2 (Figure 6c,d) [80,96]. In their study in 2013, they found that CO_2 reacts with a pre-reduced Au/CeO_2 catalyst to produce active oxygen, indicating a redox mechanism for the RWGS reaction at temperatures over 200 °C (Figure 6c) [80]. Later, in their published work in 2015, they concluded that even though it was thought that the RWGS reaction was dominated by an associative reaction pathway, some evidence of redox mechanism was seen in the RWGS reaction at temperatures above 120 °C (Figure 6d) [96].

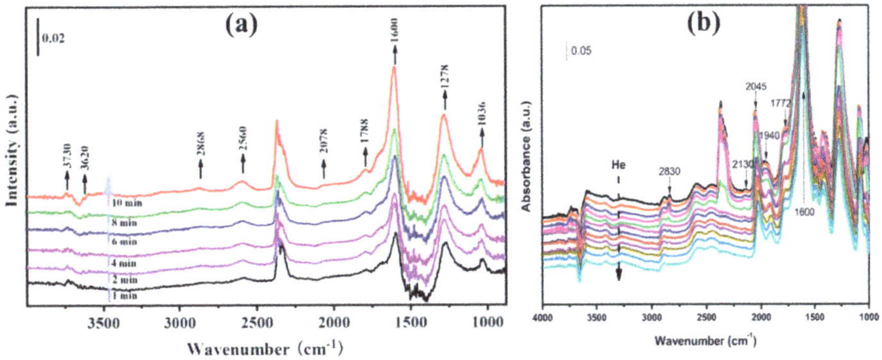

Figure 5. Operating DRIFT spectra captured after injecting a CO_2/H_2 mixture: (**a**) LN−CeO_2, reprinted with permission from [86]. Copyright Elsevier, 2022, (**b**) Pt−Ce catalyst calcined at 500 °C, reprinted with permission from [94]. Copyright Elsevier, 2016.

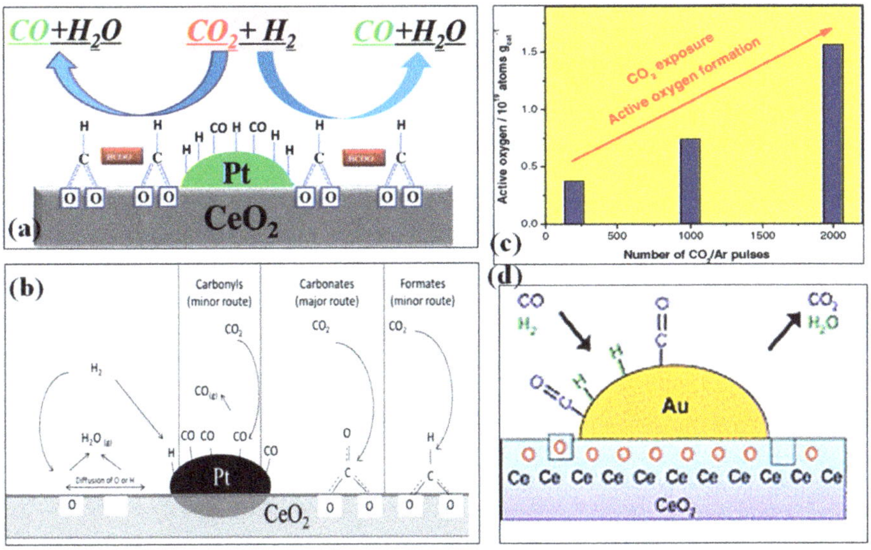

Figure 6. Model for the reaction mechanism of the RWGS reaction over (**a**) Pt−CeO_2, reprinted with permission from [94]. Copyright Elsevier, 2016, (**b**) Pt−CeO_2, reprinted with permission from [84]. Copyright Elsevier, 2004, (**c**) Au−CeO_2, reprinted with permission from [80]. Elsevier, 2013, (**d**) Au−CeO_2, reprinted with permission from [96]. Copyright Royal Society of Chemistry, 2015.

There are some reports indicating the reaction to follow neither of the two dominant pathways. Shen et al., for example, investigated the mechanism on the Ni-CeO_2 catalyst in the RWGS reaction and proposed a novel perspective into the reaction mechanism (Figure 7), arguing that CO_2 was initially adsorbed on the hydroxyl groups of Ce^{3+}–OH rather than oxygen vacancies (1, 2), resulting in the formation of the bicarbonate intermediate (*HCO_3) (3). The bicarbonate intermediate was then converted to formate (*HCOO) (4), with CO being generated by the highly active H^- in Ce–H (5) [97]. According to DFT and DRIFTS investigations, the high selectivity is attributed to the low CO affinity, but the higher catalytic activity is due to the abundance of C-H species [97].

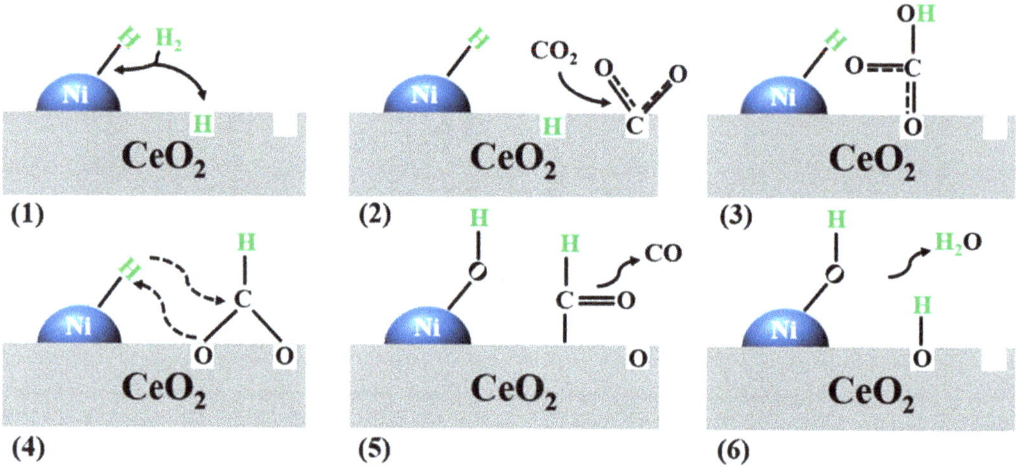

Figure 7. The proposed reaction mechanism of the RWGS reaction over Ni-CeO_2 catalyst.

Deng et al. synthesized 10Ga5Cu-CeO_2 catalyst using direct pyrolysis for photothermal catalytic RWGS reaction and identified the mechanism by in situ DRIFTS (Figure 8a) [98]. It revealed that the addition of Ga increased the synthesis of formate species, which are essential intermediates in CO_2 hydrogenation, and light irradiation promoted the breakdown of formate species to carbonyl, resulting in increased CO generation [98]. In contrast, by using Ga_2O_3 as a catalyst for the RWGS reaction after adding an amount of CeO_2 as a promoter, Zhao et al. established that adding CeO_2 to the reaction increased the development of bicarbonate species intermediates that could form H_2O and CO by interacting with H dissociated on the Ga_2O_3 surface [99]. Yang et al. provided a unique approach for boosting RWGS rate under H_2 lean conditions at low temperatures, in which they evaluated a % Cu-CeO_2 catalyst under visible light illumination, which increased catalytic activity by 30% at 250 °C (Figure 8b) [100]. Using various techniques and analysis, they discovered that, under visible light irradiation, the localized surface plasmon resonance (LSPR) created by Cu nanoparticles induces heated electron transfer to ceria, triggering desorption of bidentate formate and linear-CO transition specie [100].

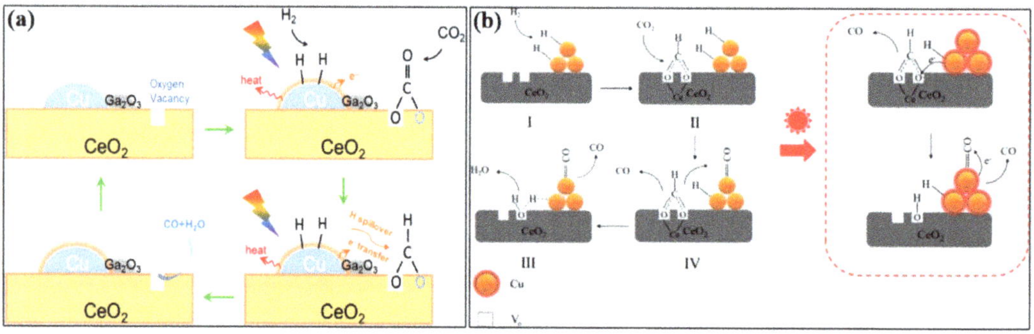

Figure 8. The associative mechanism of the RWGS reaction over (**a**) 10Ga5Cu−CeO_2, reprinted with permission from [98]. Copyright Elsevier, 2021, (**b**) Cu−CeO_2, reprinted with permission from [100]. Copyright Elsevier, 2022.

These possible alternatives indicate that there is no universal mechanism in RWGS reaction, and the reaction pathway is highly sensitive to the active metals/oxides, supports, and reaction conditions. Therefore, condensing such data into a table that succinctly explains the impact of the selectivity or functionality of active metal oxide on RWGS reaction mechanism is beneficial (Table 1).

Table 1. Dominated mechanism in RWGS reaction based on different kinds of metal-CeO_2.

Active Metal	Mechanism	Ref.
Cu	Redox or Associative	[81,91,100]
Cu5In5	Associative	[91]
Ni	Redox	[83]
Ga	Associative	[90]
Pt	Redox and/or Associative	[84,94]
Au	Redox and/or Associative	[80,96]
10Ga5Cu	Associative	[98]
Ce	Redox	[83]

3. Kinetics of RWGS Reaction

An understanding of reaction kinetics is necessary for designing a practical reactor. The majority of stated kinetic rate formulations for CO_2 hydrogenation reactions are based on Langmuir-type expressions since various reactions occur and adsorption factors influence the kinetics [101]. Interestingly, CO_2 RWGS is reported to be a faster process with a lower activation energy than the WGS reaction on many of the metal/oxides [102,103]. Recently, much research on the activity and stability of different active metals using CeO_2 support has been performed on the RWGS reaction; however, just a few kinetic and mechanistic studies have been reported on the RWGS reaction [104–106]; in most cases, the Langmuir kinetic model or power law equations have been used to explain the reaction kinetics.

Poto et al. adopted single-, dual-, and three-adsorption-site to model the kinetics of CO_2 hydrogenation to methanol over a Cu-CeO_2/ZrO_2 catalyst [107]. The findings reveal that H_2 is adsorbed and dissociated on Cu0 sites, whereas CO_2 is drawn to the oxygen vacancies formed by the CeO_2-ZrO_2 solid solution. Following that, the adsorbed H interacts selectively with the carbon atom, preferring the "formate" pathway. The CO produced by the RWGS process could desorb to the gas phase or undergo hydrogenation to yield methanol. Table 2 shows the suggested rate equations for the RWGS, where k_1 is the kinetic constant and b_i adsorption constants of i compound. The reaction rate law, first proposed by Henkel [108], was later modified by Graaf et al. [109]. The results showed that the reaction rates rise noticeably with the H_2/CO_2 concentration ratio and exponentially with temperature. They found that higher reaction rates and less H_2 consumption in the feed could be obtained from a more hydrophobic surface due to a lower H_2O adsorption and higher desorption rate from Cu^0 sites on such surfaces [107]. In order to increase the resilience of the model results, they developed a process that evaluated the sensitivity of models to the preliminary data. This method included repeatedly conducting the optimization algorithm (as a loop), with recently obtained results serving as the initial guesses. Based on the examination, it was determined that the kinetic model satisfied the physicochemical restrictions [107]. To speed up reactions and decrease the demand for H_2 in the feed, the authors suggested a more hydrophobic surface which may exhibit faster desorption from Cu^0 sites and weaker H_2O adsorption [107]. Unfortunately, very few investigations have looked into the reaction rate of M-CeO_2 catalysts in RWGS reaction; hence, the following studies will compare the reaction rates of several catalytic systems in order to gain insight into the kinetics of the RWGS reaction and the factors that affect the rate of CO_2 conversion into CO.

Lalinde et al., studied the RWGS reaction kinetics using an ordered-mesoporous $Ni\text{-}Al_2O_3$ catalyst [110]. Within the studied temperature range, they found that the rate equation for the RWGS reaction (Table 2) also comprises the WGS reaction, owing to the low equilibrium constant ($K_{eq,RWGS}$ = 0.03 and 0.1 at 320 °C and 420 °C, respectively,) and formation of CO being the rate-determining step (RDS) [110]. In their proposed reaction rate the parameters are: $k_2 = k_{CO} \cdot k_{CO2}$ as rate constant, k_β as the theoretical gas phase equilibrium compared to the equilibrium on the catalyst surface, and p_i as partial pressure of species i in gas mixture (bar). The experimental results and the local concentrations predicted by models were in good agreement [110]. Moreover, the results were consistent with another similar work by Kopyscinski et al. [111]. The reaction orders and the RDS can change depending on reaction parameters [112]. For example, reaction orders involving P_{H2} and P_{CO2} alter as the partial pressures of the gases change [104,105,113]. Ginés et al. investigated the reaction rate on the $CuO\text{-}ZnO/Al_2O_3$ catalyst as well as establishing a relationship between the reaction orders for H_2 and CO_2 in the RWGS reaction [105]. They confirmed that over a P_{CO2}:P_{H2} ratio of less than 1:3, the reaction rate was strongly sensitive to P_{CO2} (order of 1.1) and irrespective to H_2 (0 order), probably owing to surface deconstruction that favors CO_2 dissociation. Whereas, at a moderate ratio of partial pressures (>1:3), they suggested that the reaction rate is reliant on both gases with order of 0.3 for P_{CO2} and 0.8 for P_{H2}. The terms in their proposed reaction rate (provided in Table 2) are as follows: X denotes CO_2 conversion and L_o indicates the amounts of active sites in a fresh sample [105]. Kim et al. found that the RWGS reaction kinetics with Pt-reducible supports, e.g., $Pt\text{-}TiO_2$ and $Pt\text{-}Al_2O_3$ synthesized by impregnation method, were compatible with the redox mechanism [79,114]. The suggested reaction rate, as presented in Table 2, was developed using C_t as the concentration of active sites in unused catalysts [79]. Using a commercial $Pt\text{-}Al_2O_3$ catalyst this time, Jadhav et al. studied the RWGS reaction kinetics and impact of reaction parameters on catalyst efficiency, and discovered that increasing pressure increased CO_2 conversion value [115]. The reactor was tested in varied reaction settings to see how the temperature and H_2/CO_2 ratio in the feed affected reaction rate. It was realized that as the H_2 concentration in the feed increased (the partial pressure of hydrogen increased) a higher CO_2 conversion and reaction rate were achieved. Moreover, at the same P_{H2}/P_{CO2}, as the temperature increased from 573 to 673 K, the reaction rate rose from 0.009 to 0.044 mol/h/g [115]. In addition, the conversion values calculated in this study were lower than those obtained by Kim et al. using impregnated $Pt\text{-}Al_2O_3$ [79,115]. Ghodoosi et al. used a redox kinetic model over the $Fe\text{-}Mo/Al_2O_3$ catalyst for the RWGS reaction in a fixed-bed reactor to develop their rate model to describe the reaction kinetics [116]. They looked at the influence of feed gas temperature and discovered that, at high feed temperatures, the temperature drop in the reactor rises. The reason is, with increasing gas mixture temperature in feed, the rate of the reaction increases and because of the endothermic nature of RWGS reaction, it consumes more heat, causing the temperature to decrease [116]. In order to assess the proposed reactor model, three sets of experimental data were used. The outcomes demonstrated a good consistency between the model predictions and experimental findings under non-equilibrium conditions, with an average deviation of 2.7%, and an average absolute inaccuracy of 5.6%, proving that the suggested model can effectively predict the operation of the fixed-bed reactor for the catalytic RWGS reaction [116]. In another study, Chen et al. employed atomic layer epitaxy (ALE) as an alternate approach to synthesize homogenous Cu nanoparticles supported on SiO_2 [117]. The RWGS pathway using $ALE\text{-}Cu/SiO_2$ catalysts proceeded via a formate transition species, and the CO_2 and CO adsorption sites for the forward and reverse mechanisms were independent [117]. Najari et al. synthesized $Fe\text{-}K/\gamma\text{-}Al_2O_3$ catalysts via the impregnation method to perform the RWGS reaction in a lab-scale fixed-bed reactor [118]. Based on their observations, they suggested the use of the Artificial Bee Colony (ABC) method to predict kinetic values and utilize Differential Evolution (DE) optimization methods based on the Langmuir–Hinshelwood–Hougen–Watson (LHHW) mechanism to meet the kinetic requirements [118]. The ABC method outperformed the DE algorithm in forecasting total hydrocarbon selectivity, with an error rate of 6.3% compared to 32.9%. As a result, the ABC model was used to analyze the distribution of the products. Results showed that at 573 K and 1 MPa, a maximum selectivity of 73.21% for hydrocarbons can be reached with a 0.85% inaccuracy compared to the experimental result of 72.59% [118]. The proposed reaction rate (presented in Table 2), is similar to their other works [61,119]. As per their evaluations, the kinetic parameters obtained by using the ABC algorithm were roughly in line with those published in earlier investigations, taking into account the different catalytic characteristics and reaction conditions [118]. More research is required to fully comprehend how critical factors affect reaction rate and distribution of products. Using the same catalyst, but the two-step incipient wetness impregnation technique, Brübach et al. developed a new LHHW type kinetic rate statement for RWGS reaction [120]. The best performing model was derived under the assumption of H-assisted CO_2 and H-assisted CO dissociation mechanisms, supported by recent mechanistic studies. The statistical analysis demonstrated a significant association between the inhibition coefficients and reaction rate constants, as well as some lack of certainty in the estimated parameters. Since LHHW expressions have a mathematical structure, it is unfortunate that this seems to be inevitable. At a similar actual flow rate, pressure changes from 10 to 20 bar resulted in a considerable increase in CO_2 conversion. Nevertheless, they believed this was primarily due to an increase in residency time. Increasing the temperatures accelerated the reaction rate and pushed the CO selectivity to higher values for the same CO_2 conversion rate. A similar impact was seen when the H_2/CO_2 ratio was increased [120].

Table 2. Proposed reaction rate for some catalysts in RWGS reaction.

Catalyst	RWGS Reaction Rate	Assumption	Ref.
$Cu\text{-}CeO_2/ZrO_2$	$r_2 = \frac{k_2 b_{CO_2} p_{CO_2} p_{H_2} (1 - \frac{p_{CO} p_{H_2O}}{p_{CO_2} p_{H_2} k_2^{eq}})}{(1 + b_{CO} p_{CO} + b_{CO_2} p_{CO_2})(p_{H_2}^{0.5} + b_{H_2O/H_2} p_{H_2O})}$	- Dual-site LHHW mechanism	[107]
$Ni\text{-}Al_2O_3$	$r_{rWGS} = \frac{k_2 \cdot p_{CO_2} \cdot k_{H_2}^{0.5} \cdot p_{H_2}^{0.5} \cdot (1 - (k_\beta \cdot \frac{p_{CO} \cdot p_{H_2O}}{p_{H_2} \cdot p_{CO_2} \cdot k_{eq,rWGS}}))}{(1 + k_C \cdot p_{CO_2}^{e} \cdot k_H^{f} \cdot p_{H_2}^{G} + \sqrt{k_{H_2} \cdot p_{H_2}} + k_{OH} \cdot \frac{p_{H_2O}}{\sqrt{p_{H_2}}})^2}$	- CO_2 dissociation as RDS - The rate equation also includes WGS reaction - Langmuir–Hinshelwood mechanism	[110]
$CuO\text{-}ZnO/Al_2O_3$	$r = \frac{k_1 L_0 p_{CO_2}^0 (p_{H_2}^0 (1-X)^2 - \frac{p_{CO_2}^0 X^2}{k})}{p_{H_2}^0 (1-X) + \sqrt{k_2} p_{H_2}^0 1.5 (1-X)^{1.5} + \frac{p_{CO_2}^0 X}{k_2 k_3}}$	- CO_2 dissociation as RDS - Langmuir–Hinshelwood mechanism	[105]
$Pt\text{-}TiO_2$ $Pt\text{-}Al_2O_3$	$r = \frac{k_A k_b C_t (p_{CO_2} p_{H_2} - p_{CO} p_{H_2O}/k_{eq})}{(k_A p_{CO_2} + k_{-A} p_{CO} + k_{-b} p_{H_2O})}$	- CO_2 dissociation as RDS - Adsorption of CO and H_2O at the site of catalytic activity was not counted - At low P_{H2} ($1 < P^0_{H2}/P^0_{CO2} < 4$) the dissociation adsorption phase was not included - Langmuir–Hinshelwood and surface redox mechanisms	[79]
$Pt\text{-}Al_2O_3$	$r = \frac{C_S (k_1 k_2 p_{CO_2} p_{H_2} - k_{-1} k_{-2} p_{CO} p_{H_2O})}{k_2 p_{H_2} + k_{-1} p_{CO}}$	- In low CO_2 conversions - Surface redox mechanism and associative formate mechanism - Steady state; r = r1 = r2	[115]
$FeMo\text{-}Al_2O_3$	$-r_{CO_2} = \frac{k(p_{CO_2} - \frac{p_{CO} p_{H_2O}}{k_e p_{H_2}})}{1 + k_{H_2O} \frac{p_{H_2O}}{p_{H_2}}}$	- Redox kinetic model	[116]
$ALE\text{-}Cu/SiO_2$	$r = 2^{1/2} k_4 k_1^{1/2} k_2^{1/2} k_3 p_{H_2}^{1/2} p_{CO_2}^{1/2}$	- HCOO–2S → CO–S + OH–S is RDS - Langmuir–Hinshelwood model	[117]

Table 2. *Cont.*

Catalyst	RWGS Reaction Rate	Assumption	Ref.
Fe-K/γ-Al_2O_3	$r_{rWGS} = k_{rWGS}\frac{p_{CO_2}p_{H_2} - p_{CO}p_{H_2O}/k_{eq}}{p_{CO} + A_{rWGS,H_2O}p_{H_2O} + b_{rWGS,CO_2}p_{CO_2}}$	- LHHW mechanism - One-dimensional heterogeneous model - Steady-state conditions - CO_2 dissociation as RDS	[118]
Fe/K@γ-Al_2O_3	$r_{rWGS} = \frac{k_{rWGS}(p_{CO_2}p_{H_2}^{0.5} - \frac{p_{CO}p_{H_2O}}{k_{eq}p_{H_2}^{0.5}})}{(1 + A_{rWGS}\frac{p_{H_2O}}{p_{H_2}})^2}$	- Elementary reaction steps - LHHW mechanism - H-assisted CO_2 and CO dissociation mechanisms	[120]

4. Effect of Different Parameters on Catalyst Performance in RWGS Reaction

4.1. Preparation Methods

The method of catalyst preparation determines the metal-support interaction and morphological properties of the final catalysts that considerably impact the performance of the catalysts. The impact of different preparation procedures of 1 wt.% Ni-CeO_2, e.g., co-precipitation (CP), deposition–precipitation (DP), and impregnation (IM) approaches, on the physicochemical and catalytic characteristics in the RWGS reaction was explored by Luhui et al. [121]. The Ni-CeO_2-CP catalyst achieved the highest conversion rate in the RWGS reaction when compared to the Ni-CeO_2-DP and Ni-CeO_2-IM catalysts (Figure 9a); however, the CO selectivity followed the order: Ni-CeO_2-IM > Ni-CeO_2-CP > Ni-CeO_2-DP (Figure 9b). As confirmed by TPR analysis, an integration of numerous oxygen vacancies and broadly dispersed small NiO particles was considered to be the reason for the outstanding performance of the Ni-CeO_2-CP catalyst in terms of high activity and good selectivity. This suggests that more nickel ions were integrated into the CeO_2 lattice to develop a solid solution. The Ni-CeO2-DP catalyst has only a limited number of oxygen vacancies in comparison to the Ni-CeO2-CP catalyst, which results in low RWGS selectivity. It was proposed that the RWGS selectivity was strongly influenced by the oxygen vacancies. It is believed that the solid solution of Ce_xNi_yO is produced when the Ni^{2+} ions are inserted into the ceria lattice to substitute certain Ce^{4+} cations [121]. Oxygen vacancies are produced by the lattice distortion and charge imbalance that occur within the CeO_2 structure [122,123]. Several reports have indicated that precipitated ceria-based catalysts have distinct properties depending on the precipitants used, which significantly influence structural properties and catalytic performance [124–129]. In other work by the same group, Luhui et al. used the CP method to make a range of Ni-CeO_2 catalysts using Na_2CO_3, NaOH, as well as a combination of precipitants (Na_2CO_3:NaOH; 1:1 ratio) in order to investigate their catalytic efficacy in the RWGS reaction [130]. According to the structural characterization findings, the catalyst developed by the mixed precipitating agents (Na_2CO_3:NaOH; 1:1 ratio) exhibited the highest oxygen vacancies along with high Ni particle dispersion, resulting in the highest catalytic activity for the corresponding catalyst (Figure 9c), whereas the precipitants' catalytic selectivity for CO were ranked as: NaOH > Na_2CO_3 > Na_2CO_3:NaOH = 1:1 (Figure 9d) [130]. The technique used to synthesize the CeO_2 catalyst has a substantial impact on its structure, and the structure of the synthesized catalysts can greatly influence the catalytic performance in the CO_2 RWGS reaction [131]. Hard-template (HT), complex (CA), and precipitation strategies (PC) were used to synthesize CeO_2 catalysts with various structures, and their efficiency in the CO_2 RWGS reaction was examined by Dai et al. [131]. The Ce-HT catalyst had the greatest CO_2 RWGS reaction activity due to its porous structure (TEM), high specific surface area of 144.9 $m^2.g^{-1}$ (BET), and abundance of oxygen vacancies; Ce-HT > Ce-CA > Ce-PC is the temperature sequence in which the catalysts reduce in the presence of H_2 at low temperatures (H_2-TPR) [131]. Xiaodong et al. carried out the RWGS reaction over Pt-CeO_2 catalysts at temperatures between 200 and 500 °C under atmospheric pressure and various pretreatment conditions using the co-precipitated technique [94]. The samples were represented as PC-M-N, where PC stands for the co-precipitated 1%Pt-CeO_2 catalyst and M and N stand for the calcination and reduction temperatures of the samples, respectively, (Figure 9e) [94]. The catalyst prepared at a lower calcination temperature (PC-500-400) demonstrated a more favorable catalytic performance than the others due to its high Pt dispersion [94]. In another study, Ronda–Lloret et al. investigated the use of metal organic frameworks (MOFs) as precursors instead of merely using the traditional wet impregnation (WI) method in the production of CuO_x-CeO_2 catalysts [132]. After impregnating Cu-MOF using a ceria precursor, they flash-pyrolized (PF) the impregnated MOF applying distinctive conditions and procedures and compared the performances in the RWGS reaction with the WI synthesis technique finding that the MOF-derived catalyst outperformed the other catalyst (Figure 9f) [132]. Throughout the thermal decomposition procedure, the metal ions in MOFs are transformed into metallic or metal oxide nanoparticles, while the organic linkers produce carbonaceous formations which can function as supports and promote active phase distribution [133]. As Ronda–Lloret et al. concluded, by changing the pyrolysis environment, an oxidizing environment may be produced that prevents sintering and keeps copper oxidized during decomposition. Using air in the decomposition process causes the creation of copper oxide compounds that sinter with more difficultly than the metallic copper. This promotes the interaction with the ceria support, which improves its catalytic behavior. Thus, by utilizing air, highly dispersed CuO on CeO_2 can be created that is readily reducible and exhibits strong interactions with the ceria [132].

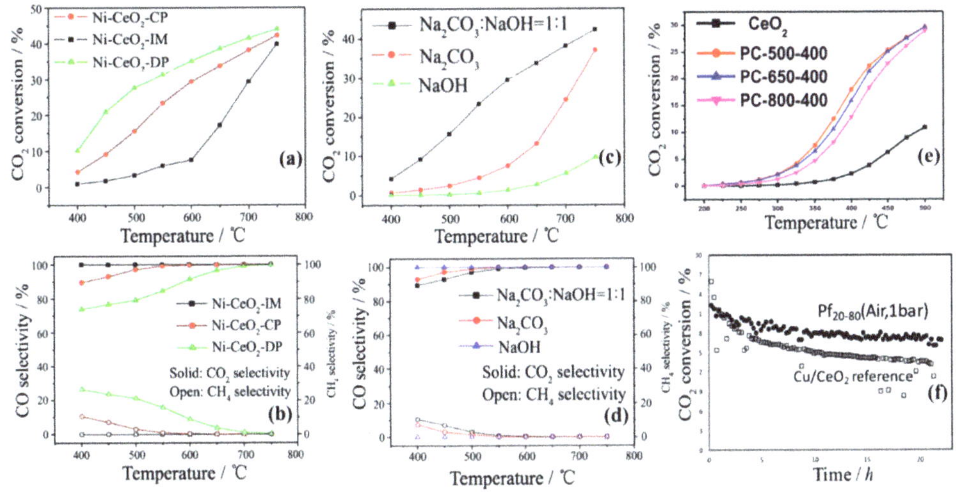

Figure 9. Effect of preparation methods on metal-CeO_2 RWGS reaction: (**a**) Ni-CeO_2 -activity, reprinted with permission from [121]. Copyright Elsevier, 2013, (**b**) Ni-CeO_2-selectivity, reprinted with permission from [121]. Copyright Elsevier, 2013, (**c**) Ni-CeO_2 -activity, reprinted with permission from [130]. Copyright Elsevier, 2013, (**d**) Ni-CeO_2-selectivity, reprinted with permission from [130]. Copyright Elsevier, 2013, (**e**) Pt-CeO_2, reprinted with permission from [94]. Copyright Elsevier, 2016, (**f**) CuO_x-CeO_2-activity, reprinted with permission from [132]. Copyright Elsevier, 2018.

4.2. *Shape and Crystal Face Effect*

The form and exposed crystal face of catalysts have a major impact on RWGS reaction activity since they may control the adsorption and desorption energies of precursors in the reaction process [134–136]. Thus, the efficiency of CeO_2 supported catalysts can be modified by conducting experiments with various morphologies. The RWGS reaction was studied by Kovacevic et al. over cerium oxide catalysts of various morphologies: cubes, rods, and particles [137]. Using TPR they found that surface oxygen is less removable in the case of nanoshapes with a high concentration of oxygen vacancies and, compared to rods and particles, cerium oxide cubes had twice more activity per surface area. The stronger intrinsic reactivity of (100) crystal planes encapsulating cubes, as opposed to less intrinsically reactive (111) facets exhibited in rods and particles, results in enhanced catalytic activity of ceria cubes in RWGS [137]. In another study, Lin et al. found that under similar conditions and the same active metal, the CeO_2(110) surface has substantially more activity than the CeO_2(111) surface, indicating that the ceria support performance is facet-dependent (Figure 10a,b) [81]. According to their study, once Cu particles are loaded onto the CeO_2-Nanorod (NR) and CeO_2-Nanosphere (NS) surfaces, the NR sample exhibits greater RWGS reaction activity. This is mostly due to the increased feasibility of CO_2 dissociative activation and the generation of active bidentate carbonate and formate intermediates over CeO_2(110) [81]. Liu et al. used RWGS to compare crystal plane reactive activity in three nano-CeO_2s with varied exposed planes [83]. The overall order of RWGS reactive efficiency of the three studied CeO_2 shapes was ceria nanocube (NC) > ceria-NR > ceria-nanooctahedra (NO), as shown in Figure 10c [83]. It is well established that oxygen vacancies formation on ceria (100) or (110) consume less energy than creating them on ceria (111) [138,139]. As a result, the ceria (100) and (110) planes seem to be more attractive choices for catalyzing processes that involve an oxygen cycle with adsorbates [83]. This could be the main reason why the ceria-NC exposed (100) plane had the best CO_2 conversion and selectivity.

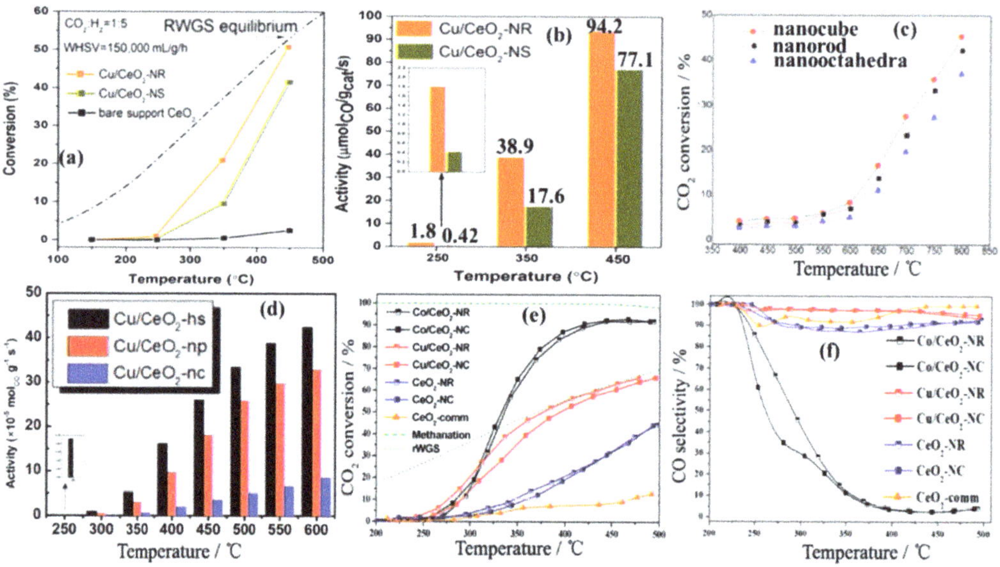

Figure 10. Effect of shape and crystal face on metal-CeO_2 RWGS reaction: (**a**,**b**) Cu−CeO_2, reprinted with permission from [81]. Copyright ACS, 2018, (**c**) CeO_2, reprinted with permission from [83]. Copyright Elsevier, 2016, (**d**) Cu−CeO_2, reprinted with permission from [88]. Copyright Elsevier, 2020, (**e**,**f**) M−CeO_2 (M = Co or Cu), reprinted with permission from [140]. Copyright MDPI, 2019.

Zhang et al. developed self-assembled CeO_2 with 3D hollow nanosphere (hs) (111), nanoparticle (np) (111), and nanocube (nc) (200) morphologies that were employed to support Cu particles [88]. Owing to the large levels of active oxygen vacancy sites, the Cu-CeO_2-hs(111) exhibited the greatest RWGS catalytic activity among the studied catalysts (Figure 10d) [88]. Konsolakis et al. looked into the influence of the active phase type and ceria nanoparticle support morphology (NR or NC) on the physicochemical characteristics and CO_2 hydrogenation capability of M-CeO_2 (M = Co or Cu) composites at 1 atm [140]. Regardless of support structure, CO_2 conversion was reported to follow the following order: Co-CeO_2 > Cu-CeO_2 > CeO_2 with the Cu-CeO_2 sample being far more selective toward CO than Co-CeO_2 (Figure 10e,f). The Co catalysts supported on NC ceria demonstrated slightly higher catalytic activity than Co supported on rod-like forms, highlighting the importance of support morphology in addition to the choice of metal element; for Cu-based samples and bare CeO_2, the pattern was the opposite [140].

Figure 11 displays the lattice characteristics of cerium oxide under various conditions based on the work of Lin et al. [81]. To assess the reduction level of support, nonthermal lattice expansion of cerium oxide was used (which represents the creation of Ce^{3+} in the bulk phase). Ceria lattice constant would rapidly drop when the reduced catalysts were exposed to reactant gas at room temperature, suggesting that the oxygen vacancies were regenerated by CO_2 breakdown on the surface of the catalyst. As marked with red squares in Figure 11, in reaction settings and in temperatures ranging from 45 to 110 °C, the CeO_2 lattice of Cu/CeO_2-NR shrank significantly with temperature. Cu/CeO_2-NS, however, had a small rise in the lattice parameter. This behavior suggested that, at low temperatures, Cu/CeO_2-NR was more favorable for CO_2 activation than Cu/CeO_2-NS [81].

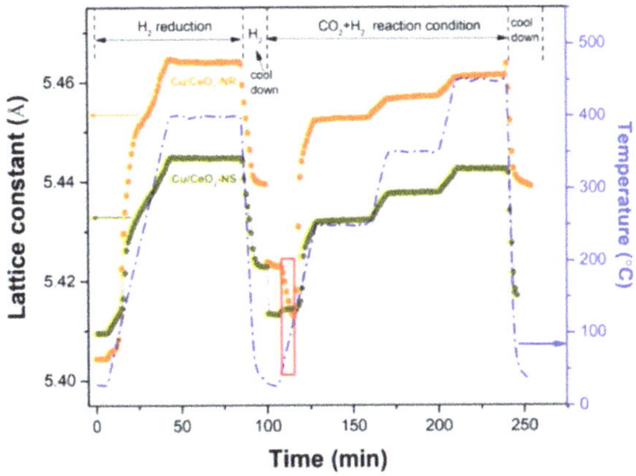

Figure 11. CeO_2 lattice constant for Cu/CeO_2-NR and Cu/CeO_2-NS catalysts, reprinted with permission from [81]. Copyright ACS, 2018.

4.3. Metal–Support Interactions

Activation of catalysts by pretreatment at high-temperature in the presence of hydrogen is often adopted to reduce the oxide nanoparticles and generate oxygen vacancies on the reducible support surface; however, such activation procedures can develop greater interactions between metal nanoparticles and the support, which has been reported to impact catalytic activity in varying ways: positively [141], negatively [142], or in some cases insignificantly [143]. The strong metal–support interaction (SMSI) phenomenon, which typically develops in metals and reducible oxides subjected to high reduction temperature, is one such case [144]. The type of the support [145], metal composition [146], and catalyst synthesis procedure can all influence metal–support interaction (MSI) [147]. According to Goguet et al., the major active site in the RWGS reaction over Pt-CeO_2 catalyst is the interface among Pt and CeO_2 and the reducible site of CeO_2, which is created by the SMSI effect of Pt and CeO_2 [84].

SMSI between Cu species and CeO_2 helps in boosting the reducibility and stability of associated catalysts, which is favorable for catalytic reduction processes [148]. The results of a study by Zhou et al. showed that the H_2 reduction at 400 °C can create oxygen vacancies and active Cu^0 species as active sites in Cu-CeO_2 catalysts [149]. The SMSI phenomenon allows electrons to move from Cu to Ce on its surface, forming the Ce^{3+}-O_v-Cu^0 and Cu^0-$CeO_{2-\delta}$ interface structures that increase the adsorption and activation of the reactant in RWGS reaction. The results suggested that the Cu-CeO_2 catalyst with 8 wt.% Cu had the best CO_2 conversion yield (Figure 12a). The full synergistic interaction between the active species via Ce^{3+}- oxygen vacancy-Cu^0 was attributed to its high catalytic activity in the RWGS process [149].

Aitbekova et al. designed the 2.6 nm Ru equally distributed on Al_2O_3, TiO_2, and CeO_2 supports and tested in a CO_2 reduction process [55]. Ru catalysts supported on TiO_2 and CeO_2 were significantly more active than those supported on Al_2O_3, but CH_4 was the predominant product in all cases. Nonetheless, they reported that moderate oxidization of the catalyst at a temperature of 230 °C followed by low temperature reduction (230 °C), named as OX-LTR, leads to the Ru particles' re-dispersion on CeO_2, giving a nearly complete switching of product selectivity from methane to CO (Figure 12b), indicating that a weaker adsorption of CO on the single RuO_x site is likely to result in increased selectivity. As they stated in their research, such re-dispersion appears only slightly in Al_2O_3- and TiO_2-supported Ru, probably due to the lower Al_2O_3 and TiO_2 and RuO_x interaction as compared to the CeO_2 support with RuO_x. Moreover, a light oxidation of the catalysts at 230 °C coupled with a high reduction temperature of 500 °C, named as OX-HTR, favored the formation of SMSI in the case of Ru-TiO_2; however, the Ru-CeO_2 catalysts (both OX-LTR and OX-HTR) exhibited fairly similar rates, implying the effect of SMSI is negligible for CeO_2-supported Ru materials under the CO_2 hydrogenation conditions investigated [55]. Similar conclusions were derived by Tauster et al. in a separate study [150].

A capping layer encircling the supported nanoparticles is commonly observed as evidence of the impact [151]. The existence of such an action, on the other hand, could be linked to charge transfer across metallic nanoparticles and the oxide support [152]. For example, Figueiredo et al. synthesized Cu_xNi_{1-x}-CeO_2 (x = 0.25, 0.35 and 60) nanoparticles for use in the RWGS reaction and investigated the SMSI influence on CO_2 dissociation reaction by exploring the nanoparticles' electrical and structural features and discovered the reactivity of nanoparticles was proportional to the Cu content on the surface with Cu-richer ones having a negative impact on the CO_2 dissociation reactivity [153]. According to their experimental results, through the reduction treatment, the SMSI effect does not actually impact nanoparticles synthesized with low Cu amounts (Figure 12c). The SMSI situation caused the support's capping layer that surrounds the nanoparticle surface to cover the catalytic active spots on the surface of the nanoparticle, leading to a decrease in the reactivity of CO_2 dissociation [153].

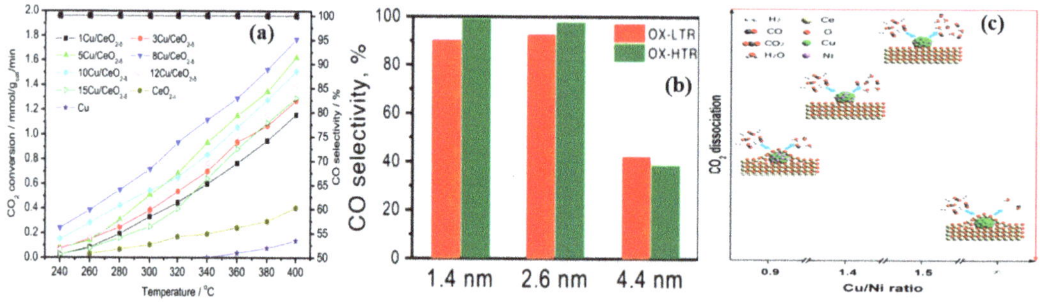

Figure 12. Effect of Effect of metal–support interactions on metal-CeO_2 in RWGS reaction: (**a**) Cu-CeO_2, reprinted with permission from [149]. Copyright Elsevier, 2020, (**b**) Ru-CeO_2, reprinted with permission from [55]. Copyright ACS, 2018, (**c**) Cu_xNi_{1-x}-CeO_2, reprinted with permission from [153]. Copyright ACS, 2020.

Looking more into the work of Zhou et al., the CO_2-TPD profiles of the xCu/$CeO_{2-\delta}$ catalysts are presented in Figure 13, and show two unique CO_2 desorption peaks for each sample [149]. The area of the CO_2 desorption peak and the number of CO_2 adsorption active sites present on the surface of the relevant catalyst are invariably connected. By increasing Cu loading, the peak regions for the ε peak exhibited a volcano pattern, reaching their highest on the 8Cu/$CeO_{2-\delta}$ catalyst. This was explained in their work by the fact that increasing Cu loading (8%) led to a greater number of Cu- $CeO_{2-\delta}$ junctions and encouraged strong MSI. Nonetheless, at larger Cu loadings (>8%), a portion of the surface active sites may be coated by an extreme amount of Cu species [149]. When it comes to the peak φ, the peak regions initially start to increase, up to Cu loading of 10%, and then continue to decrease. This was explained by the fact that at lower Cu loadings up to 10%, the catalysts' high specific surface areas enable CO_2 molecules to adsorb on their surfaces, whereas at larger Cu loadings (above 10%), the catalysts' active sites are reduced as a result of the evident decrease in specific surface areas [149].

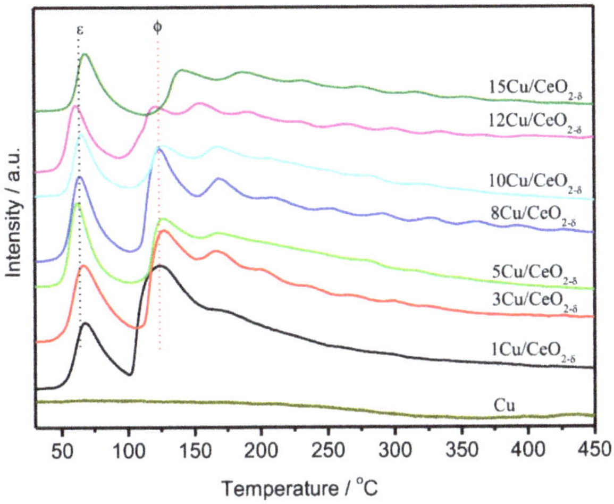

Figure 13. CO_2-TPD analysis of the $Cu/CeO_{2-\delta}$ catalysts, reprinted with permission from [149]. Copyright Elsevier, 2020.

4.4. Active Metal Loading

The influence of the catalysts' composition has been investigated in many studies. Lloret et al. designed catalysts with two different Cu concentrations and two different quantities of ceria precursor, aiming to have two distinct molar ratios of 20Cu:80Ce and 40Cu:60Ce with two different decomposition methods: pyrolysis (P), and PF (Figure 14a) [132]. It was revealed that samples with a higher Cu level have weak catalytic activity, whereas catalysts with a lower Cu content have better catalytic behavior in the RWGS process [132]. It is interesting to note that the ceria crystal size is smaller in the catalyst with higher ceria content (lower Cu), and copper dispersion is reduced when there is a large copper loading. According to the TPR results, hydrogen consumption was higher and peaked at lower temperatures in the sample with less Cu. A high Cu content needed the degradation of a greater quantity of HKUST-1, creating a more reducing environment. Therefore, if higher Cu content is there, additional MOFs are needed to break down to adjust to the reducing environment [132]. The impact of metal nanoparticle concentration on CeO_2-supported Pt and Ru catalysts with metal contents of 1, 5, and 10% on CeO_2 has been investigated by Einakchi et al. [154]. Comparing Figure 14b,c, raising the metal loading from 1 wt.% to 10 wt.% clearly degrades the activity of Pt catalysts, while it seems to favor the RWGS activity of Ru catalysts. Ru and Pt catalysts are likely to have lower metal dispersion at 10 wt.% when compared to metal loadings of 1 wt.%. The high catalytic performance of Pt-CeO_2 was found to be linked with metal loading and particularly sensitive to metal dispersion, with 1 wt.% Pt displaying the optimum catalytic performance. Unlike Pt-CeO_2, no correlation was identified between Ru catalyst RWGS activity and metal loading (Ru dispersion); nonetheless, 5 wt.% Ru metal was proven to be the best loading for Ru-CeO_2 catalysts [154]. In another study, Wang et al. prepared Co-CeO_2 catalysts with different cobalt concentrations (0, 1, 2, 5, 10%) using the CP method employed in the RWGS reaction [54]. The findings revealed that the sample with 2% Co on CeO_2 support showed highly dispersed Co_3O_4 on CeO_2 surface displaying a strong MSI that resulted in an outstanding RWGS catalytic efficiency in terms of activity, CO selectivity, and minimal carbon deposition. Nevertheless, bulk Co_3O_4 with bigger particle size generated in catalysts having high Co content (5% and 10%) lead to considerably higher carbon deposition and enhanced by-product CH_4 generation throughout the process. Their results suggested that for the RWGS reaction, widely dispersed Co which is reduced from highly distributed Co_3O_4 on CeO_2 support, ought to be a major active material, whereas solid Co that has a large particle size could be the main active component for methanation as well as carbon deposition (Figure 14d,e) [54]. The same group in another study investigated the effect of the content of cobalt supported on CeO_2 prepared by a colloidal solution combustion technique to form mesoporous catalysts (Co-CeO_2-M) and examined their activity and selectivity toward RWGS reaction and then compared the optimum Co amount sample with the same catalyst prepared by IM and CP [155]. The catalytic analyses revealed that the mesoporous 5% and 10%Co-CeO_2 catalyst had high activity in the RWGS reaction (Figure 14f); however, 10%Co-CeO_2 was less selective to CO formation than the 5%Co-CeO_2 one. Nonetheless, both had good stability over a 10-h period at 600 °C. Moreover, the activity and selectivity of 5% Co-CeO_2-M was higher than the 5% Co-CeO_2-IM and 5%Co-CeO_2-CP catalysts. They concluded that the superior catalytic performance of the 5%Co-CeO_2-M catalyst was owing to its unique mesoporous configuration, in which the Co particle is dispersed throughout the pore wall and is in close contact to small CeO_2 particles [155]. When defining the optimal catalyst in terms of activity and selectivity, metal dispersion is not the only factor to consider; the nature of the support also plays an important role. According to research by Jurkovic et al. on various supports for Cu-based catalysts, the supports with the greatest Cu dispersion were Al_2O_3 (77.7%) followed by ZrO_2 (73.6%), CeO_2 (67.6%), TiO_2 (66.3%), and SiO_2 (36.2%) [156]. Nevertheless, the alumina support was found to have the highest reported catalytic activity, followed by ceria, titania, silica, and zirconia. Ceria was ranked by TPR as the second-best support among the studied group, while having the third-best Cu dispersion, most likely because of its reducibility and capacity to hold oxygen [156]. Moreover, according to the literature, there is a direct correlation between the catalysts' activity and acid-base properties [157]. According to Pino et al., the synergistic interaction of Ni and La_2O_3 on a La_2O_3 doped Ni-CeO_2 catalyst boosted catalytic activity because of the creation of a basic site and Ni dispersion improvement [158]. Therefore, it can also be concluded that improving basicity of a catalyst might facilitate CO_2 adsorption [74]. According to the research of Jurkovic et al., Al_2O_3 is a frequently used irreducible support with good performance, where its modest acidity could be a likely contributing factor [74,156].

4.5. Metal Size Effect

The RWGS reaction is sensitive to the structure of the supported catalysts, and the size of the attached metal active sties that influence the adsorption, intermediate formation and desorption of the products [159]. Reducing the metal particle size may improve MSI by generating a larger metal–support interface, leading to positively increasing the RWGS reaction activity. Indeed, due to the SMSI affect, more oxygen atoms should be attached to the metal surface when the particle size decreases [160]. For example, Li et al. developed a 5% Ir-CeO_2 catalyst including an Ir particle size of around 1 nm, and a 0.7% Ir-CeO_2 catalyst with atomic dispersion of Ir [161]. Even though the dominant product of CH_4 was produced by an Ir-CeO_2 catalyst with large Ir particles (>2.5 nm), CO was produced mainly by the 5% and 0.7% Ir-CeO_2 catalysts when Ir particle size was less than 1 nm, and the catalytic activity per mole of Ir increased (Figure 15a). They also discovered that due of the intense interaction with CeO_2, 1 nm Ir particles and atomically dispersed Ir get partially oxidized, but large Ir particles, more than 2.5 nm, are mostly reduced. As a result, they concluded that the primary active site for a RWGS reaction is partly oxidized Ir that engages extensively with CeO_2 support regardless of Ir particles or atomically dispersed Ir atoms [161]. Metal active sites distributed on an atomic level add more to the CO product than metal clusters at a 3D level [162]. In a study, Zhao et al. produced Pt-CeO_2 catalysts with various Pt sizes to test the influence of size on CO selectivity in the RWGS reaction [163]. Three Pt-CeO_2 catalysts were produced using CeO_2 nanorods, including atomically dispersed Pt species as well as Pt clusters or particles of two different sizes (2.1 and 5.2). According to Figure 15 b, within the temperature range of 200 to 450 °C, the atomically distributed Pt-CeO_2 catalyst (shown as Pt-CeO_2-AA-350) exhibited almost 100% CO selectivity, whereas CO selectivity using Pt nanoparticles (5Pt-CeO_2-IMP-350 catalyst with 5.2 nm particle size) reduced noticeably at elevated reaction temperature. Large-sized Pt species demonstrated high CO adsorption, resulting in the production of CH_4 by further hydrogenation [163].

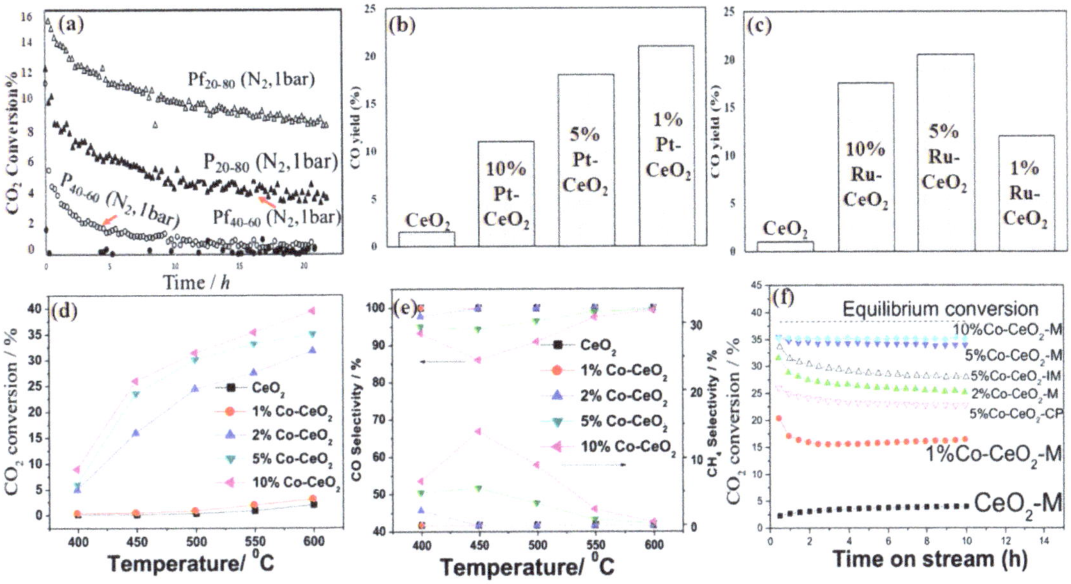

Figure 14. Effect of metal loading on metal-CeO_2 RWGS reaction: (**a**) CuO_x-CeO_2, adapted with permission from [132]. Copyright Elsevier, 2018, (**b**,**c**) CO yield at 450 °C, (**b**) Pt-CeO_2, (**c**) Ru-CeO_2, (**d**,**e**) activity and selectivity of Co-CeO_2, reprinted with permission from [54]. Copyright Elsevier, 2017, (**f**) Co-CeO_2, reprinted with permission from [155]. Copyright Elsevier, 2018.

Lu et al. developed mesoporous CeO_2 (surface area = 100 m^2 g^{-1}) as well as NiO-CeO_2 with large surface areas, narrow pore size dispersion, and homogeneous mesopores (intercrystallite voids) [53]. According to the results, with increasing temperature and NiO quantity, the CO_2 conversion rate in RWGS reaction increased. As for CO selectivity, when less than 3 wt.% NiO was used, NiO particles monodispersed in mesoporous CeO_2 resulting in a complete CO_2 conversion to CO which was irrespective of temperature. For more than 3.5 wt.%, due to NiO particle aggregation, 100% CO selectivity was improbable below 700 °C (Figure 15c) [53]. Wang et al. studied the effect of CeO_2 on RWGS in comparison to In_2O_3 [164]. The surface areas of the In_2O_3–CeO_2 catalysts enhanced compared to pure In_2O_3; as the CeO_2 content increased, the size of the In_2O_3 particles in the In_2O_3–CeO_2 samples reduced and the dispersion of In_2O_3 particles in the In_2O_3–CeO_2 increased. Besides, additional oxygen vacancies were formed (Figure 15d), which in turn enhanced the dissociative hydrogen adsorption and increased the quantity of bicarbonate species produced by activated CO_2 adsorption [164].

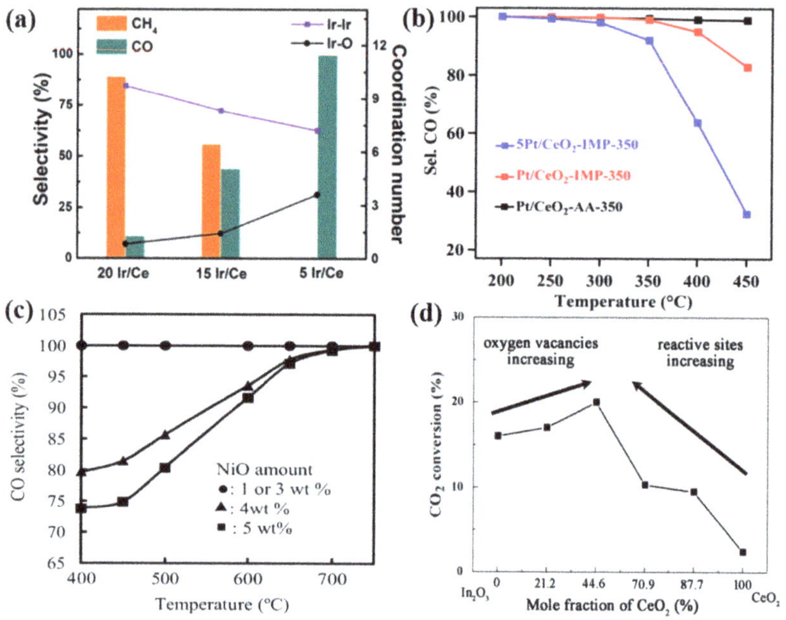

Figure 15. Effect of metal size CeO_2 support on RWGS reaction: (**a**) Ir-CeO_2, reprinted with permission from [161], Copyright Wiley, 2017, (**b**) Pt-CeO_2, reprinted with permission from [163]. Copyright Elsevier, 2021, (**c**) NiO-CeO_2, reprinted with permission from [53]. Copyright Elsevier, 2014, (**d**) In_2O_3–CeO_2, reprinted with permission from [164]. Copyright Elsevier, 2016.

4.6. Effect of Adding CeO_2 as a Reducible Transition Metal Oxide Promoter

The inclusion of a cerium oxide as a promoter can affect CO_2 adsorption and activation, as well as the activity and selectivity of the RWGS process [165,166]. Yang et al. showed in their work how adding ceria to alumina support (CeO_2-Al_2O_3) helped in lowering acidity of Ni-based catalysts, which also minimized carbon deposition [167]. Figure 10a shows the catalytic performance of all samples versus temperature in terms of CO_2 conversion. Moreover, introduction CeO_2 to the unpromoted Ni-Al catalyst also improved its reducibility. As it is shown in Figure 16a, ceria improved unpromoted Ni-Al catalyst reducibility in that the Ni-CeAl catalyst had a reduction percentage (RP) of ~93%, compared to 86.5% for the reference Ni-Al catalyst. The authors concluded that their findings were mostly due to two aspects occurring together: (1) CeO_2, as a promoter here, reduced the Ni-Al_2O_3 connection, leading to an increase in Ni particle reducibility, owing to the generated Ni-promoter interaction, and (2) because of its intrinsic redox capabilities, CeO_2 offered additional oxygen mobility to the catalysts [96]. In another study by Lee et al., a set of Pt-CeO_2-TiO_2 catalysts were impregnated with different support combinations ranging from 0 to 20% to evaluate the influence of varied CeO_2/TiO_2 ratio on catalytic activity during RWGS reaction [82]. Accordingly (Figure 16b), increasing CeO_2 loading improved the catalytic activity of Pt-impregnated catalysts, the Pt-20%CeO_2-TiO_2 sample showing the highest CO_2 conversion. Based on their analysis, by substituting TiO_2 with CeO_2, the lattice and pore configuration changed in favor of more CO_2 conversion in RWGS reaction (the mechanism is explained in Figure 3b) [82]. With the aim of improving adsorption and activation of CO_2 on Ga_2O_3, Zhao et al. used CeO_2 as a promoter with an optimum Ga:Ce ratio of 99:1 and employed RWGS reaction as a test reaction and indicator for the catalytic performance of the resultant samples [99]. The results showed the positive performance of CeO_2-Ga_2O_3 (Figure 16c) due to the fact that the inclusion of CeO_2 increases the production of bicarbonate species in CO_2 adsorption, which is thermodynamically more advantageous [99].

The same catalytic system was studied by Dai et al., where the gel sol-gel process was used to create a variety of Ga_2O_3-CeO_2 composite oxide catalysts with various Ga_2O_3 and CeO_2 ratios [90]. When compared to pure Ga_2O_3 and pure CeO_2, the composite oxide catalysts had smaller particles and showed high CO selectivity in the RWGS process. It was discovered that Ga_2O_3 has distinct reaction intermediates from CeO_2 and Ga_2O_3-CeO_2, making it easier to create methane in high H_2 conditions, whereas CeO_2 promotes CO selectivity. In a Ga_2O_3 to CeO_2 ratio of 3:1, composite oxide showed the greatest activity (Figure 16d). This is mostly due to the creation of the $Ga_xCe_yO_z$ solid solution phase and the development of additional active sites that result in an increased number of oxygen vacancies, which facilitate CO_2 adsorption and activation. Moreover, it was found that GaCe composite oxides have a more homogeneous mesoporous structure and a greater pore volume, making mass transport in reactions easier [90]. In contrast, there are a few cases in which once ceria was introduced to a catalytic system, CO selectivity improved but CO_2 conversion was slightly decreased [168]. Galvita et al., for example, developed a Fe_2O_3–CeO_2 composite and discovered that incorporating ceria to iron oxide increased solid solution stability but reduced CO production capabilities [169].

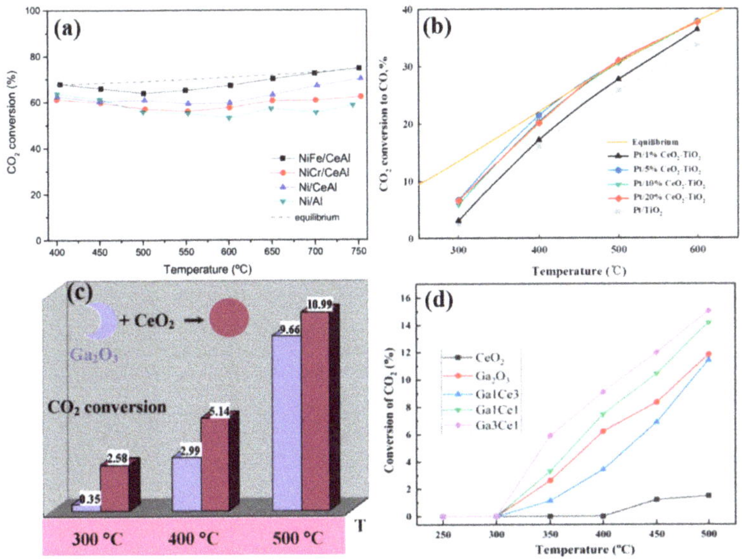

Figure 16. Effect of adding promoters on RWGS reaction: (**a**) Ni-CeAl, reprinted with permission from [167]. Copyright Elsevier, 2018, (**b**) Pt-CeTi at 400 °C, (**c**) CeO_2-Ga_2O_3, reprinted with permission from [99]. Copyright Elsevier, 2012, (**d**) $Ga_2O_3-CeO_2$, reprinted with permission from [90]. Copyright Wiley, 2022.

On the other hand, when a non-reducible transition metal oxide is used as a promoter for M-CeO_2 catalysts, more oxygen vacancies can emerge during the reduction process [170]. For example, more thermally stable support can be developed by a mixed framework of Al_2O_3-CeO_2 which offers a broad surface for optimum active phase dispersion and enables the development of oxygen vacancies on the surface throughout the catalytic reaction to improve catalytic performance [171]. Zonetti et al. and Wenzel et al. showed that adding Zr to the CeO_2 lattice improved its ability to create oxygen vacancies as well as its thermal stability, which is a desirable feature of catalytic systems [172,173].

4.7. Bimetallic Effect

Reports indicate that bimetals, where a second metal is introduced along with the primary active metal, can be used to help boost the catalytically active phase [174]. Evidently in many systems, bimetallic compounds have outperformed their individual components [175]. Furthermore, the generation of metal carbide (coke precursor) could be prevented due to the electrical effect caused by metal–metal interactions, resulting in less deactivation [176]. Yang et al. showed how the inclusion of a second element (Cr or Fe) can positively affect the reducibility of monometallic Ni-based catalysts (Ni-CeAl) [167]. As it is displayed in Figure 17a, the addition of Fe to the Ni-CeAl catalyst system can increase the reducibility of Ni- to 95% compared to 93% for the Ni-CeAl catalyst [167]. Chen et al. synthesized a Cu-Fe bimetallic phase loaded on CeO_2 and evaluated its performance for RWGS reaction at temperatures ranging from 450 to 750 °C at 1 atm (Figure 17b). The efficiency of the iron-containing copper-based catalyst was greatly increased over that of the catalyst without iron, and CO_2 conversion nearly approached theoretical levels. The bimetallic CeO_2-supported catalyst was shown to have high selectivity, stability, with no secondary reactions, and no carbon deposition on the catalyst surface after the process [177]. In contrast, Li et al. used CuIn bimetallic catalysts for the RWGS reaction, demonstrating that the promotional impact of In on Cu is support dependent [91]. The CO_2 conversion of the CuIn-ZrO_2 catalyst was higher by far than the Cu-ZrO_2 catalyst; however, the CO_2 conversion of CuIn-CeO_2 was considerably lower than Cu-CeO_2 (Figure 17c). The cause of the support-sensitivity of RWGS activity was further discovered through systematic analysis. Cu and In combined to form CuIn alloys on the ZrO_2 support, which allowed CO_2 to be activated by oxygen vacancies from partially reduced In_2O_3, whereas, Cu and In were found as metallic Cu and In_2O_3 on the CeO_2 support, respectively. The addition of In prevented Cu dispersion and the development of oxygen vacancies on CeO_2, resulting in lower RWGS activity [91]. The cata lytic activity of 5 wt.% mono- and bimetallic Ru- and Pt-based catalysts supported on CeO_2 was compared as a function of temperature as shown in Figure 17d [154]. The bimetallic activity is higher than that of the individual Ru activities due to the synergy between Ru, Pt, and CeO_2. Moreover, the Fe-promotional impact in the Ru90Fe10-CeO_2 catalyst is most noticeable at lower temperatures, whereas there is no difference in catalytic activity above 450 °C when compared to Ru-CeO_2 [154].

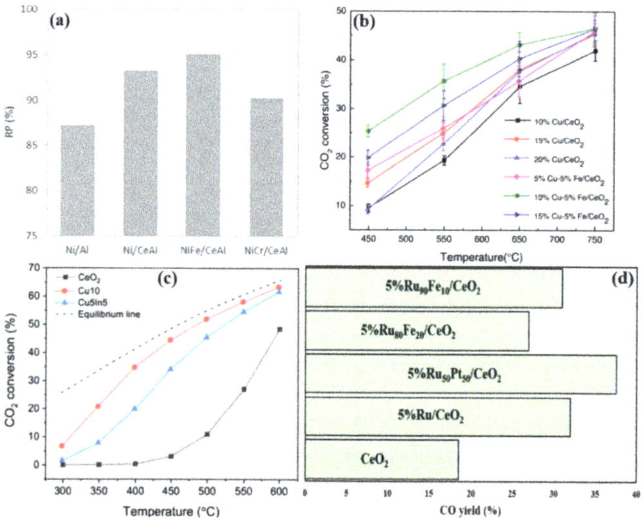

Figure 17. Effect of bimetallic on metal-CeO_2 in RWGS reaction: (**a**) NiFe-CeO_2, reprinted with permission from [167]. Copyright Elsevier, 2018, (**b**) CuFe-CeO_2, reprinted with permission from [177]. Copyright Elsevier, 2021, (**c**) CuIn-CeO_2, reprinted with permission from [91]. Copyright ACS, 2022, (**d**) CO yield for RuPt/CeO_2 and RuFe/CeO_2 at 600 °C.

The following table (Table 3) provides a summary of M-CeO_2 catalysts used in RWGS reaction. The synthesis methods and relevant results are provided for comparing their performance.

Table 3. Conditions and catalytic performance in RWGS reaction on different kinds of metal-CeO_2.

Catalyst	Synthesizing Method	Temperature (°C)	Conversion (%)	Selectivity (%)	Ref.
Pt-CeO_2	Co-Precipitation	300	6.7	-	[94]
2%Pt-CeO_2	Commercial	290	21.7	~100	[178]
1 wt.%Pt-CeO_2	Polyol	500	~24	~100	[154]
3.2%PtCo-CeO_2	Incipient Wetness Impregnation	300	9.1	92.3	[179]
5 wt.%Ru-CeO_2	Polyol	500	~25	~100	[154]
5%Ru-CeO_2	Polyol	350	~16	~31	[180]
Ru50Pt50-CeO_2	Polyol	500	~28	~100	[154]
RuNi-CeZ	Wet Impregnation	350	53	93	[181]
5%Ru/Sm-CeO_2	Polyol	350	~16	~69	[180]
FeNi-CeZr	Wet Impregnation	350	13	60	[181]
PtCo-CeO_2	Incipient Wetness Impregnation	300	9.1	92.31	[179]
0.5Pd10Cu-CeO_2	Precipitation–Impregnation	270	12	-	[182]
1Pd10Cu-CeO_2	Precipitation–Impregnation	270	17.8	-	[182]
2Pd10Cu-CeO_2	Precipitation–Impregnation	270	11.3	-	[182]
PdNi-CeO_2	Incipient Wetness Impregnation	300	2.5	37.5	[183]
10Cu-CeO_2	Precipitation-Impregnation	270	6.4	-	[182]
Cu-CeO_2	Space-Confined	300	~18	~100	[184]
CeO_2-NC	Hydrothermal–Incipient Wetness Impregnation	700	27.8	~100	[83]
CeO_2-NR	Hydrothermal–Incipient Wetness Impregnation	700	23.8	~100	[83]
CeO_2-NO	Hydrothermal–Incipient Wetness Impregnation	700	19.8	~100	[83]
CeO_2-HT	Hard-Template	580	15.9	~100	[131]
Cu/CeO_2	Wet Impregnation	380	52	95	[140]
$CuCeO_x$	Hard Template	400	33	~100	[51]
CuO_x-CeO_2	Wet Impregnation	400	10	~100	[132]
1 wt.%Cu-CeO_2	Combustion	600	~70	~100	[28]
5.60 wt.%Cu-CeO_2-hs	Hydrothermal–Impregnation	600	~50	~100	[88]
5 wt.%Cu-CeO_2-nr	Hydrothermal–incipient wetness impregnation	450	~50	-	[81]
5 wt.%Cu-CeO_2-ns	Microemulsion-incipient wetness impregnation	450	~40	-	[81]
0.25 mole%Cu-CeO_2	Hard Template–Impregnation	400	31.34	100	[185]
5Cu/48CeO_2/ZSM	Physical Mixing	600	~68	100	[186]
Fe-CeO_2	Hard Template–Impregnation	340	3.3	~100	[185]
Mn-CeO_2	Hard Template–Impregnation	340	3.3	~100	[128]
Co-CeO_2	Hard Template–Impregnation	340	9.3	~100	[128]
10%Co-CeO_2	Colloidal Solution Combustion	300	3.8	39.4	[155]
10Cu5Fe-CeO_2	Impregnation	750	42	100	[177]
5Cu5In-CeO_2	Impregnation	500	45	100	[91]
10Cu-CeO_2	Impregnation	500	50	100	[91]
10 wt.%Co-CeO_2	Co-precipitation–Impregnation	500	28	~91	[187]
1%K10% Co-CeO_2	Co-precipitation–Impregnation	500	31	100	[187]
1%Ni-CeO_2	Co-precipitation	600	~35	100	[188]
1%Ni-CeO_2	Co-precipitation	400	~4.5	~90	[130]
Ni-CeO_2	Co-precipitation	700	37.5	~100	[121]
Ni-CeO_2	Deposition–Precipitation	700	41.7	~100	[121]
Ni-CeO_2	Impregnation	700	29.2	~100	[121]
Ni-$CeZrO_x$	Impregnation	700	46.1	97.3	[189]
Ni/CeO_2-Al_2O_3	Wet Impregnation	750	59	94	[167]
Ni-$CeZrO_x$	Precipitation-Co-precipitation	550	48	87.5	[172]
1%NiO-CeO_2/SBA-15	Calcination	450	~2.5	100	[53]
6%Ni/Al_2O_3-CeO_2	Combustion	750	~63	~90	[190]
RuFe-CeO_2	Polyol	800	47.5	~100	[180]
0.7%Ir-CeO_2	Adsorption–Precipitation	300	2.9	>99	[161]
5%Ir-CeO_2	Adsorption–Precipitation	300	6.8	>99	[161]
20%Ir-CeO_2	Adsorption–Precipitation	300	8.8	12	[161]
In_2O_3-CeO_2	Co-precipitation	500	20.4	~100	[164]
Ga_2O_3-CeO_2	Physical Mixing	400	5.14	~100	[99]

5. Summary and Outlook

The RWGS is a potential process with a promising prospect for large-scale CO_2 to CO conversion in the near future, when integrated with large-scale technologies for sustainable H_2 production. If the RWGS reaction is to play a significant role in lowering CO_2 levels in the atmosphere, a suitable catalyst made of earth-abundant elements is desirable. CeO_2 support has been shown to aid in increasing CO_2 catalytic activity, and the following key features make CeO_2 an attractive choice:

- CeO_2 supports are renowned for their acid-base properties, and high oxygen mobility and stability, which improve RWGS reaction activity when added to a suitable active metal catalytic system.
- Surface defects (Ce^{3+} and oxygen vacancies) enable metal particles to attach well to the ceria support, resulting in increased metal dispersion.
- A Ce^{4+}/Ce^{3+} redox pair with strong activity can significantly improve CO_2 dissociative activation.
- Sintering processes and coke formation may be hampered by a strong metal–support interaction (SMSI) leading to an enhancement in catalyst stability.
- CeO_2 can be produced by a variety of methods to provide suitable design and technical features generating appropriate metal–support interactions that can further be tuned by synthesis methods and catalyst pretreatment techniques.
- CeO_2 support could influence the reaction mechanism by changing the adsorption energy of key intermediates with MSI.

The mechanism of the RWGS reaction involving CeO_2 as a support, as well as an active site, have been thoroughly debated, since understanding these concepts is critical for the rational design of highly efficient and selective RWGS catalysts. Although the RWGS reaction mechanism is still a controversial topic, the reaction routes on different catalysts working under different reaction parameters may be clarified in the near future due to the advancements in in situ and *operando* spectroscopic technologies, including in situ (*or near ambient pressure (NAP)*) XPS and in situ XRD, in combination with theoretical approaches involving DFT. Identifying the deactivation pathway may also help in the development of stable catalysts for the RWGS reaction.

In this context, developing a more structurally and thermally stable CeO_2 that can achieve high CO selectivity at high CO_2 conversion rates with less carbon deposition to prevent catalyst deactivation is critical. A careful approach is needed to develop and activate CeO_2 based catalysts and fine-tune their properties by regulating MSI to generate the desired effects. The literature shows the implementation of a range of parameters that influence CeO_2 catalyst performance in RWGS reaction, such as the addition of a suitable second metal in bimetallic systems, adaptation of apt catalyst preparation methods to optimize active metal size, addition of CeO_2 as a promoter for metal-supported catalysts, tuning CeO_2 crystal planes for oxide catalysts etc. Recently, transition metal carbides have been seen as an appealing and suitable option for commercial applications in the RWGS reaction due to their interesting features comparable to those of noble metals such as their dual capabilities for H_2 dissociation and C-O bond regulation, and possibility to act similarly to reducible oxides. Even though numerous materials have been investigated, no commercial-scale catalyst is available for RWGS reaction and further progress and scientific understanding is required before an industrial level of implementation can be thought of. So far, only a few studies are available on the kinetic model and reaction rate determination on M-CeO_2 catalysts in RWGS reaction, and more dedicated future studies are required to fulfill this critical gap; it is necessary to expand this study further, to reactor development and large-scale expansion. The engineering of an industrial RWGS unit requires a comprehensive knowledge of reaction kinetics, involving internal/external mass transport impacts, along with the RWGS thermodynamic equilibrium.

Author Contributions: Conceptualization, P.E. and A.K.; methodology, P.E. and A.K.; writing—original draft preparation, P.E.; writing—review and editing, A.K. and M.K.; supervision, A.K. and M.K.; project administration, A.K.; funding acquisition, A.K. and M.K. All authors have read and agreed to the published version of the manuscript.

Funding: This research was funded by Qatar National Research Fund, grant number NPRP8-509-2-209 and NPRP10-0107-170119.

Data Availability Statement: Not applicable.

Acknowledgments: The work was made possible by a grant from the Qatar National Research Fund under the National Priorities Research Program award numbers NPRP8-509-2-209 and NPRP10-0107-170119. Its contents are solely the responsibility of the authors and do not necessarily represent the official views of QNRF. The authors would also like to gratefully acknowledge the support of Graduate Teaching/Research Assistantship (GTRA) from Qatar University in conducting this research.

Conflicts of Interest: The authors declare no conflict of interest.

Abbreviations

ABC	Artificial Bee Colony
ALE	Atomic Layer Epitaxy
BET	Brunauer–Emmett–Teller
CA	Complex
CP	Co-Precipitation
DE	Differential Evolution
DFT	Density Functional Theory
DP	Deposition-Precipitation
DRIFTS	Diffuse Reflectance Infrared Fourier-Transform Spectroscopy
EDX	Energy Dispersive X-ray
FCC	Face-Centered Cubic
FTS	Fischer–Tropsch Synthesis
HT	Hard-template
HTR	High Temperature Reduction
IM	Impregnation
LHHW	Langmuir–Hinshelwood–Hougen–Watson
LN	Liquid Nitrogen
LSPR	Localized Surface Plasmon Resonance
LTR	Low Temperature Reduction
M	Mesoporous
MOF	Metal Organic Framework
MSI	Metal–Support Interaction
NAP	Near Ambient Pressure
NC	Nanocube
NO	Nanooctahedra
NR	Nanorod
NS	Nanospheres
OSC	Oxygen Storage Capacity
OX	Oxidation
P	Pyrolysis
PC	Precipitation
PF	Flash Pyrolized
RDS	Rate-Determining Step
RM	Reverse Microemulsion
RP	Reduction Percentage
RWGS	Reverse Water Gas Shift
SEM	Scanning Electron Microscopy
SMSI	Strong Metal–Support Interaction
TEM	Transmission Electron Microscopy
TPR	Temperature Programmed Reduction

WGS	Water Gas Shift
WI	Wet Impregnation
XPS	X-ray Photoelectron Spectroscopy
XRD	X-ray Diffraction

References

1. Yang, X.; Su, X.; Chen, X.; Duan, H.; Liang, B.; Liu, Q.; Liu, X.; Ren, Y.; Huang, Y.; Zhang, T. Promotion effects of potassium on the activity and selectivity of Pt/zeolite catalysts for reverse water gas shift reaction. *Appl. Catal. B Environ.* **2017**, *216*, 95–105. [CrossRef]
2. Ebrahimi, P.; Kumar, A.; Khraisheh, M. Thermodynamic assessment of effect of ammonia, hydrazine and urea on water gas shift reaction. *Int. J. Hydrogen Energy* **2020**, *47*, 3237–3247. [CrossRef]
3. Sasol ECOFT and Deutsche Aircraft join forces to accelerate power-to-liquid (PTL) aiming for carbon-neutral flight. *Focus Catal.* **2022**, *2022*, 4. [CrossRef]
4. Liu, X.; Pajares, A.; Matienzo, D.D.C.; de la Piscina, P.R.; Homs, N. Preparation and characterization of bulk MoXC catalysts and their use in the reverse water-gas shift reaction. *Catal. Today* **2020**, *356*, 384–389. [CrossRef]
5. Centi, G.; Perathoner, S. Opportunities and prospects in the chemical recycling of carbon dioxide to fuels. *Catal. Today* **2009**, *148*, 191–205. [CrossRef]
6. Aresta, M. *Carbon Dioxide as Chemical Feedstock*; John Wiley & Sons: Hoboken, NJ, USA, 2010.
7. Ueckerdt, F.; Bauer, C.; Dirnaichner, A.; Everall, J.; Sacchi, R.; Luderer, G. Potential and risks of hydrogen-based e-fuels in climate change mitigation. *Nat. Clim. Chang.* **2021**, *11*, 384–393. [CrossRef]
8. Olah, G.A.; Goeppert, A.; Prakash, G.S. Chemical recycling of carbon dioxide to methanol and dimethyl ether: From greenhouse gas to renewable, environmentally carbon neutral fuels and synthetic hydrocarbons. *J. Org. Chem.* **2009**, *74*, 487–498. [CrossRef]
9. Sterner, M. *Bioenergy and Renewable Power Methane in Integrated 100% Renewable Energy Systems. Limiting Global Warming by Transforming Energy Systems: Limiting Global Warming by Transforming Energy Systems*; Kassel University Press GmbH: Hesse, Germany, 2009; Volume 14.
10. Pastor-Pérez, L.; Baibars, F.; Le Sache, E.; Arellano-Garcia, H.; Gu, S.; Reina, T.R. CO_2 valorisation via reverse water-gas shift reaction using advanced Cs doped Fe-Cu/Al_2O_3 catalysts. *J. CO2 Util.* **2017**, *21*, 423–428. [CrossRef]
11. Galván, C.Á.; Schumann, J.; Behrens, M.; Fierro, J.L.G.; Schlögl, R.; Frei, E. Reverse water-gas shift reaction at the Cu/ZnO interface: Influence of the Cu/Zn ratio on structure-activity correlations. *Appl. Catal. B Environ.* **2016**, *195*, 104–111. [CrossRef]
12. Sheehan, S.W. Electrochemical methane production from CO_2 for orbital and interplanetary refueling. *Iscience* **2021**, *24*, 102230. [CrossRef]
13. Xu, X.D.; Moulijn, J.A. Mitigation of CO2 by chemical conversion: Plausible chemical reactions and promising products. *Energy Fuels* **1996**, *10*, 305–325.
14. Kwak, J.H.; Kovarik, L.; Szanyi, J.N. Heterogeneous catalysis on atomically dispersed supported metals: CO_2 reduction on multifunctional Pd catalysts. *ACS Catal.* **2013**, *3*, 2094–2100. [CrossRef]
15. Yasuda, T.; Uchiage, E.; Fujitani, T.; Tominaga, K.-I.; Nishida, M. Reverse water gas shift reaction using supported ionic liquid phase catalysts. *Appl. Catal. B Environ.* **2018**, *232*, 299–305. [CrossRef]
16. Wender, I. Reactions of synthesis gas. *Fuel Process. Technol.* **1996**, *48*, 189–297. [CrossRef]
17. ExxonMobil unveils tech for methanol-to-SAF. *Focus Catal.* **2022**, *2022*, 6. [CrossRef]
18. Kumar, A.; Mohammed, A.A.; Saad, M.A.; Al-Marri, M.J. Effect of nickel on combustion synthesized copper/fumed-SiO_2 catalyst for selective reduction of CO_2 to CO. *Int. J. Energy Res.* **2022**, *46*, 441–451. [CrossRef]
19. De Miranda, P.E.V. Chapter 5.3.3—Application of Hydrogen by Use of Chemical Reactions of Hydrogen and Carbon Dioxide. In *Science and Engineering of Hydrogen-Based Energy Technologies*; de Miranda, P.E.V., Ed.; Academic Press: Cambridge, MA, USA, 2019; pp. 279–289.
20. Stangeland, K.; Kalai, D.; Li, H.; Yu, Z. CO_2 methanation: The effect of catalysts and reaction conditions. *Energy Procedia* **2017**, *105*, 2022–2027. [CrossRef]
21. Nityashree, N.; Price, C.; Pastor-Perez, L.; Manohara, G.; Garcia, S.; Maroto-Valer, M.M.; Reina, T. Carbon stabilised saponite supported transition metal-alloy catalysts for chemical CO_2 utilisation via reverse water-gas shift reaction. *Appl. Catal. B Environ.* **2020**, *261*, 118241. [CrossRef]
22. Dias, Y.R.; Perez-Lopez, O.W. Carbon dioxide methanation over Ni-Cu/SiO_2 catalysts. *Energy Convers. Manage.* **2020**, *203*, 112214. [CrossRef]
23. Elsernagawy, O.Y.; Hoadley, A.; Patel, J.; Bhatelia, T.; Lim, S.; Haque, N. Thermo-economic analysis of reverse water-gas shift process with different temperatures for green methanol production as a hydrogen carrier. *J. CO2 Util.* **2020**, *41*, 101280. [CrossRef]
24. Kaiser, P.; Unde, R.B.; Kern, C.; Jess, A. Production of liquid hydrocarbons with CO_2 as carbon source based on reverse water-gas shift and Fischer-Tropsch synthesis. *Chem. Ing. Tech.* **2013**, *85*, 489–499. [CrossRef]
25. York, A.P.; Xiao, T.C.; Green, M.L.; Claridge, J.B. Methane oxyforming for synthesis gas production. *Catal. Rev.* **2007**, *49*, 511–560. [CrossRef]
26. Unde, R.B. *Kinetics and Reaction Engineering Aspects of Syngas Production by the Heterogeneously Catalysed Reverse Water Gas Shift Reaction*; Universitaet Bayreuth: Bayreuth, Germany, 2012.

27. Wu, H.; Chang, Y.; Wu, J.; Lin, J.; Lin, I.; Chen, C. Methanation of CO_2 and reverse water gas shift reactions on Ni/SiO_2 catalysts: The influence of particle size on selectivity and reaction pathway. *Catal. Sci. Technol.* **2015**, *5*, 4154–4163. [CrossRef]
28. Ebrahimi, P.; Kumar, A.; Khraisheh, M. Combustion synthesis of copper ceria solid solution for CO_2 conversion to CO via reverse water gas shift reaction. *Int. J. Hydrogen Energy* 2022, *in press*. [CrossRef]
29. Ammal, S.C.; Heyden, A. Origin of the unique activity of Pt/TiO_2 catalysts for the water–gas shift reaction. *J. Catal.* **2013**, *306*, 78–90. [CrossRef]
30. Ebrahimi, P.; Kumar, A.; Khraisheh, M. A review of recent advances in water-gas shift catalysis for hydrogen production. *Emergent Mater.* **2020**, *3*, 881–917. [CrossRef]
31. Boaro, M.; Colussi, S.; Trovarelli, A. Ceria-based materials in hydrogenation and reforming reactions for CO_2 valorization. *Front. Chem.* **2019**, *7*, 28. [CrossRef]
32. Ebrahimi, P.; Kumar, A.; Khraisheh, M. Analysis of combustion synthesis method for Cu/CeO_2 synthesis by integrating thermodynamics and design of experiments approach. *Results Eng.* **2022**, *15*, 100574. [CrossRef]
33. Ishito, N.; Hara, K.; Nakajima, K.; Fukuoka, A. Selective synthesis of carbon monoxide via formates in reverse water–gas shift reaction over alumina-supported gold catalyst. *J. Energy Chem.* **2016**, *25*, 306–310. [CrossRef]
34. Trovarelli, A. Catalytic properties of ceria and CeO_2-containing materials. *Catal. Rev.* **1996**, *38*, 439–520. [CrossRef]
35. Yeung, C.M.; Tsang, S.C. A study of co-precipitated bimetallic gold catalysts for water–gas shift reaction. *Catal. Commun.* **2008**, *9*, 1551–1557.
36. Chang, K.; Zhang, H.; Cheng, M.-J.; Lu, Q. Application of ceria in CO_2 conversion catalysis. *ACS Catal.* **2019**, *10*, 613–631. [CrossRef]
37. Castaño, M.G.; Reina, T.R.; Ivanova, S.; Centeno, M.; Odriozola, J.A. Pt vs. Au in water–gas shift reaction. *J. Catal.* **2014**, *314*, 1–9. [CrossRef]
38. Moreira, M.N.; Ribeiro, A.M.; Cunha, A.F.; Rodrigues, A.E.; Zabilskiy, M.; Djinović, P.; Pintar, A. Copper based materials for water-gas shift equilibrium displacement. *Appl. Catal. B Environ.* **2016**, *189*, 199–209. [CrossRef]
39. Porosoff, M.D.; Chen, J.G. Trends in the catalytic reduction of CO_2 by hydrogen over supported monometallic and bimetallic catalysts. *J. Catal.* **2013**, *301*, 30–37. [CrossRef]
40. Younis, A.; Chu, D.; Li, S. Cerium oxide nanostructures and their applications. *Funct. Nanomater.* **2016**, *3*, 53–68.
41. Yao, S.; Xu, W.; Johnston-Peck, A.; Zhao, F.; Liu, Z.; Luo, S.; Senanayake, S.; Martínez-Arias, A.; Liu, W.; Rodriguez, J. Morphological effects of the nanostructured ceria support on the activity and stability of CuO/CeO_2 catalysts for the water-gas shift reaction. *PCCP* **2014**, *16*, 17183–17195. [CrossRef]
42. Sayle, D.C.; Maicaneanu, S.A.; Watson, G.W. Atomistic models for CeO_2 (111), (110), and (100) nanoparticles, supported on yttrium-stabilized zirconia. *J. Am. Chem. Soc.* **2002**, *124*, 11429–11439. [CrossRef]
43. Jiang, Y.; Adams, J.B.; Van Schilfgaarde, M. Density-functional calculation of CeO_2 surfaces and prediction of effects of oxygen partial pressure and temperature on stabilities. *J. Chem. Phys.* **2005**, *123*, 064701. [CrossRef]
44. Chen, Y.; Hu, P.; Lee, M.-H.; Wang, H. Au on (1 1 1) and (1 1 0) surfaces of CeO_2: A density-functional theory study. *Surf. Sci.* **2008**, *602*, 1736–1741. [CrossRef]
45. Sayle, T.X.; Cantoni, M.; Bhatta, U.M.; Parker, S.C.; Hall, S.R.; Möbus, G.; Molinari, M.; Reid, D.; Seal, S.; Sayle, D.C. Strain and architecture-tuned reactivity in ceria nanostructures; enhanced catalytic oxidation of CO to CO_2. *Chem. Mater.* **2012**, *24*, 1811–1821. [CrossRef]
46. Liu, X.; Zhou, K.; Wang, L.; Wang, B.; Li, Y. Oxygen vacancy clusters promoting reducibility and activity of ceria nanorods. *J. Am. Chem. Soc.* **2009**, *131*, 3140–3141. [CrossRef] [PubMed]
47. Si, R.; Flytzani-Stephanopoulos, M. Shape and crystal-plane effects of nanoscale ceria on the activity of Au-CeO_2 catalysts for the water–gas shift reaction. *Angew. Chem.* **2008**, *120*, 2926–2929. [CrossRef]
48. Stubenrauch, J.; Brosha, E.; Vohs, J. Reaction of carboxylic acids on CeO_2 (111) and CeO_2 (100). *Catal. Today* **1996**, *28*, 431–441. [CrossRef]
49. Mei, D. First-principles characterization of formate and carboxyl adsorption on stoichiometric CeO2 (111) and CeO2 (110) surfaces. *J. Energy Chem.* **2013**, *22*, 524–532. [CrossRef]
50. Senanayake, S.D.; Mullins, D.R. Redox pathways for HCOOH decomposition over CeO_2 surfaces. *J. Phys. Chem. C* **2008**, *112*, 9744–9752. [CrossRef]
51. Zhou, G.; Dai, B.; Xie, H.; Zhang, G.; Xiong, K.; Zheng, X. CeCu composite catalyst for CO synthesis by reverse water–gas shift reaction: Effect of Ce/Cu mole ratio. *J. CO2 Util.* **2017**, *21*, 292–301. [CrossRef]
52. Reddy, B.M.; Thrimurthulu, G.; Katta, L. Design of efficient CexM1− xO2− δ (M= Zr, Hf, Tb and Pr) nanosized model solid solutions for CO oxidation. *Catal. Lett.* **2011**, *141*, 572–581. [CrossRef]
53. Lu, B.; Kawamoto, K. Preparation of mesoporous CeO_2 and monodispersed NiO particles in CeO_2, and enhanced selectivity of NiO/CeO2 for reverse water gas shift reaction. *Mater. Res. Bull.* **2014**, *53*, 70–78. [CrossRef]
54. Wang, L.; Liu, H.; Chen, Y.; Yang, S. Reverse water–gas shift reaction over co-precipitated Co–CeO_2 catalysts: Effect of Co content on selectivity and carbon formation. *Int. J. Hydrogen Energy* **2017**, *42*, 3682–3689. [CrossRef]
55. Aitbekova, A.; Wu, L.; Wrasman, C.J.; Boubnov, A.; Hoffman, A.S.; Goodman, E.D.; Bare, S.R.; Cargnello, M. Low-temperature restructuring of CeO_2-supported Ru nanoparticles determines selectivity in CO_2 catalytic reduction. *J. Am. Chem. Soc.* **2018**, *140*, 13736–13745. [CrossRef] [PubMed]

56. Cheng, Z.; Sherman, B.J.; Lo, C.S. Carbon dioxide activation and dissociation on ceria (110): A density functional theory study. *J. Chem. Phys.* **2013**, *138*, 014702. [CrossRef] [PubMed]
57. Lu, X.; Wang, W.; Wei, S.; Guo, C.; Shao, Y.; Zhang, M.; Deng, Z.; Zhu, H.; Guo, W. Initial reduction of CO_2 on perfect and O-defective CeO_2 (111) surfaces: Towards CO or COOH? *RSC Adv.* **2015**, *5*, 97528–97535. [CrossRef]
58. Wu, K.; Sun, L.D.; Yan, C.H. Recent Progress in Well-Controlled Synthesis of Ceria-Based Nanocatalysts towards Enhanced Catalytic Performance. *Adv. Energy Mater.* **2016**, *6*, 1600501. [CrossRef]
59. Trovarelli, A.; Llorca, J. Ceria catalysts at nanoscale: How do crystal shapes shape catalysis? *ACS Catal.* **2017**, *7*, 4716–4735. [CrossRef]
60. Ma, Y.; Gao, W.; Zhang, Z.; Zhang, S.; Tian, Z.; Liu, Y.; Ho, J.C.; Qu, Y. Regulating the surface of nanoceria and its applications in heterogeneous catalysis. *Surf. Sci. Rep.* **2018**, *73*, 1–36. [CrossRef]
61. Saeidi, S.; Najari, S.; Fazlollahi, F.; Nikoo, M.K.; Sefidkon, F.; Klemeš, J.J.; Baxter, L.L. Mechanisms and kinetics of CO_2 hydrogenation to value-added products: A detailed review on current status and future trends. *Renew. Sustain. Energy Rev.* **2017**, *80*, 1292–1311. [CrossRef]
62. Wang, W.; Wang, S.; Ma, X.; Gong, J. Recent advances in catalytic hydrogenation of carbon dioxide. *Chem. Soc. Rev.* **2011**, *40*, 3703–3727. [CrossRef]
63. Porosoff, M.D.; Yan, B.; Chen, J.G. Catalytic reduction of CO_2 by H_2 for synthesis of CO, methanol and hydrocarbons: Challenges and opportunities. *Energy Environ. Sci.* **2016**, *9*, 62–73. [CrossRef]
64. Kattel, S.; Liu, P.; Chen, J.G. Tuning selectivity of CO_2 hydrogenation reactions at the metal/oxide interface. *J. Am. Chem. Soc.* **2017**, *139*, 9739–9754. [CrossRef]
65. Su, X.; Yang, X.; Zhao, B.; Huang, Y. Designing of highly selective and high-temperature endurable RWGS heterogeneous catalysts: Recent advances and the future directions. *J. Energy Chem.* **2017**, *26*, 854–867. [CrossRef]
66. Trovarelli, A. *Catalysis by Ceria and Related Materials*; World Scientific: Singapore, 2002; Volume 2.
67. Montini, T.; Melchionna, M.; Monai, M.; Fornasiero, P. Fundamentals and catalytic applications of CeO_2-based materials. *Chem. Rev.* **2016**, *116*, 5987–6041. [CrossRef]
68. Xie, S.; Wang, Z.; Cheng, F.; Zhang, P.; Mai, W.; Tong, Y. Ceria and ceria-based nanostructured materials for photoenergy applications. *Nano Energy* **2017**, *34*, 313–337. [CrossRef]
69. Devaiah, D.; Reddy, L.H.; Park, S.-E.; Reddy, B.M. Ceria–zirconia mixed oxides: Synthetic methods and applications. *Catal. Rev.* **2018**, *60*, 177–277. [CrossRef]
70. Su, X.; Xu, J.; Liang, B.; Duan, H.; Hou, B.; Huang, Y. Catalytic carbon dioxide hydrogenation to methane: A review of recent studies. *J. Energy Chem.* **2016**, *25*, 553–565. [CrossRef]
71. Frontera, P.; Macario, A.; Ferraro, M.; Antonucci, P. Supported catalysts for CO_2 methanation: A review. *Catalysts* **2017**, *7*, 59. [CrossRef]
72. Zhang, M.; Zijlstra, B.; Filot, I.A.; Li, F.; Wang, H.; Li, J.; Hensen, E.J. A theoretical study of the reverse water-gas shift reaction on Ni (111) and Ni (311) surfaces. *Can. J. Chem. Eng.* **2020**, *98*, 740–748. [CrossRef]
73. Fornero, E.L.; Chiavassa, D.L.; Bonivardi, A.L.; Baltanás, M.A. Transient analysis of the reverse water gas shift reaction on Cu/ZrO_2 and $Ga_2O_3/Cu/ZrO_2$ catalysts. *J. CO2 Util.* **2017**, *22*, 289–298. [CrossRef]
74. Zhu, M.; Ge, Q.; Zhu, X. Catalytic reduction of CO_2 to CO via reverse water gas shift reaction: Recent advances in the design of active and selective supported metal catalysts. *Trans. Tianjin Univ.* **2020**, *26*, 172–187. [CrossRef]
75. Lin, W.; Stocker, K.M.; Schatz, G.C. Mechanisms of hydrogen-assisted CO_2 reduction on nickel. *J. Am. Chem. Soc.* **2017**, *139*, 4663–4666. [CrossRef]
76. González-Castaño, M.; Dorneanu, B.; Arellano-García, H. The reverse water gas shift reaction: A process systems engineering perspective. *React. Chem. Eng.* **2021**, *6*, 954–976.
77. Kim, K.-J.; Lee, Y.-L.; Na, H.-S.; Ahn, S.-Y.; Shim, J.-O.; Jeon, B.-H.; Roh, H.-S. Efficient waste to energy conversion based on CO-CeO_2 catalyzed water-gas shift reaction. *Catalysts* **2020**, *10*, 420.
78. Mathew, T.; Saju, S.; Raveendran, S.N. Survey of Heterogeneous Catalysts for the CO_2 Reduction to CO via Reverse Water Gas Shift. In *Engineering Solutions for CO_2 Conversion*; Reina, T.R., Arellano-Garcia, H., Odriozola, J.A., Eds.; Wiley: Hoboken, NJ, USA, 2021; Chapter 12; pp. 281–316.
79. Kim, S.S.; Lee, H.H.; Hong, S.C. A study on the effect of support's reducibility on the reverse water-gas shift reaction over Pt catalysts. *Appl. Catal. A Gen.* **2012**, *423*, 100–107.
80. Wang, L.-C.; Khazaneh, M.T.; Widmann, D.; Behm, R.J. TAP reactor studies of the oxidizing capability of CO_2 on a Au/CeO_2 catalyst—A first step toward identifying a redox mechanism in the Reverse Water–Gas Shift reaction. *J. Catal.* **2013**, *302*, 20–30.
81. Lin, L.; Yao, S.; Liu, Z.; Zhang, F.; Li, N.; Vovchok, D.; Martinez-Arias, A.; Castañeda, R.; Lin, J.; Senanayake, S.D. In situ characterization of Cu/CeO_2 nanocatalysts for CO_2 hydrogenation: Morphological effects of nanostructured ceria on the catalytic activity. *J. Phys. Chem. C* **2018**, *122*, 12934–12943.
82. Lee, S.M.; Eom, H.; Kim, S.S. A study on the effect of CeO_2 addition to a Pt/TiO_2 catalyst on the reverse water gas shift reaction. *Environ. Technol.* **2021**, *42*, 182–192.
83. Liu, Y.; Li, Z.; Xu, H.; Han, Y. Reverse water–gas shift reaction over ceria nanocube synthesized by hydrothermal method. *Catal. Commun.* **2016**, *76*, 1–6.

84. Goguet, A.; Meunier, F.C.; Tibiletti, D.; Breen, J.P.; Burch, R. Spectrokinetic investigation of reverse water-gas-shift reaction intermediates over a Pt/CeO_2 catalyst. *J. Phys. Chem. B* **2004**, *108*, 20240–20246.
85. Zhu, X.; Qu, X.; Li, X.; Liu, J.; Liu, J.; Zhu, B.; Shi, C. Selective reduction of carbon dioxide to carbon monoxide over Au/CeO_2 catalyst and identification of reaction intermediate. *Chin. J. Catal.* **2016**, *37*, 2053–2058.
86. Lu, B.; Zhang, T.; Zhang, L.; Xu, Y.; Zhang, Z.; Wu, F.; Li, X.; Luo, C. Promotion effects of oxygen vacancies on activity of Na-doped CeO_2 catalysts for reverse water gas shift reaction. *Appl. Surf. Sci.* **2022**, *587*, 152881.
87. Wang, M.; Shen, M.; Jin, X.; Tian, J.; Li, M.; Zhou, Y.; Zhang, L.; Li, Y.; Shi, J. Oxygen vacancy generation and stabilization in CeO_2−x by Cu introduction with improved CO_2 photocatalytic reduction activity. *ACS Catal.* **2019**, *9*, 4573–4581. [CrossRef]
88. Zhang, Y.; Liang, L.; Chen, Z.; Wen, J.; Zhong, W.; Zou, S.; Fu, M.; Chen, L.; Ye, D. Highly efficient Cu/CeO_2-hollow nanospheres catalyst for the reverse water-gas shift reaction: Investigation on the role of oxygen vacancies through in situ UV-Raman and DRIFTS. *Appl. Surf. Sci.* **2020**, *516*, 146035. [CrossRef]
89. Jacobs, G.; Davis, B.H. Reverse water-gas shift reaction: Steady state isotope switching study of the reverse water-gas shift reaction using in situ DRIFTS and a Pt/ceria catalyst. *Appl. Catal. A Gen.* **2005**, *284*, 31–38. [CrossRef]
90. Dai, H.; Zhang, A.; Xiong, S.; Xiao, X.; Zhou, C.; Pan, Y. The Catalytic Performance of $Ga_2O_3-CeO_2$ Composite Oxides over Reverse Water Gas Shift Reaction. *ChemCatChem* **2022**, *14*, e202200049. [CrossRef]
91. Li, M.; My Pham, T.H.; Ko, Y.; Zhao, K.; Zhong, L.; Luo, W.; Züttel, A. Support-Dependent Cu–In Bimetallic Catalysts for Tailoring the Activity of Reverse Water Gas Shift Reaction. *ACS Sustain. Chem. Eng.* **2022**, *10*, 1524–1535. [CrossRef]
92. Meunier, F.C.; Tibiletti, D.; Goguet, A.; Reid, D.; Burch, R. On the reactivity of carbonate species on a Pt/CeO_2 catalyst under various reaction atmospheres: Application of the isotopic exchange technique. *Appl. Catal. A Gen.* **2005**, *289*, 104–112. [CrossRef]
93. Kalamaras, C.M.; Americanou, S.; Efstathiou, A.M. "Redox" vs "associative formate with–OH group regeneration" WGS reaction mechanism on Pt/CeO_2: Effect of platinum particle size. *J. Catal.* **2011**, *279*, 287–300. [CrossRef]
94. Chen, X.; Su, X.; Liang, B.; Yang, X.; Ren, X.; Duan, H.; Huang, Y.; Zhang, T. Identification of relevant active sites and a mechanism study for reverse water gas shift reaction over Pt/CeO_2 catalysts. *J. Energy Chem.* **2016**, *25*, 1051–1057. [CrossRef]
95. Juarez, R.; Parker, S.F.; Concepcion, P.; Corma, A.; Garcia, H. Heterolytic and heterotopic dissociation of hydrogen on ceria-supported gold nanoparticles. Combined inelastic neutron scattering and FT-IR spectroscopic study on the nature and reactivity of surface hydrogen species. *Chem. Sci.* **2010**, *1*, 731–738. [CrossRef]
96. Wang, L.; Widmann, D.; Behm, R.J. Reactive removal of surface oxygen by H_2, CO and CO/H_2 on a Au/CeO_2 catalyst and its relevance to the preferential CO oxidation (PROX) and reverse water gas shift (RWGS) reaction. *Catal. Sci. Technol.* **2015**, *5*, 925–941. [CrossRef]
97. Shen, H.; Dong, Y.; Yang, S.; He, Y.; Wang, Q.; Cao, Y.; Wang, W.; Wang, T.; Zhang, Q.; Zhang, H. Identifying the roles of $Ce^{3+}-OH$ and Ce−H in the reverse water-gas shift reaction over highly active Ni-doped CeO_2 catalyst. *Nano Res.* **2022**, *15*, 5831–5841. [CrossRef]
98. Deng, B.; Song, H.; Peng, K.; Li, Q.; Ye, J. Metal-organic framework-derived Ga-Cu/CeO_2 catalyst for highly efficient photothermal catalytic CO_2 reduction. *Appl. Catal. B Environ.* **2021**, *298*, 120519. [CrossRef]
99. Zhao, B.; Pan, Y.-X.; Liu, C.-J. The promotion effect of CeO_2 on CO_2 adsorption and hydrogenation over Ga_2O_3. *Catal. Today* **2012**, *194*, 60–64. [CrossRef]
100. Yang, Z.; Zeng, M.; Wang, K.; Yue, X.; Chen, X.; Dai, W.; Fu, X. Visible light-assisted thermal catalytic reverse water gas reaction over Cu-CeO_2: The synergistic of hot electrons and oxygen vacancies induced by LSPR effect. *Fuel* **2022**, *315*, 123186. [CrossRef]
101. Wolf, A.; Jess, A.; Kern, C. Syngas Production via Reverse Water-Gas Shift Reaction over a Ni-Al_2O_3 Catalyst: Catalyst Stability, Reaction Kinetics, and Modeling. *Chem. Eng. Technol.* **2016**, *39*, 1040–1048. [CrossRef]
102. Pekridis, G.; Kalimeri, K.; Kaklidis, N.; Vakouftsi, E.; Iliopoulou, E.; Athanasiou, C.; Marnellos, G. Study of the reverse water gas shift (RWGS) reaction over Pt in a solid oxide fuel cell (SOFC) operating under open and closed-circuit conditions. *Catal. Today* **2007**, *127*, 337–346. [CrossRef]
103. Spencer, M. On the activation energies of the forward and reverse water-gas shift reaction. *Catal. Lett.* **1995**, *32*, 9–13. [CrossRef]
104. Ernst, K.-H.; Campbell, C.T.; Moretti, G. Kinetics of the reverse water-gas shift reaction over Cu (110). *J. Catal.* **1992**, *134*, 66–74. [CrossRef]
105. Ginés, M.; Marchi, A.; Apesteguía, C. Kinetic study of the reverse water-gas shift reaction over CuO/ZnO/Al_2O_3 catalysts. *Appl. Catal. A Gen.* **1997**, *154*, 155–171. [CrossRef]
106. Osaki, T.; Narita, N.; Horiuchi, T.; Sugiyama, T.; Masuda, H.; Suzuki, K. Kinetics of reverse water gas shift (RWGS) reaction on metal disulfide catalysts. *J. Mol. Catal. A Chem.* **1997**, *125*, 63–71. [CrossRef]
107. Poto, S.; van Berkel, D.V.; Gallucci, F.; d'Angelo, M.F.N. Kinetic modelling of the methanol synthesis from CO_2 and H_2 over a CuO/CeO_2/ZrO_2 catalyst: The role of CO_2 and CO hydrogenation. *Chem. Eng. J.* **2022**, *435*, 134946. [CrossRef]
108. Henkel, T. Modellierung Von Reaktion und Stofftransport in Geformten Katalysatoren am Beispiel der Methanolsynthese. Ph.D. Thesis, Technische Universität München, München, Germany, 2011.
109. Graaf, G.; Stamhuis, E.; Beenackers, A. Kinetics of low-pressure methanol synthesis. *Chem. Eng. Sci.* **1988**, *43*, 3185–3195. [CrossRef]
110. Lalinde, J.A.H.; Roongruangsree, P.; Ilsemann, J.; Baeumer, M.; Kopyscinski, J. CO_2 methanation and reverse water gas shift reaction. Kinetic study based on in situ spatially-resolved measurements. *Chem. Eng. J.* **2020**, *390*, 124629. [CrossRef]

111. Kopyscinski, J.; Schildhauer, T.J.; Vogel, F.; Biollaz, S.M.; Wokaun, A. Applying spatially resolved concentration and temperature measurements in a catalytic plate reactor for the kinetic study of CO methanation. *J. Catal.* **2010**, *271*, 262–279. [CrossRef]
112. Daza, Y.A.; Kuhn, J.N. CO_2 conversion by reverse water gas shift catalysis: Comparison of catalysts, mechanisms and their consequences for CO_2 conversion to liquid fuels. *RSC Adv.* **2016**, *6*, 49675–49691. [CrossRef]
113. Campbell, C.T.; Ernst, K.-H. *Forward and Reverse Water—Gas Shift Reactions on Model Copper Catalysts: Kinetics and Elementary Steps*; ACS Publications: Washington, DC, USA, 1992.
114. Kim, S.S.; Lee, H.H.; Hong, S.C. The effect of the morphological characteristics of TiO_2 supports on the reverse water–gas shift reaction over Pt/TiO_2 catalysts. *Appl. Catal. B Environ.* **2012**, *119*, 100–108. [CrossRef]
115. Jadhav, S.G.; Vaidya, P.D.; Bhanage, B.M.; Joshi, J.B. Kinetics of reverse water-gas shift reaction over Pt/Al_2O_3 catalyst. *Can. J. Chem. Eng.* **2016**, *94*, 101–106. [CrossRef]
116. Ghodoosi, F.; Khosravi-Nikou, M.R.; Shariati, A. Mathematical Modeling of Reverse Water-Gas Shift Reaction in a Fixed-Bed Reactor. *Chem. Eng. Technol.* **2017**, *40*, 598–607. [CrossRef]
117. Chen, C.S.; Wu, J.H.; Lai, T.W. Carbon dioxide hydrogenation on Cu nanoparticles. *J. Phys. Chem. C* **2010**, *114*, 15021–15028. [CrossRef]
118. Najari, S.; Gróf, G.; Saeidi, S.; Gallucci, F. Modeling and optimization of hydrogenation of CO_2: Estimation of kinetic parameters via Artificial Bee Colony (ABC) and Differential Evolution (DE) algorithms. *Int. J. Hydrogen Energy* **2019**, *44*, 4630–4649. [CrossRef]
119. Najari, S.; Saeidi, S.; Gróf, G.; Keil, F.J.; Rodrigues, A.E. Kinetic parameters estimation via dragonfly algorithm (DA) and comparison of cylindrical and spherical reactors performance for CO_2 hydrogenation to hydrocarbons. *Energy Convers. Manage.* **2020**, *226*, 113550. [CrossRef]
120. Brübach, L.; Hodonj, D.; Pfeifer, P. Kinetic Analysis of CO_2 Hydrogenation to Long-Chain Hydrocarbons on a Supported Iron Catalyst. *Ind. Eng. Chem. Res.* **2022**, *61*, 1644–1654. [CrossRef]
121. Luhui, W.; Hui, L.; Yuan, L.; Ying, C.; Shuqing, Y. Influence of preparation method on performance of Ni-CeO_2 catalysts for reverse water-gas shift reaction. *J. Rare Earths* **2013**, *31*, 559–564.
122. Li, Y.; Zhang, B.; Tang, X.; Xu, Y.; Shen, W. Hydrogen production from methane decomposition over Ni/CeO_2 catalysts. *Catal. Commun.* **2006**, *7*, 380–386. [CrossRef]
123. Yisup, N.; Cao, Y.; Feng, W.-L.; Dai, W.-L.; Fan, K.-N. Catalytic oxidation of methane over novel Ce–Ni–O mixed oxide catalysts prepared by oxalate gel-coprecipitation. *Catal. Lett.* **2005**, *99*, 207–213. [CrossRef]
124. Huang, Y.; Wang, A.; Li, L.; Wang, X.; Su, D.; Zhang, T. "Ir-in-ceria": A highly selective catalyst for preferential CO oxidation. *J. Catal.* **2008**, *255*, 144–152. [CrossRef]
125. Li, L.; Song, L.; Wang, H.; Chen, C.; She, Y.; Zhan, Y.; Lin, X.; Zheng, Q. Water-gas shift reaction over CuO/CeO_2 catalysts: Effect of CeO_2 supports previously prepared by precipitation with different precipitants. *Int. J. Hydrogen Energy* **2011**, *36*, 8839–8849. [CrossRef]
126. Yongyi, H.; Qibiao, L.; Yongzhao, W.; Yongxiang, Z. Selective catalytic dehydration of 1, 4-butanediol to 3-buten-1-ol over CeO_2 with different morphology. *Chin. J. Catal.* **2010**, *31*, 619–622.
127. Dong, X.-F.; Zou, H.-B.; Lin, W.-M. Effect of preparation conditions of CuO–CeO_2–ZrO_2 catalyst on CO removal from hydrogen-rich gas. *Int. J. Hydrogen Energy* **2006**, *31*, 2337–2344. [CrossRef]
128. Xue, L.; Zhang, C.; He, H.; Teraoka, Y. Promotion effect of residual K on the decomposition of N_2O over cobalt–cerium mixed oxide catalyst. *Catal. Today* **2007**, *126*, 449–455. [CrossRef]
129. Mei, Y.; Meisheng, C.; Zhang, N.; Zhiqi, L.; Huang, X. Characterization of CeO_2-ZrO_2 mixed oxides prepared by two different co-precipitation methods. *J. Rare Earths* **2013**, *31*, 251–256.
130. Luhui, W.; Hui, L.; Yuan, L.; Ying, C.; Shuqing, Y. Effect of precipitants on Ni-CeO_2 catalysts prepared by a co-precipitation method for the reverse water-gas shift reaction. *J. Rare Earths* **2013**, *31*, 969–974.
131. Dai, B.; Cao, S.; Xie, H.; Zhou, G.; Chen, S. Reduction of CO_2 to CO via reverse water-gas shift reaction over CeO_2 catalyst. *Korean J. Chem. Eng.* **2018**, *35*, 421–427. [CrossRef]
132. Ronda-Lloret, M.; Rico-Francés, S.; Sepúlveda-Escribano, A.; Ramos-Fernandez, E.V. CuOx/CeO_2 catalyst derived from metal organic framework for reverse water-gas shift reaction. *Appl. Catal. A: Gen.* **2018**, *562*, 28–36. [CrossRef]
133. Kim, T.K.; Lee, K.J.; Cheon, J.Y.; Lee, J.H.; Joo, S.H.; Moon, H.R. Nanoporous metal oxides with tunable and nanocrystalline frameworks via conversion of metal–organic frameworks. *J. Am. Chem. Soc.* **2013**, *135*, 8940–8946. [CrossRef]
134. Liu, L.; Fan, F.; Bai, M.; Xue, F.; Ma, X.; Jiang, Z.; Fang, T. Mechanistic study of methanol synthesis from CO_2 hydrogenation on Rh-doped Cu (111) surfaces. *Mol. Catal.* **2019**, *466*, 26–36. [CrossRef]
135. Lykaki, M.; Pachatouridou, E.; Carabineiro, S.A.; Iliopoulou, E.; Andriopoulou, C.; Kallithrakas-Kontos, N.; Boghosian, S.; Konsolakis, M. Ceria nanoparticles shape effects on the structural defects and surface chemistry: Implications in CO oxidation by Cu/CeO_2 catalysts. *Appl. Catal. B: Environ.* **2018**, *230*, 18–28. [CrossRef]
136. Zheng, X.; Li, Y.; Zhang, L.; Shen, L.; Xiao, Y.; Zhang, Y.; Au, C.; Jiang, L. Insight into the effect of morphology on catalytic performance of porous CeO_2 nanocrystals for H_2S selective oxidation. *Appl. Catal. B Environ.* **2019**, *252*, 98–110. [CrossRef]
137. Kovacevic, M.; Mojet, B.L.; van Ommen, J.G.; Lefferts, L. Effects of morphology of cerium oxide catalysts for reverse water gas shift reaction. *Catal. Lett.* **2016**, *146*, 770–777. [CrossRef]
138. Esch, F.; Fabris, S.; Zhou, L.; Montini, T.; Africh, C.; Fornasiero, P.; Comelli, G.; Rosei, R. Electron localization determines defect formation on ceria substrates. *Science* **2005**, *309*, 752–755. [CrossRef]

139. Skorodumova, N.; Baudin, M.; Hermansson, K. Surface properties of CeO_2 from first principles. *Phys. Rev. B* **2004**, *69*, 075401. [CrossRef]
140. Konsolakis, M.; Lykaki, M.; Stefa, S.; Carabineiro, S.A.; Varvoutis, G.; Papista, E.; Marnellos, G.E. CO_2 hydrogenation over nanoceria-supported transition metal catalysts: Role of ceria morphology (nanorods versus nanocubes) and active phase nature (Co versus Cu). *Nanomaterials* **2019**, *9*, 1739. [CrossRef] [PubMed]
141. Kim, M.-S.; Chung, S.-H.; Yoo, C.-J.; Lee, M.S.; Cho, I.-H.; Lee, D.-W.; Lee, K.-Y. Catalytic reduction of nitrate in water over Pd–Cu/TiO_2 catalyst: Effect of the strong metal-support interaction (SMSI) on the catalytic activity. *Appl. Catal. B Environ.* **2013**, *142*, 354–361. [CrossRef]
142. Liotta, L.; Longo, A.; Macaluso, A.; Martorana, A.; Pantaleo, G.; Venezia, A.; Deganello, G. Influence of the SMSI effect on the catalytic activity of a Pt (1%)/$Ce_{0.6}Zr_{0.4}O_2$ catalyst: SAXS, XRD, XPS and TPR investigations. *Appl. Catal. B Environ.* **2004**, *48*, 133–149. [CrossRef]
143. Bertella, F.; Concepción, P.; Martínez, A. The impact of support surface area on the SMSI decoration effect and catalytic performance for Fischer-Tropsch synthesis of Co-Ru/TiO_2-anatase catalysts. *Catal. Today* **2017**, *296*, 170–180. [CrossRef]
144. Tauster, S.; Fung, S.; Garten, R.L. Strong metal-support interactions. Group 8 noble metals supported on titanium dioxide. *J. Am. Chem. Soc.* **1978**, *100*, 170–175. [CrossRef]
145. Hosokawa, S.; Taniguchi, M.; Utani, K.; Kanai, H.; Imamura, S. Affinity order among noble metals and CeO_2. *Appl. Catal. A: Gen.* **2005**, *289*, 115–120. [CrossRef]
146. Ivanova, A.; Slavinskaya, E.; Gulyaev, R.; Zaikovskii, V.; Stonkus, O.; Danilova, I.; Plyasova, L.; Polukhina, I.; Boronin, A. Metal–support interactions in Pt/Al_2O_3 and Pd/Al_2O_3 catalysts for CO oxidation. *Appl. Catal. B Environ.* **2010**, *97*, 57–71. [CrossRef]
147. Strobel, R.; Pratsinis, S.E. Flame synthesis of supported platinum group metals for catalysis and sensors. *Platin. Met. Rev.* **2009**, *53*, 11–20. [CrossRef]
148. Zou, H.; Dong, X.; Lin, W. Selective CO oxidation in hydrogen-rich gas over CuO/CeO_2 catalysts. *Appl. Surf. Sci.* **2006**, *253*, 2893–2898. [CrossRef]
149. Zhou, G.; Xie, F.; Deng, L.; Zhang, G.; Xie, H. Supported mesoporous Cu/CeO_2-δ catalyst for CO_2 reverse water–gas shift reaction to syngas. *Int. J. Hydrogen Energy* **2020**, *45*, 11380–11393. [CrossRef]
150. Tauster, S. Strong metal-support interactions. *Acc. Chem. Res.* **1987**, *20*, 389–394. [CrossRef]
151. Matte, L.P.; Kilian, A.S.; Luza, L.; Alves, M.C.; Morais, J.; Baptista, D.L.; Dupont, J.; Bernardi, F. Influence of the CeO_2 support on the reduction properties of Cu/CeO_2 and Ni/CeO_2 nanoparticles. *J. Phys. Chem. C* **2015**, *119*, 26459–26470. [CrossRef]
152. Thill, A.S.; Kilian, A.S.; Bernardi, F. Key role played by metallic nanoparticles on the ceria reduction. *J. Phys. Chem. C* **2017**, *121*, 25323–25332. [CrossRef]
153. Figueiredo, W.T.; Escudero, C.; Perez-Dieste, V.; Ospina, C.A.; Bernardi, F. Determining the Surface Atomic Population of $Cu_x Ni_{1-x}/CeO_2$ ($0 < x \leq 1$) Nanoparticles during the Reverse Water–Gas Shift (RWGS) Reaction. *J. Phys. Chem. C* **2020**, *124*, 16868–16878.
154. Einakchi, R. Metal Nanoparticles Over Active Ionic-Conductive Supports for the Reverse Water Gas Shift Reaction. Master's Dissertation, Université d'Ottawa/University of Ottawa, Ottawa, ON, Canada, 2016.
155. Wang, L.; Liu, H. Mesoporous Co-CeO_2 catalyst prepared by colloidal solution combustion method for reverse water-gas shift reaction. *Catal. Today* **2018**, *316*, 155–161. [CrossRef]
156. Jurković, D.L.; Pohar, A.; Dasireddy, V.D.; Likozar, B. Effect of copper-based catalyst support on reverse water-gas shift reaction (RWGS) activity for CO_2 reduction. *Chem. Eng. Technol.* **2017**, *40*, 973–980. [CrossRef]
157. Bouarab, R.; Akdim, O.; Auroux, A.; Cherifi, O.; Mirodatos, C. Effect of MgO additive on catalytic properties of Co/SiO_2 in the dry reforming of methane. *Appl. Catal. A Gen.* **2004**, *264*, 161–168. [CrossRef]
158. Pino, L.; Vita, A.; Cipitì, F.; Laganà, M.; Recupero, V. Hydrogen production by methane tri-reforming process over Ni–ceria catalysts: Effect of La doping. *Appl. Catal. B Environ.* **2011**, *104*, 64–73. [CrossRef]
159. Chen, X.; Su, X.; Duan, H.; Liang, B.; Huang, Y.; Zhang, T. Catalytic performance of the Pt/TiO_2 catalysts in reverse water gas shift reaction: Controlled product selectivity and a mechanism study. *Catal. Today* **2017**, *281*, 312–318. [CrossRef]
160. Liu, L.; Corma, A. Metal catalysts for heterogeneous catalysis: From single atoms to nanoclusters and nanoparticles. *Chem. Rev.* **2018**, *118*, 4981–5079. [CrossRef] [PubMed]
161. Li, S.; Xu, Y.; Chen, Y.; Li, W.; Lin, L.; Li, M.; Deng, Y.; Wang, X.; Ge, B.; Yang, C. Tuning the selectivity of catalytic carbon dioxide hydrogenation over iridium/cerium oxide catalysts with a strong metal–support interaction. *Angew. Chem.* **2017**, *129*, 10901–10905. [CrossRef]
162. Chen, X.; Chen, Y.; Song, C.; Ji, P.; Wang, N.; Wang, W.; Cui, L. Recent advances in supported metal catalysts and oxide catalysts for the reverse water-gas shift reaction. *Front. Chem.* **2020**, *8*, 709. [CrossRef] [PubMed]
163. Zhao, Z.; Wang, M.; Ma, P.; Zheng, Y.; Chen, J.; Li, H.; Zhang, X.; Zheng, K.; Kuang, Q.; Xie, Z.-X. Atomically dispersed Pt/CeO_2 catalyst with superior CO selectivity in reverse water gas shift reaction. *Appl. Catal. B Environ.* **2021**, *291*, 120101. [CrossRef]
164. Wang, W.; Zhang, Y.; Wang, Z.; Yan, J.-M.; Ge, Q.; Liu, C.-J. Reverse water gas shift over In $2O_3$–CeO_2 catalysts. *Catal. Today* **2016**, *259*, 402–408. [CrossRef]

165. Ro, I.; Sener, C.; Stadelman, T.M.; Ball, M.R.; Venegas, J.M.; Burt, S.P.; Hermans, I.; Dumesic, J.A.; Huber, G.W. Measurement of intrinsic catalytic activity of Pt monometallic and Pt-MoOx interfacial sites over visible light enhanced PtMoOx/SiO_2 catalyst in reverse water gas shift reaction. *J. Catal.* **2016**, *344*, 784–794. [CrossRef]
166. Pettigrew, D.; Trimm, D.; Cant, N. The effects of rare earth oxides on the reverse water-gas shift reaction on palladium/alumina. *Catal. Lett.* **1994**, *28*, 313–319. [CrossRef]
167. Yang, L.; Pastor-Pérez, L.; Gu, S.; Sepúlveda-Escribano, A.; Reina, T. Highly efficient Ni/CeO_2-Al_2O_3 catalysts for CO_2 upgrading via reverse water-gas shift: Effect of selected transition metal promoters. *Appl. Catal. B Environ.* **2018**, *232*, 464–471. [CrossRef]
168. Dorner, R.W.; Hardy, D.R.; Williams, F.W.; Willauer, H.D. Effects of ceria-doping on a CO_2 hydrogenation iron–manganese catalyst. *Catal. Commun.* **2010**, *11*, 816–819. [CrossRef]
169. Galvita, V.V.; Poelman, H.; Bliznuk, V.; Detavernier, C.; Marin, G.B. CeO_2-modified Fe_2O_3 for CO_2 utilization via chemical looping. *Ind. Eng. Chem. Res.* **2013**, *52*, 8416–8426. [CrossRef]
170. Yan, B.; Zhao, B.; Kattel, S.; Wu, Q.; Yao, S.; Su, D.; Chen, J.G. Tuning CO_2 hydrogenation selectivity via metal-oxide interfacial sites. *J. Catal.* **2019**, *374*, 60–71. [CrossRef]
171. Rahmani, F.; Haghighi, M.; Estifaee, P. Synthesis and characterization of Pt/Al_2O_3–CeO_2 nanocatalyst used for toluene abatement from waste gas streams at low temperature: Conventional vs. plasma–ultrasound hybrid synthesis methods. *Microporous Mesoporous Mater.* **2014**, *185*, 213–223. [CrossRef]
172. Zonetti, P.C.; Letichevsky, S.; Gaspar, A.B.; Sousa-Aguiar, E.F.; Appel, L.G. The $Ni_xCe_{0.75}Zr_{0.25-x}O_2$ solid solution and the RWGS. *Appl. Catal. A Gen.* **2014**, *475*, 48–54. [CrossRef]
173. Wenzel, M.; Rihko-Struckmann, L.; Sundmacher, K. Continuous production of CO from CO_2 by RWGS chemical looping in fixed and fluidized bed reactors. *Chem. Eng. J.* **2018**, *336*, 278–296. [CrossRef]
174. Yang, L.; Pastor-Pérez, L.; Villora-Picó, J.J.; Gu, S.; Sepúlveda-Escribano, A.; Reina, T.R. CO_2 valorisation via reverse water-gas shift reaction using promoted Fe/CeO_2-Al_2O_3 catalysts: Showcasing the potential of advanced catalysts to explore new processes design. *Appl. Catal. A Gen.* **2020**, *593*, 117442. [CrossRef]
175. Ray, K.; Sengupta, S.; Deo, G. Reforming and cracking of CH_4 over Al_2O_3 supported Ni, Ni-Fe and Ni-Co catalysts. *Fuel Proces. Technol.* **2017**, *156*, 195–203. [CrossRef]
176. Theofanidis, S.A.; Galvita, V.V.; Poelman, H.; Marin, G.B. Enhanced carbon-resistant dry reforming Fe-Ni catalyst: Role of Fe. *Acs Catal.* **2015**, *5*, 3028–3039. [CrossRef]
177. Chen, L.; Wu, D.; Wang, C.; Ji, M.; Wu, Z. Study on Cu-Fe/CeO_2 bimetallic catalyst for reverse water gas shift reaction. *J. Environ. Chem. Eng.* **2021**, *9*, 105183. [CrossRef]
178. Goguet, A.; Meunier, F.; Breen, J.; Burch, R.; Petch, M.; Ghenciu, A.F. Study of the origin of the deactivation of a Pt/CeO_2 catalyst during reverse water gas shift (RWGS) reaction. *J. Catal.* **2004**, *226*, 382–392. [CrossRef]
179. Kattel, S.; Yu, W.; Yang, X.; Yan, B.; Huang, Y.; Wan, W.; Liu, P.; Chen, J.G. CO_2 Hydrogenation over Oxide-Supported PtCo Catalysts: The Role of the Oxide Support in Determining the Product Selectivity. *Angew. Chem. Int. Ed.* **2016**, *55*, 7968–7973. [CrossRef]
180. Panaritis, C.; Edake, M.; Couillard, M.; Einakchi, R.; Baranova, E.A. Insight towards the role of ceria-based supports for reverse water gas shift reaction over RuFe nanoparticles. *J. CO2 Util.* **2018**, *26*, 350–358. [CrossRef]
181. Le Saché, E.; Pastor-Perez, L.; Haycock, B.J.; Villora-Picó, J.J.; Sepulveda-Escribano, A.; Reina, T.R. Switchable catalysts for chemical CO_2 recycling: A step forward in the methanation and reverse water–Gas shift reactions. *ACS Sustain. Chem. Eng.* **2020**, *8*, 4614–4622. [CrossRef]
182. Choi, E.J.; Lee, Y.H.; Lee, D.-W.; Moon, D.-J.; Lee, K.-Y. Hydrogenation of CO_2 to methanol over Pd–Cu/CeO_2 catalysts. *Mol. Catal.* **2017**, *434*, 146–153. [CrossRef]
183. Porosoff, M.D.; Yang, X.; Boscoboinik, J.A.; Chen, J.G. Molybdenum carbide as alternative catalysts to precious metals for highly selective reduction of CO_2 to CO. *Angew. Chem. Int. Ed.* **2014**, *53*, 6705–6709. [CrossRef] [PubMed]
184. Yang, S.-C.; Pang, S.H.; Sulmonetti, T.P.; Su, W.-N.; Lee, J.-F.; Hwang, B.-J.; Jones, C.W. Synergy between ceria oxygen vacancies and Cu nanoparticles facilitates the catalytic conversion of CO_2 to CO under mild conditions. *ACS Catal.* **2018**, *8*, 12056–12066. [CrossRef]
185. Dai, B.; Zhou, G.; Ge, S.; Xie, H.; Jiao, Z.; Zhang, G.; Xiong, K. CO_2 reverse water-gas shift reaction on mesoporous M-CeO_2 catalysts. *Can. J. Chem. Eng.* **2017**, *95*, 634–642. [CrossRef]
186. Vovchok, D.; Zhang, C.; Hwang, S.; Jiao, L.; Zhang, F.; Liu, Z.; Senanayake, S.D.; Rodriguez, J.A. Deciphering dynamic structural and mechanistic complexity in Cu/CeO_2/ZSM-5 catalysts for the reverse water-gas shift reaction. *ACS Catal.* **2020**, *10*, 10216–10228. [CrossRef]
187. Wang, L.; Liu, H.; Chen, Y.; Zhang, R.; Yang, S. K-Promoted Co–CeO_2 Catalyst for the Reverse Water–Gas Shift Reaction. *Chem. Lett.* **2013**, *42*, 682–683. [CrossRef]
188. Luhui, W.; Zhang, S.; Yuan, L. Reverse water gas shift reaction over Co-precipitated Ni-CeO_2 catalysts. *J. Rare Earths* **2008**, *26*, 66–70.
189. Sun, F.-M.; Yan, C.-F.; Guo, C.-Q.; Huang, S.-L. Ni/Ce–Zr–O catalyst for high CO_2 conversion during reverse water gas shift reaction (RWGS). *Int. J. Hydrogen Energy* **2015**, *40*, 15985–15993. [CrossRef]
190. Belekar, R. Suppression of coke formation during reverse water-gas shift reaction for CO_2 conversion using highly active Ni/Al_2O_3-CeO_2 catalyst material. *Phys. Lett. A* **2021**, *395*, 127206. [CrossRef]

catalysts

Review

A Review of Biomass-Derived Heterogeneous Catalysts for Biodiesel Production

Somasundaram Chandra Kishore [1,†], Suguna Perumal [2,†], Raji Atchudan [3,*,†], Ashok K. Sundramoorthy [4,†], Muthulakshmi Alagan [5,†], Sambasivam Sangaraju [6] and Yong Rok Lee [3,*]

1 Department of Biomedical Engineering, Saveetha School of Engineering, Saveetha Institute of Medical and Technical Sciences (SIMATS), Saveetha University, Chennai 602105, India
2 Department of Chemistry, Sejong University, Seoul 143747, Republic of Korea
3 School of Chemical Engineering, Yeungnam University, Gyeongsan 38541, Republic of Korea
4 Department of Prosthodontics, Saveetha Dental College and Hospitals, Saveetha Institute of Medical and Technical Sciences, Poonamallee High Road, Velappanchavadi, Chennai 600077, India
5 Center for Environmental Management Laboratory, National Institute of Technical Teachers Training and Research, Chennai 600113, India
6 National Water and Energy Center, United Arab Emirates University, Al Ain 15551, United Arab Emirates
* Correspondence: atchudanr@yu.ac.kr (R.A.); yrlee@yu.ac.kr (Y.R.L.)
† These authors contributed equally to this work.

Abstract: The scientific community is being forced to consider alternative renewable fuels such as biodiesel as a result of the sharp increases in the price of petroleum and the increased demand for petroleum-derived products. Transesterification is a technique used to create biodiesel where a variety of edible oils, non-edible oils, and animal fats are used. For this, either a homogeneous or heterogeneous catalyst is utilized. An appropriate catalyst is chosen based on the quantity of free fatty acid content in the oil. The main distinction between homogeneous and heterogeneous catalysts is that compared to the heterogeneous catalyst, the homogeneous catalyst is not affected by the quantity of free fatty acids in the oil. Early methods of producing biodiesel relied on homogeneous catalysts, which have drawbacks such as high flammability, toxicity, corrosion, byproducts such as soap and glycerol, and high wastewater output. The majority of these issues are solved by heterogeneous catalysts. Recent innovations use novel heterogeneous catalysts that are obtained from biomass and biowaste resources. Numerous researchers have documented the use of biomass-derived heterogeneous catalysts in the production of high-quality, pure biodiesel as a potentially greener manufacturing method. The catalysts were significantly altered through conventional physical processes that were both cost- and energy-effective. The present review is intended to analyze catalysts from biowaste for making biodiesel at a minimal cost. The most recent methods for creating diverse kinds of catalysts—including acidic, basic, bifunctional, and nanocatalysts—from various chemicals and biomass are highlighted in this review. Additionally, the effects of various catalyst preparation methods on biodiesel yield are thoroughly explored.

Keywords: biomass; heterogeneous catalyst; biodiesel

1. Introduction

In recent years, nanomaterials with different architectures and supports have been playing a key role in industrial catalysis [1]. Because of worries about the destructive effects of fossil fuel usage, such as climate change, global warming, and their detrimental effects on human health, biomass is now the most popular renewable energy source, and its utilization is growing [2,3]. In reality, biomass has enormous potential for use as a source of renewable energy, with the ability to produce biofuels for heat, power, and transportation. Rising bioenergy endorsement as part of agricultural policies has resulted in a different and more fruitful utilization of land and has greatly aided in the advancement of economic possibilities for rural areas, as well as a ramp-up in farmer income, a reduction in rural–urban migration, and environmental and cultural preservation. Recent technological advancements in biomass exploitation have led to the development of promising methods for efficiently generating heat, power, and liquid biofuels out of these widely available and affordable natural resources [4]. A blend of monoalkyl esters made from renewable bioresources known as biodiesel is of particular relevance since it is quickly becoming a significant transportation fuel. Due to its lower pollution levels, biodiesel has been promoted as a cleaner alternative to gasoline and diesel (primarily in terms of CO_2 emissions). A multitude of air pollutants, including particulate matter, sulfur, hydrocarbons, and carbon monoxide, can be considerably reduced by using biodiesel. The advantages of biodiesel for the environment are driving up production. However, production costs are still quite high compared to diesel fuel made from petroleum. With the addition of a solid heterogeneous catalyst, biodiesel production could lower its cost and become economically competitive with diesel. As a result, recent significant research efforts have been made to identify the appropriate catalysts.

Biomass is any organic material that originates in some way from the photosynthetic process. The definition of biomass varies because of the variability of the components, the purpose, and the place of origin. However, in a broader sense, biomass is a collection of organically derived materials, including those from plants such as crops, trees, shrubs, and algae as well as any other materials made of an organic matrix, with the exception of plastics made from petrochemical and fossil resources [5]. The most significant biomass sources include sewage [6], algae [7], aquatic crops [8], agricultural and forestry leftovers [9], animal residues [10], and waste products from the wood-processing sector [11]. They also fall under the biomass classification particularly if they cannot be reused in later processing such as municipal solid waste and waste discharges from human activities. Biomass has a remarkably diversified chemical makeup. For instance, grains and livestock waste are mostly made up of starch, whereas residues of plant origin are primarily made up of cellulose, hemicellulose, and lignin in different amounts. Obviously, various chemical characteristics come from diverse chemical structures. Figure 1 depicts the schematic diagram for the synthesis of biodiesel from biomass.

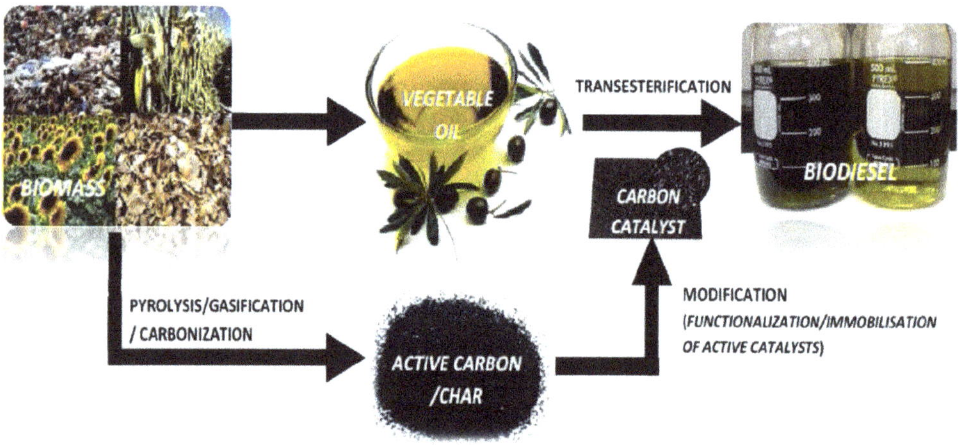

Figure 1. Schematic diagram for the synthesis of biodiesel by transesterification. Reprinted with permission from Ref. [12]. 2022, Elsevier.

Fatty acid methyl esters (FAME) are the main component of the renewable transportation fuel known as biodiesel [13,14]. These FAME are normally prepared by the transesterification of vegetable and animal fats. The demand for biodiesel is rising both domestically and internationally. Concerns over greenhouse gas (GHG) emissions and climate change, a need for renewable/sustainable energy sources, and a desire to build domestic and more reliable fuel supplies are the main drivers behind this. Although the terms "first generation" and "second generation" biofuels are used often, they have no legal or regulatory significance [15,16]. The phrase "1st Generation" often refers to biofuels made from readily accessible, edible feedstocks utilizing proven conversion techniques. The majority of current biofuels are categorized as first generation. This comprises biodiesel created by the transesterification of triglycerides present in animal fat and vegetable oils and ethanol made through the fermentation of sugars with its source in maize, sugar cane, or sorghum, etc. The phrase "2nd Generation" can be used to describe biofuels made from cutting-edge, non-food feedstocks or via the use of modern processing technologies. The non-edible triglycerides and lignocellulose are two examples of advanced feedstocks that include algae and jatropha. Catalytic hydroprocessing of triglycerides to create sustainable fuel and thermal conversion such as pyrolysis and gasification of lignocellulose are two examples of contemporary processing technology [17,18].

Direct usage and mixing of raw oils [19], microemulsion [20], thermal cracking [21], pyrolysis [22], and transesterification [23] are some of the current processes for making biodiesel. Transesterification is by far the simplest and most economical process for producing biodiesel. Transesterification, also known as alcoholysis, is the process by which fats or oils react with an alcohol in the presence of a catalyst to produce glycerol and fatty acid methyl ester [24]. We can utilize methanol, ethanol, butanol, and amyl alcohol, among other forms of alcohol. Nevertheless, methanol is frequently employed since it is less expensive, has a short chain, reacts rapidly, and disperses readily in the reaction media [25]. Although many different FAME species can be found in biodiesel fuel made by transesterification of triglycerides, most fuels are typically dominated by a limited number of species such as lauric acid, Myristic acid, myristoleic acid, palmitic acid, palmitoleic acid, stearic acid, oleic acid, linoleic acid, linolenic acid, arachidic acid, gondoic acid, behenic acid, and erucic acid [26]. The most efficient technique for producing biodiesel is thought to be enzymatic transesterification. However, the highly expensive cost of the catalyst and the slower reaction rate prevent its wider use [27]. Figure 2 illustrates the mechanism for the formation of biodiesel. To produce three moles of fatty acid methyl ester and one mole of glycerol, the process needs three moles of methanol and one mole of triglyceride. Monoglycerides and diglycerides are produced as intermediates in three sequential reversible processes that result from this. Following the reaction, the glycerol is separated by settling or centrifugation, and then it is purified for various applications. Before being employed as a diesel fuel, the methyl ester fraction is purified.

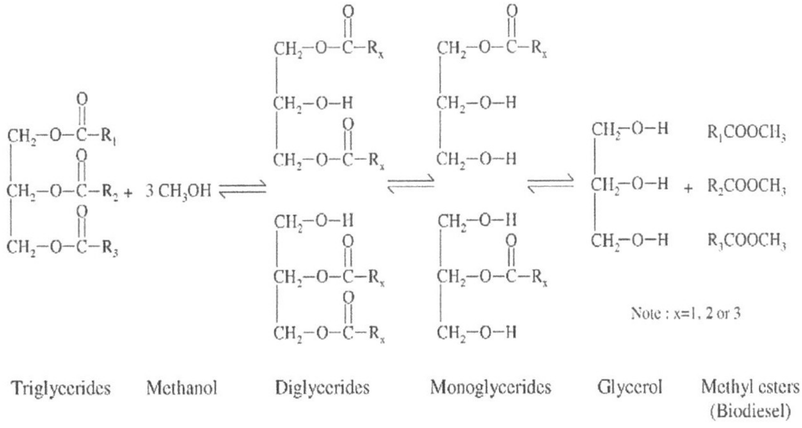

Figure 2. Formation of FAME from triglyceride [28].

1.1. Homogeneous Catalysis

At the moment, homogenous base catalysts are often used on an industrial basis to produce biodiesel [29]. The benefits of a homogeneous base catalyst, such as sodium hydroxide (NaOH) and potassium hydroxide (KOH), include high catalytic activity, faster reaction times, moderate working conditions, and readily accessible and inexpensive raw materials. Free fatty acids (FFA) and water can greatly affect the homogeneous base catalyst process [30]. Additionally, the production of soaps as a byproduct of neutralization and saponification would hinder the extraction and purification process, create a significant amount of wastewater, and increase operating costs [31]. This requirement renders this catalyst unfavorable to the environment.

For feedstock with a high FFA concentration, such as used cooking oil, unrefined vegetable oils, and animal fats, homogenous acid catalysts such as H_2SO_4, HCl, and H_3PO_4 are appropriate. The lack of sensitivity to the existence of FFA and water, the capacity to catalyze both transesterification and esterification processes, and the absence of soap byproduct production are some of its main benefits. The main obstacle to the widespread use of this catalyst, however, is its delayed response time. According to reports, base-catalyzed transesterification has a conversion rate that is around 4000 times slower than acid-catalyzed transesterification [32]. Homogeneous acid catalysts are also extremely acidic and reactive by nature. In a homogenous operation, product isolation and purification requires a number of stages, generates a lot of effluents, and increases operational costs. The recovery and recycling of homogeneous catalysts are challenging, impractical, costly, and need additional processing [33].

1.2. Heterogeneous Catalysis

In recent years, research in using solid or heterogeneous catalysts in the synthesis of biodiesel has significantly grown [34]. The phases or states of heterogeneous catalysts differ from those of reactants. These are just the kinds of catalysts that usually generate active sites during a reaction with their reactants [35]. Due to the fact that it takes place in a three-phase system with a solid heterogeneous catalyst and two immiscible liquid phases such as methanol and oil, the heterogeneously catalyzed methanolysis process is exceedingly complicated. Along with methanolysis, other processes including the saponification of glycerides and methyl esters and the catalyst-assisted neutralization of FFAs also take place. The main drawbacks of this catalysis comprise higher oil/alcohol ratios and higher temperatures than in homogeneous catalysis. It is simpler to extract the catalysts from the product afterward since they are not either destroyed or decomposed in the reaction mixture. Additionally, since the recovered catalyst may be utilized again in the process, less catalyst is needed overall, reducing the expenses. The heterogeneous-based operation has a number of advantages, such as being noncorrosive, ease of separation, and having extended catalyst endurance [36,37]. Examples of heterogeneous catalysts including alkaline earth metals such as MgO and SrO are commonly employed. The most popular application of CaO is as a transesterification catalyst. As a heterogeneous base catalyst, Al is employed in the form of oxides such as Al_2O_3, which are supported by CaO and MgO. Numerous naturally occurring calcium sources, including eggshell, mollusk shell, and bones, are frequently employed as the starting materials for the manufacture of catalysts because they can both reduce wastes and yield catalysts with a high level of cost competitiveness. Several catalysts are on sale for basic-catalyzed reactions, including metal oxide, mixed oxide, and hydrotalcite [38]. On the contrary side, catalysts for acidic activity include various forms of carbon, zeolites, ion-exchange resin, and transition metal oxides [39]. Nevertheless, the three-phase system's existence in a heterogeneous environment will cause a diffusion issue that will hinder the reaction [40]. The mass transfer performance is constrained by three highly immiscible phases such as solid catalyst, alcohol, and oil, which slow the pace of reaction. These solid catalysts generated from non-renewable resources and ecologically unfriendly sources suffer from a smaller number of active sites, microporosity, leaching, being poisonous, and costly [41]. Therefore, the catalyst must possess a larger pore diameter and greater specific surface area that support the hydrophobicity and external catalytic sites in them to make a superior solid acid catalyst.

1.3. Enzymatic Catalysis

A lipase-catalyzed approach has been devised and has been the subject of substantial research over the past 10 years to reduce the issues related to the use of acid and/or alkali for the synthesis of biodiesel [42]. The enzyme-based biodiesel process has the following benefits: (1) a more straightforward manufacturing process; (2) less energy consumption; (3) higher purity of the byproduct glycerol; (4) no soap development in the system; and (5) simple separation and recycling of immobilized enzymes [43]. The enzymatic pathway for the manufacturing of biodiesel is now viewed as an ecofriendly substitute that is increasingly feasible as new, more effective, and less expensive enzymes are now being evolved. However, the enzyme catalyst has some limitations, primarily linked to lower reaction rates, increased prices, and loss of activity or enzyme inhibition. The enzymatic catalyst may be regenerated to reduce enzyme costs, which improves the effectiveness of recovering biodiesel. Additionally, genetic engineering and the identification of high alcohol-tolerant microorganisms can be used to improve enzyme catalytic activity [44]. Introducing enzymes that can catalyze both the esterification of FFA and the transesterification of TGs will make it much easier to overcome the significant challenge posed by the existence of FFA in the precursor material that prevents transesterification using conventional techniques. Waste oils and fats include FFA that can be totally turned into biodiesel. In reality, as the FFA are immediately esterified enzymatically into FAME, the enzyme-catalyzed transesterification is more appropriate for usage on FFA-rich feedstocks including waste oils, greases, beef tallow, and lard [27].

1.4. Non-Catalyzed Process

Non-catalytic approaches have been examined for the production of biodiesel in order to overcome challenges related to traditional catalyzed (trans)esterification. Saka and Kusdiana, who explored the synthesis of biodiesel from rapeseed oil using methanol under the supercritical state in 2001, discovered non-catalytic (trans)esterification [45]. If reactions are conducted at temperatures of 239 °C and pressures 8.09 MPa above the critical point of methanol, as demonstrated in the reactions they have conducted, large yields may be reached in a short amount of time even without the use of a catalyst. Molecules have high kinetic energy and density under supercritical conditions of the alkyl donor, increasing chemical reactivity. Protic solvents have lower dielectric constants because they easily break hydrogen bonds, making them miscible with oils and fats. The authors also listed additional advantages of supercritical methanol transesterification, including easier product purification and an environmentally beneficial method. The severity of the processing conditions affecting operational safety and energy needs is the key issue with this strategy. According to the literature that is currently accessible, this technique of catalyst-free alcoholysis reactions at high temperature and pressure settings allows for strong phase solubility and less mass-transfer constraints. Since the reaction rate is drastically accelerated in the supercritical state, the reaction is guaranteed to be completed in a shorter amount of time while requiring less laborious segregation and purification processes [46].

Additionally, the presence of alcohol, which performs a double function as a reactant and a catalyst, amplifies the absence of a catalyst. Moreover, compared to the standard alkali-catalyzed process, the supercritical method (SCM) has been claimed to be more tolerant of the accompanying water and free fatty acids in the feedstock [47]. Consequently, the method is considerably more successful against different kinds of vegetable oils, which has increased interest in research considering the application of this method in an unconventional way.

2. Common Methodology for the Synthesis of Heterogeneous Catalyst

2.1. Hydrothermal

With the advantage of the economical synthesis method and ease of optimization of time and temperature, hydrothermal synthesis is an excellent method for producing bulk metal catalysts with high purity. Teo et al. studied the effectiveness of a catalyst for transesterifying crude Jatropha curcas oil (JCO), which has a high fatty acid content and was evaluated in the current study. At various synthesis times, calcium methoxide ($Ca(OCH_3)_2$) has been produced effectively using a cost-effective hydrothermal technique [48]. CaO was utilized as a precursor because it is widely available, reasonably priced, and ecologically safe. On the surface of CaO, $Ca(OCH_3)_2$ can form, and its active basic surface is exceptionally developed. Three different morphologies of $Ca(OCH_3)_2$ catalysts were observed, including (a) spherical particles with non-uniform sizes, (b) well-organization plate-like structures with irregular surfaces, and (c) a collection of small plate-like designs with even surfaces. The relationship between the biodiesel yield and the catalysts' synthesis time, surface area, and shape was investigated. At ideal experimental parameters of 2 h reaction time, $Ca(OCH_3)_2$ was able to sustain 86% of FAME content after a 5th cycle. The optimized condition was the temperature (65 °C), 1 molar ratio (methanol: oil), and a catalyst loading (2 wt.%). For the transesterification process of non-edible Jatropha curcas oil for biodiesel synthesis, $Ca(OCH_3)_2$ was proved to be the better solid heterogeneous catalyst.

Using acid and basic catalysts, the transesterification of triglycerides from waste oilseed fruits with methanol has been explored in heterogeneous and homogeneous systems. Macario and his co-workers performed hydrothermal synthesis to synthesize the acid catalysts MCM-41 and ITQ-6 [49]. Ionic exchange was used to load potassium onto various materials to create acid-base catalysts (obtaining K-MCM-41, K-ITQ-6), as shown in Figure 3. After 24 h of reaction at 180 °C, K-ITQ-6 catalysts produced the maximum levels of triglycerides conversion and biodiesel output. This catalyst is deactivated by potassium leaching, yet it is possible and simple to regenerate and reuse it. Soltani and his co-workers prepared a highly recyclable mesoporous SO_3H-GO@TiO_2 catalyst that has both Lewis and Brönsted acid sites for one-pot catalyzing the esterification of cheap acidic feedstocks, such as palm fatty acid distillate (PFAD), which contains a significant amount of water (3 wt.%) and FFAs > 85 wt.% [50]. The mesoporous sulfonated GO@TiO_2 catalyst was synthesized utilizing a two-step hydrothermal microwave-assisted technique that included post-sulfonation treatment and homogenous TiO_2 nanoparticle dispersion inside the GO mesopore framework [50]. After examining the effects of several GO:Ti ratios on textural characteristics, the ratio of 1:1 was chosen as the best ratio. The outcomes showed that the catalyst had high acidity with a mix of Brönsted and Lewis acid sites, as well as favorable textural qualities. Additionally, the post-sulfonation treatment had no effect on the crystalline structure of the as-prepared mesoporous GO@TiO_2 material. Furthermore, ten repeated esterification processes using the synthesized mesoporous SO_3H-GO@TiO_2 catalyst demonstrated good stability without further interventions.

An alternate method for creating activated carbon in the nanometer range from biomass with a large surface area is hydrothermal carbonization (HTC). In their work, RF Abdullah et al. synthesized empty fruit bunch-based activated carbon (EFBHAC) using the HTC method [51]. K_2CO_3 and $Cu(NO_3)_2$ were added to the activated carbon to create a bifunctional nanocatalyst for the tandem esterification and transesterification of waste cooking oil (WCO). As a consequence of hydrolysis, dehydration, decarboxylation, aromatization, and recondensation during the HTC process, the generated EFBHAC had a BET surface area (4056.17 m^2 g^{-1}), pore volume (0.827 cm^3 g^{-1}), and pore diameter (5.42 nm). The maximum biodiesel yield of 97.1% was obtained by impregnating K_2CO_3 and $Cu(NO_3)_2$, over EFBHAC, which resulted in high levels of basicity (9.21 mmol g^{-1}) and acidity (31.41 mmol g^{-1}). The optimized conditions were 5 wt.% catalyst loading, 12:1 methanol-to-oil molar ratio, and 70 °C for 2 h. After the fifth cycle, more than 80% of biodiesel was generated, demonstrating its high reusability. These findings demonstrate the outstanding potential of waste materials to make bifunctional nanocatalysts that can increase biodiesel output and have commercial viability.

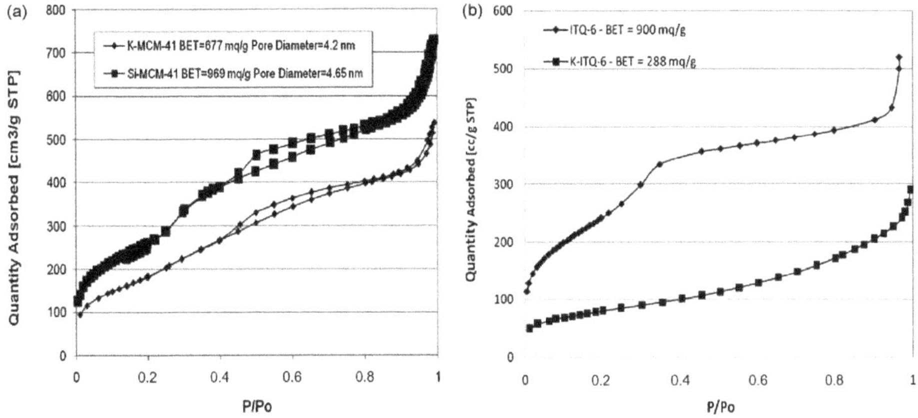

Figure 3. N_2 adsorption isotherms of catalyst (**a**) before and (**b**) after ionic exchange. Reprinted with permission from Ref. [49]. 2022, Elsevier.

In order to produce biodiesel from the transesterification of soybean oil with methanol, Mo-KIT-6 catalyst precursors produced by direct hydrothermal synthesis utilizing various Si/Mo molar ratios (10, 20, and 30) were reported by Cardoso and his group [52]. The factorial design revealed that employing the 10 Mo-KIT-6 catalysts and an alcohol/oil ratio of 20/1 at 150 °C for 3 h are the ideal conditions for optimizing biodiesel production. However, when employing the 20 Mo-KIT-6 catalysts with a 15:1 alcohol-to-oil ratio, the biodiesel output approaches its maximal potential while also benefiting from the use of less methanol, which reduces the energy required for the separation of unreacted alcohol.

2.2. Carbonization

Based on the report by Atelge, in his work, a heterogeneous catalyst made of activated carbon derived from used coffee grounds was produced utilizing KOH as an activator by the carbonization method [53]. The synthesized catalyst was characterized and assessed in terms of biodiesel and hydrogen generation efficiency, with the maximum iodine number being achieved at 600 °C carbonization temperature. Three-weight percent catalyst loading, a 9:1 methanol-to-waste cooking oil molar ratio, a reaction temperature of 90 °C, and a reaction duration of 120 min were found to be the ideal transesterification reaction conditions. The yield of biodiesel under ideal reaction conditions was 91.57%. In order to verify the catalyst's capacity to be reused, the selected superior catalyst was successfully utilized five times. For both the transesterification and the dehydrogenation processes, the catalyst performance was nearly identical to that of the initial cycle following regeneration. Furthermore, 19.15 and 15.48 kJ/mol were estimated as the activation energies of the transesterification and methanolysis processes, respectively.

Through a transesterification process, Fernández et al. produced biodiesel using biochar and coffee husk ashes as heterogeneous catalysts [54]. The raw coffee husks were carbonized, activated, and charred to produce the catalysts. The findings demonstrated the existence of several inorganic compounds combined with the turbostratic structure, primarily K- and Ca-containing phases (in the case of the biochar samples). The synthesized biochar sample at 700 °C demonstrates a biodiesel conversion of 66%; the activated biochar catalyst prepared at the same temperature exhibited a higher biodiesel conversion (74%), which explains its outstanding specific surface area. Coffee husk ashes had the highest catalytic effectiveness (biodiesel conversion of 93%), which is consistent with their greater Ca and K salt levels when compared to biochar samples.

For the formation of biodiesel, incomplete carbonization of shrimp shell with loading KF on the resulting material and activation at a specified temperature were used by Yang and his co-workers to create a high-performance and ecologically friendly shrimp shell catalyst [55]. The transesterification of rapeseed oil with methanol was used to study the catalytic performance. The outcomes showed that the ideal preparation conditions were ideal for carbonization at 450 °C, loading KF at 25 wt.%, and activation at 250 °C. The chemical reaction using shrimp shell as the catalyst was conducted at 65 °C with a 2.5 wt.% of catalyst loading, 9:1 of a methanol/rapeseed oil molar ratio, and 3 h reaction duration to achieve the desired conversion. The catalyst shrimp shell had a porous framework morphology, and the active sites created by the interaction of the shrimp shell that had not fully carbonized with KF during the activation phase were what gave it its catalytic activity for transesterification. It was discovered that the shrimp shell has significant catalytic activity and environmentally favorable qualities, offering the possibility of using it as a heterogeneous base catalyst in the biodiesel synthesis method.

2.3. Calcination

Jayakumar et al. performed a calcination method to develop a low-cost, pollution-less, and very efficient heterogeneous catalytic synthesis for the conversion of Jatropha curcas oil (JCO) into biodiesel using waste animal (ox) bone [56]. The calcined ox bone catalyst analysis clearly demonstrated that there was a significant amount of catalytic activity for the production of biodiesel. According to the reaction time (1–4 h) and temperature, biodiesel yields at a 500 rpm agitation speed, a 5% catalyst loading rate (w/w) of oil, and a 12:1 methanol-oil ratio (molar) were monitored (313–338 °C). It was discovered that 338 °C was the ideal temperature for achieving the highest (96.82%) biodiesel output. With a frequency factor of 7.03×10^6 h^{-1}, the activation needed 38.55 kJ mol^{-1} of energy (Ea). The experiments on reusability showed that the calcined animal bone catalyst was very stable for up to three cycles with FAME > 90%, which was considerably reduced to 61% in the fourth cycle. The findings of this analysis revealed the potential for using JCO and calcined ox bone catalyst as inexpensive and easily accessible unwanted waste materials that may be utilized as feedstock for the market-scale production of biodiesel to meet the potential needs of the community.

The potential of using discarded rohu fish (Labeo rohita) scale as a high-performance, recyclable, and inexpensive heterogeneous catalyst for the synthesis of biodiesel from soybean oil is examined in the research performed by Chakraborty et al. [57]. X-ray diffraction and thermogravimetric analysis showed that calcining hydroxyapatite over 900 °C for two hours might convert a sizeable amount of β-tri-calcium phosphate. A fibrous layer with a porous structure was seen in SEM morphological investigations of the calcined scale, and a BET surface area of 39 m^2/g was determined. The ideal parametric parameters were 6.27:1 (methanol/oil molar ratio), 997.42 °C (the calcination temperature), and the catalyst concentration (1.01 wt.% of oil), leading to the best FAME production of 97.73%, which was found using response surface methodology. Reusability tests depict that the developed catalyst might be used up to six more times, opening up a possible route for the production of biodiesel.

It was reported by Nair et al. that waste frying oil can be converted into biodiesel utilizing a heterogeneous catalyst made of powdered calcined clamshell (Mereterix mereterix) [58]. Clamshells calcined for 2.5 and 3.5 h at 1173 °C showed varying catalytic activity. A catalyst with increased activity and a quicker transesterification reaction time was produced after 3.5 h of calcination. According to XRF analysis, the catalyst is mostly composed of calcium (97%) as well as some minor elements such as Na, Si, Fe, Sr, Al, Mn, and S. Figure 4 represents the XRD spectrum of the catalyst before and after the reaction. The calcined clamshell's surface area was determined to be 2.6 m^2 g^{-1}, and 3.0 g of catalyst and 6.03:1 (methanol to oil molar ratio) at 333 °C in 3 h were determined to be the best values for key variables. On the basis of variable optimization, high yields (>89%) and conversions (>97%) were attained.

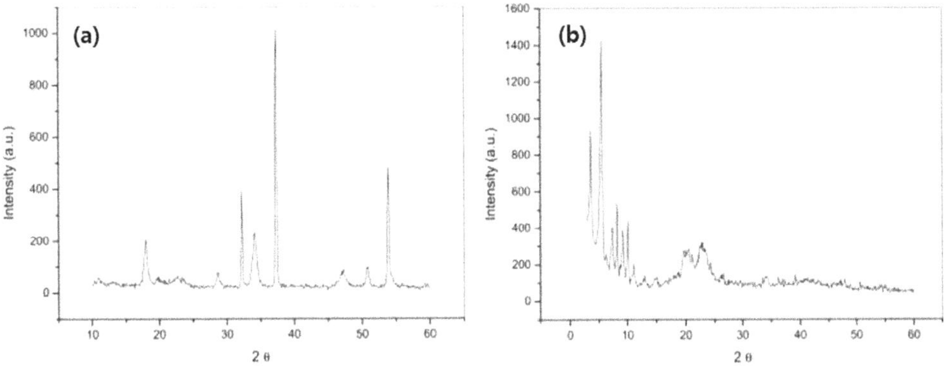

Figure 4. XRD of clamshell (**a**) calcined at 1173 K for 3.5 h and (**b**) after use. Reprinted with permission from Ref. [58]. 2022, Elsevier.

2.4. Pyrolysis

According to the experiment performed by Li and his co-researchers, the catalyst used was rice husk that had been pyrolyzed and sulfonated using strong sulfuric acid [59]. The catalyst was then applied to catalyze the transesterification and esterification of WCO to produce biodiesel. The conversion of FFAs to FAME in the presence of the catalyst achieved 98.17% after 3 h and 87.57% after 15 h. In comparison, the standard solid acid catalyst Amberlyst-15 only produced FFA conversion and FAME yields of 95.25 and 45.17%, respectively. As a result, the produced catalyst displayed a high catalytic activity for both transesterification and esterification to occur simultaneously.

The transformation of potato peel as a possible source of bio-oil and green heterogeneous catalysts has been reported by Daimary and his co-workers using pyrolysis [60]. The thermochemical conversion's biochar waste was recycled as a clean, sustainable, and recyclable source for heterogeneous catalysts. The alkali and alkaline-rich biochar was reclaimed through calcination and converted into bio-based mixed metal oxide and carbonates. Reasonably high bio-char (29.50%) was observed to be generated at 500 °C, which was also the optimal operating temperature for pyrolysis. The high potassium concentration (36.54%) in both the oxide and carbonate forms of the produced catalyst contributed to its high catalytic activity. The optimal parameters of temperature, methanol-to-oil ratio, duration, and catalyst loading were used to achieve the highest oil conversion (97.50%) possible. Utilizing an integrated process for catalyst synthesis, food waste was converted into high-value products including biochar and bio-oil.

In another work by Bhatia and the team, WCO was subjected to transesterification using a heterogeneous catalyst made by a pyrolyzing waste cork (Quercus suber), as shown in Figure 5 [61]. The findings of the experiment reveal that the optimum FAME conversion (98%) was achieved at alcohol:oil ratios of 25:1, catalyst loadings of 1.5% w/v, and temperatures of 65 °C for heterogeneous catalysts produced at 600 °C. WCO is used to generate biodiesel, which is mainly made up of FAME in the ascending order C18:1, C18:2, C16:0, C18:0, C20:0. The cetane number (CN) 50.56, higher heating value (HHV) 39.5, kinematic viscosity 3.9, and density 0.87 were determined as the properties of the generated biodiesel.

Hazm et al. described the anaerobic pyrolysis of waste polyethylene terephthalate food containers to produce plastic waste char (PWC), which was then transformed into an acidic solid catalyst for the biodiesel production process [62]. Such a strategy may enable the development of biofuels while also offering a potential answer to the PW environmental problem. Here, plastic waste was carbonized at 600 °C to produce a carbon precursor, which was then subjected to three sulfonation ratios of sulfuric acid (1:10, 1:15, and 1:20), resulting in a series of solid acid sulfonated carbon catalysts, PWC-SO_3H (a), (b), and (c). The PWC-SO_3H catalysts that were created were thermally stable up to 375 °C. A mesoporous textural morphology with non-uniform dispersion of the sulfonate group was discovered via surface morphology screening. The esterification of PFAD with methanol then assessed the effectiveness of the catalysts in producing biodiesel. The best conversion of PFAD-derived biodiesel (96.9%) was produced by the PWC-SO_3H (b) catalyst with an impregnation ratio (1:15) under the ideal experimental parameters of 5 wt.% catalysts at 110 °C for 2 h and an 18:1 PFAD molar ratio with methanol. According to research on recyclability, the PWC-SO3H (b) catalyst may be reused for four subsequent reactions while still having a high level of catalytic activity.

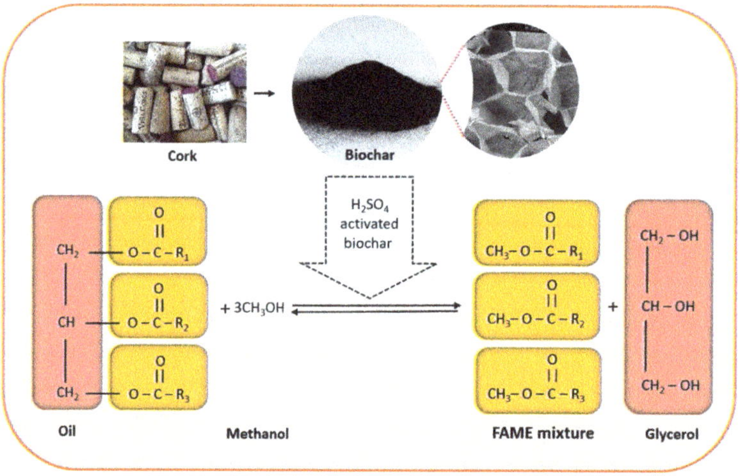

Figure 5. Pictorial representation of transesterification of waste cooking oil by using a heterogeneous catalyst prepared by pyrolysis of waste cork (Quercus suber). Reprinted with permission from Ref. [61]. 2022, Elsevier.

2.5. Impregnation

The investigation performed by Mahesh et al. proposed wet impregnation to synthesize a heterogeneous catalyst (KBr/CaO) using conventional calcium oxide and potassium bromide [63]. Tests on this solid catalyst for transesterification of used cooking oil were conducted. Techniques such as FTIR, XRD, and SEM were used to analyze the produced catalyst. Regression models revealed that the ideal conditions were ideal for a 12:1 methanol:oil ratio, a 3 wt.% catalyst loading, and a 1.8 h reaction time.

The work performed by Alsharif et al. concentrated on developing a heterogeneous catalyst for the transesterification process by wet impregnating lithium onto TiO_2 [64]. With varying quantities of Li (20, 30, 40 wt.%) and at varying calcination temperatures, a series of Li/TiO_2 was produced (450, 600, 750). The most effective catalysts, with a transesterification yield of 98%, were determined to be the Li/TiO_2 catalysts with 30% w/w Li and calcined at 600 °C. The highest catalyst performance was obtained using a methanol-to-oil ratio of 24:1, a 5 wt.% catalyst loading, and reaction temperatures of 55 °C for three hours. According to the kinetic studies, the transesterification process was acceptable with the zero-order hypothesis. However, with each subsequent usage, the reusability lowers.

Wang et al. reported the synthesis of biodiesel from soybean oil, very stable heterogeneous catalysts synthesized by the method of impregnation where the carbonates' suspension solutions were over the electric furnace dust (EFD) [65]. The single impregnation with K_2CO_3 suspension solution can only enhance its catalytic performance in the initial phase, whereas in the process of impregnating with various suspension solutions of $CdCO_3$, $BaCO_3$, $CaCO_3$, and $SrCO_3$ and being calcined at high temperature did not enhance the catalytic performance of the EFD powder. A kinetic study using Na_2CO_3@EFD was conducted, and it was noticed that the values 32.81 kJ/mol (activation energy) and 3760.85 /min (frequency factor) were both lesser than those for the solid base catalysts that have been previously reported. This suggested that the as-prepared Na_2CO_3@EFD catalyst was a potential candidate for industrial use in biomass transformation to energy.

2.6. Sol-Gel

Using the sol-gel process, Shimada and his group created four catalysts that altered the molar ratios [66]. They produced two zinc aluminates (Al:Zn = 1:0.1 and Al:Zn = 1:1) and two potassium aluminates (Al:K = 1:0.1 and Al:K = 1:1). The heterogeneous catalyst underwent one transesterification and five recycling processes while displaying an amorphous structure. Comparable to homogeneous catalysts, the reaction time was less than that of heterogeneous catalysts. Because the catalyst may be recycled, this may lower the expenses associated with producing biodiesel. With strong catalytic activity (97.58% ester concentration and 92.94% yield) and effective biodiesel production, the novel catalyst also generated wastewater and glycerin with less residue.

Salinas and his co-workers synthesized La_2O_3-ZrO_2-mixed oxides with different La_2O_3 contents (0, 1, 2, 3, and 5 wt.%) and calcined them at 600 °C using the sol-gel technique [67]. The synthesis of biodiesel from canola oil through a transesterification process served as the benchmark for catalytic activity. According to the characterization results, the development of La_2O_3 monolayers and the degree of m-ZrO_2 basicity significantly impact the generation of biodiesel. The 3% La_2O_3-ZrO_2 catalyst produced a higher conversion of biodiesel (56% at 4 h) in the presence of basic sites and the creation of a monolayer of La_2O_3.

Pure αFe_2O_3 and $\alpha Fe_2O_{3\,1-x}/ZnO_x$ nanoparticles were synthesized by Maleki et al. using the sol-gel method [68]. The sonochemical technique was used to examine the ability of pure αFe_2O_3 and $\alpha Fe_2O_{3\,1-x}/ZnO_x$ nanoparticles to produce biodiesel from canola oil. x = 0, 40, and 80 weight percent. The dispersion of ZnO is well integrated into the structure of Fe_2O_3, improving the basicity of the nanocatalyst by producing oxygen vacancy, according to the CO_2-TPD and Raman data. The catalytic performance of ZnO was optimally included when x = 47.24 wt.%. In addition, the optimized nanocatalyst had dimensions of 40 nm and a form of a needle or hexagon. The reaction was optimized, and the highest biodiesel output was 94.21%. However, after seven successive cycles of reusing the best catalysts, findings indicate that Fe_2O_3/ZnO nanoparticles performed better catalytically and were more stable than pure Fe_2O_3. The findings of the engine test show that canola biodiesel's high oxygen content significantly improves fuel combustion conditions, especially in B30 (30% biodiesel/70% diesel).

2.7. Co-Solvent Preparation

In this research, Pham et al. examined the optimal conditions for biodiesel synthesis with 10 wt.% isopropanol (IPA) as a co-solvent and the transesterification kinetics at various IPA concentrations [69] The transesterification process was assigned in a heterogeneous state, which includes both FAME and glycerol [GL] phases in the IPA quantity of 5 to 25 wt.% as the GL phase developed and isolated from the FAME phase. The correlation coefficient (r) between the rate constant and IPA amount was 0.97 ($p < 0.05$), indicating that the transesterification rate increased as IPA quantity increased. The following were the ideal transesterification parameters for producing biodiesel from canola oil with 10 wt.% IPA: 1.0 wt.% KOH, a 6:1 molar ratio of MeOH to oil, and a 30 °C reaction temperature. Phase separation following the reaction was able to be facilitated, and waste was decreased by transesterification using 10% IPA. Since IPA recovered at a rate of >96% during the FAME step, IPA was able to be recycled. These findings showed that IPA was a more effective co-solvent for producing biodiesel from feedstocks derived from plant oils.

A homogeneous reaction procedure employing acetone as a co-solvent for the transesterification of vegetable oils with methanol and the inclusion of a potassium hydroxide catalyst have been proposed by Okitsu and his team [70]. The impacts of the primary factors, such as the quantities of acetone and KOH, the molar ratio of methanol to oil, and the reaction temperature, on FAME production were examined. The ideal parameters were 25 wt.% acetone and 1 wt.% KOH catalyst in oil, 4.5 wt.% methanol in oil, and a reaction temperature of 25 °C. After 30 min, the transformation of vegetable oil to FAME under these circumstances approached 98%.

The research by Akkarawatkhoosith et al. focuses on a quick and efficient method for concurrently raising gasoline quality and biodiesel synthesis. To achieve this, co-solvent technology was used [71]. The transesterification of ethanol and palm oil in a microtube reactor at supercritical conditions served as the basis for experiments. The isopropanol was added to the process and served as a co-solvent to improve the homogeneity of the mixture as well as a source of reactant to synthesize isopropyl esters. Considering reaction temperature, pressure, residence time, and the ethanol-to-oil molar ratio, large biodiesel yields were produced after the addition of isopropanol at milder parameters than those described in the literature. Biodiesel quality was improved, especially the cloud and pour points, which were superior to those produced using more traditional methods.

3. Biodiesel Production Using Heterogeneous Catalysts

A number of feedstocks were used to produce biodiesel. The choice of feedstock is crucial to the production of biodiesel since it affects a variety of variables, including chemical composition, yield, and biodiesel quality. The availability, pricing, amount of oil, and physical and chemical characteristics of the feedstocks can all be taken into consideration. Both edible and non-edible vegetable oils are used extensively in the production of biodiesel. Vegetable oils' primary component, triglyceride, controls the chemical and physical properties of these feedstocks, as well as the quality of biodiesel. Figure 6 represents the flowchart for the production of biodiesel using the heterogeneous catalyst.

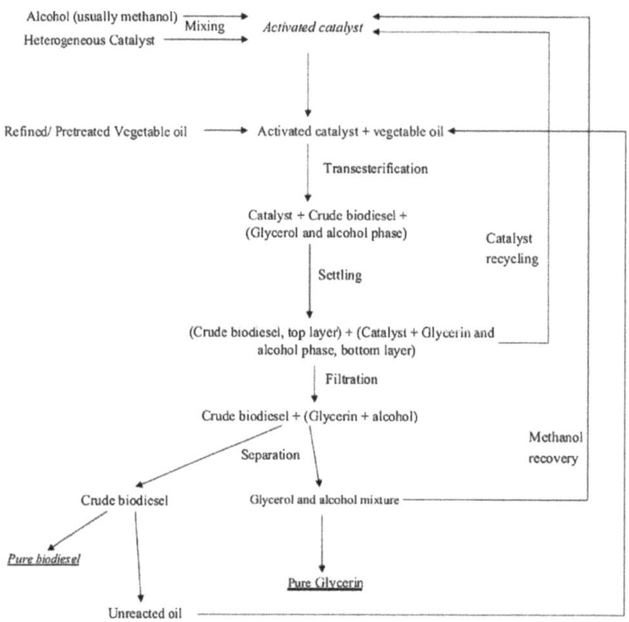

Figure 6. Flowchart of biodiesel synthesis using heterogeneous catalyst. Reprinted with permission from Ref. [12]. 2022, Elsevier.

3.1. Edible Plant Oils

Because edible oils are used in the manufacturing of biodiesel and the food business, two significant industries, there is a growing need and demand for them. Numerous edible plant oils, including coconut, rapeseed, palm, canola, peanut, olive, mustard, soybean, and sunflower, have been used in the production of biodiesel. As a result of these types of feedstocks' low FFA and water content, as well as their simple conversion method, first-generation biodiesel are produced using them. However, rising production costs and questions about food versus fuel are the primary drawbacks of edible feedstocks that have prompted many researchers to look for more dependable alternatives. Crops containing edible oil require growing agricultural areas and are weather-dependent. Additionally, moving these commodities from their growing site to the plant location may raise the overall cost of making biodiesel. Therefore, producing biodiesel from food oils is not worthwhile and indicates the need for greater study into using less expensive and more friendly feedstocks.

The study by Bambase et al. looked at the usage of OH-impregnated CaO as a heterogeneous catalyst in the conversion of refined coconut oil into crude biodiesel [72]. The effects of NaOH content (10–20% (*w*/*v*) solution), calcination temperature (550–700 °C), and calcination duration were investigated using the wet impregnation technique (2–5.5 h). Using OH-impregnated CaO as a catalyst, refined coconut oil was transesterified with methanol at 60 °C to produce biodiesel. The results indicate that when 20% (*w*/*v*) NaOH solution was employed during impregnation with calcination at 600 °C for 2 h, a 66.36% conversion to biodiesel could be accomplished in 10 min. Tetrahydrofuran was added as a co-solvent, which further boosted conversion to 81.70%. This is equivalent to the 85.98% conversion rate attained when employing NaOH as a catalyst under identical reaction conditions.

The comparative physicochemical and catalytic analyses of metal oxide MgO catalysts in a transesterification process using rapeseed are presented by Szkudlarek and his group [73]. It was thoroughly assessed how the Si/Al ratio in the catalytic material affected their catalytic characteristics in the process under study. Additionally, the impact of the zeolite's ZSM-5 form on the catalytic reactivity of catalysts made of magnesium oxide was looked at. The impregnation approach was used to create a number of MgO/ZSM-5 catalysts in order to accomplish the major objectives of this study. The MgO catalyst based on ZSM-5 with the greatest silica-to-alumina ratio had the best activity in the investigated process, and 10% MgO/ZSM-5 (Si/Al = 280), which demonstrated the greatest value of higher fatty acid methyl esters (94.6%) and a high yield of triglyceride conversion (92.9%), was the best active catalyst combination in the transesterification process. The alkalinity, methanol-specific sorption characteristics, and large specific surface area of this system relative to the other MgO-based catalysts under investigation all contribute to its high activity.

Wild olive (Olea oleaster) plant oil was selected as a potential feedstock for effective biodiesel synthesis by Islam and his co-workers [74]. Various concentrations of 12-tungstophosphonic acid (TPA) were added to mixed oxides of aluminium and chromium to produce biodiesel from inexpensive local wild olive oil. Transesterification utilizing a Cr-Al mixed oxide catalyst coated with TPA allowed the biodiesel process to be completed in a single step. The optimal parameters of the oil/methanol molar ratio of 1:21, reaction temperature of 80 °C, catalyst dose of 4 wt.%, reaction periods of 5 h, and stirring speed of 600 rpm led to the highest 93% biodiesel production. The reusability tests, which were interesting, revealed that the synthesized catalyst maintained outstanding stability and reusability for up to five runs in the biodiesel synthesis from wild olive oil. Additionally, the synthetic biodiesel complies with all applicable international criteria.

3.2. Non-Edible Plant Oils

Because they include some harmful elements, non-edible resource oils are not fit for human consumption. Therefore, it is anticipated that the use of non-edible vegetable oils will assist biodiesel to be more cost-competitive than the use of food-grade vegetable oils. Non-edible vegetable oils provide a number of benefits over edible ones when used as diesel fuel, including mobility, quick availability, renewability, a greater rate of combustion, lower sulfur, and aromatic content, as well as a better rate of biodegradability. However, a significant disadvantage is that the majority of non-edible oils have high levels of free fatty acids (FFAs), which raises the price of producing biodiesel. Vegetable oil contains a variety of fatty acids, depending on the plant's species and environmental factors that affect its growth. Non-edible oilseed crops come in a variety of forms, including the Jatropha tree (Jatropha curcas), Karanja (Pongamia pinnata), mahua (Madhuca indica), castor bean seed (Ricinus communis), neem (Azadirachta indica), rubber seed tree (Hevea brasiliensis), tobacco seed (Nicotiana tabacum), and rice bran, etc. Over the past few years, a great deal of research has been done on the manufacture of biodiesel from various non-edible oilseed crops.

By hydrothermally treating Jatropha oil in an autoclave reactor with TiO_2/CaO base and SO_4/TiO_2 acid catalysts, biodiesel from Jatropha oil has been produced by Pratika and their team [75]. Utilizing H_2SO_4, the acid catalyst was synthesized by sulfating TiO_2. By loading TiO_2 on CaO, the basic catalyst was prepared. This study produced the greatest basicity and acidity catalysts from TiO_2/CaO and SO_4/TiO_2 catalysts, respectively. They were used for the esterification of Jatropha oil to lower its free fatty acid content and the transesterification of esterified Jatropha oil to produce biodiesel. The ST-2-400 catalyst among the SO_4/TiO_2 catalysts had an ideal total acidity of 6.93 mmol NH_3/g. With a catalyst weight of 3%, an oil-to-methanol ratio of 1:15, and a reaction period of 90 min, this catalyst was capable of lowering the FFA level of Jatropha oil from 1.22% to 0.59%. The TiO_2/CaO catalyst with the highest total basicity, TC-20-800, had a total basicity of 28.67 mmol HCl/g. Using 3% of TC-20-800 catalyst and 65 °C for 90 min, Jatropha oil was converted into biodiesel by esterification. Approximately 79.68% of biodiesel could be produced with the TC-20-800 catalyst.

The goal of the investigation carried out by Pandiangan et al. was to learn more about the best alcohol for converting rubber seed oil into biodiesel [76]. In this investigation, three distinct kinds of short-chain alcohols, including methanol, ethanol, and 2-propanol, were used for the transesterification of rubber seed oil. Each alcohol was utilized in a 3:1 ratio with the oil, and transesterification was performed at 70 °C for 6 h with a 10% MgO/SiO_2 catalyst present. Gas chromatography–mass spectrometry (GC-MS) examination of the transesterification products was used to verify the conversion of fatty acids in the oil into esters. The findings indicated that methanol, ethanol, and 2-propanol are the alcohols that are most reactive, with percentages of oil conversion to products of 90.1, 73.3, and 63.2%, respectively. These findings suggest that methanol is the best alcohol for transesterifying rubber seed oil.

The study by Khan and his team describes the synthesis and use of praseodymium-doped mussel shell-based CaO catalysts for the transesterification process that produces biodiesel using castor oil [77]. Wet chemistry was successfully employed to create Pr-CaO catalysts, which were then used as a stable, effective, and reasonably priced heterogeneous catalyst for the transesterification of castor oil and methanol to produce ecologically friendly and green biodiesel. In contrast to the yield of FAME produced by undoped, calcined CaO, which was roughly 80%, a FAME yield of 87.42% was obtained using a 7% Pr-CaO mixed oxide catalyst under ideal reaction conditions of 2.5 wt.% catalysts, an oil/methanol ratio of 8:1, and a temperature of 65 °C.

3.3. Waste Cooking Oil

Rezania et al. looked at the possibility of using $LaPO_4$ foam as a heterogeneous catalyst to convert highly acidic WCO into biodiesel [78]. Under the study's circumstances, a 5:1 methanol-to-oil molar ratio and a reaction temperature of 90 °C resulted in a 91% FAME yield in 120 min using 2.5 wt.% $LaPO_4$. The outcomes showed that $LaPO_4$ foam might function well as a heterogeneous catalyst with outstanding catalytic efficiency (>90%), facile isolation, and great stability. The research was performed by Mohadesi et al. to find out if WCO could be converted to biodiesel using a clay/CaO heterogeneous catalyst [79]. Clay was used for creating the catalyst, which was then calcined using calcium oxide. Natural soil components such as clay include a lot of amorphous silica. High surface area silica with an amorphous structure was created by processing calcium oxide and heating at 800 °C. The response surface approach assessed the impact of five parameters—reaction temperature, catalyst concentration, oil-to-methanol volume ratio, toluene concentration, and reaction time—on the biodiesel's purity (RSM). The conversion rate was 97.16% under ideal circumstances, which included a temperature of 54.97 °C, a catalyst concentration of 9.6 wt.%, an oil-to-methanol volume ratio of 1.94 vol:vol, a toluene concentration of 16.13 wt.%, and a reaction duration of 74.32 min. The catalyst preparation may be utilized as a reliable and affordable catalyst for the manufacture of biodiesel because the catalyst recovery test results revealed that it could be utilized up to five times.

The technique for converting WCO to biodiesel by transesterification was examined by Kataria and their team [80]. For the best biodiesel generation, the base-catalyzed transesterification under various reactant ratios—such as the molar ratio of alcohol to oil and the mass ratio of catalyst to oil—was examined. The ideal ratio and amount of zinc-doped calcium oxide were found to be 12:1 and 5 wt.%, respectively, for base-catalyzed transesterification of WCO. The calorific value, flash point, and density of the generated biodiesel were evaluated and contrasted with those of regular diesel.

3.4. Animal Fats

Animal fat wastes (AFWs), a non-edible feedstock, have recently gained wide acceptance as an alternative to vegetable oils in biodiesel production. They are inexpensive, preserve the environment, and improve the grade of the resulting biodiesel fuel (low NO_x emissions, high cetane number, and oxidative stability). AFWs are an excellent feedstock for making biodiesel as a consequence.

Mutreja et al. reported MgO impregnated with KOH as a heterogeneous catalyst for the transesterification of mutton fat with methanol [81]. With 4 wt.% of MgO-KOH-201 (MgO impregnated with 20 wt.% of KOH), the mutton fat (fat) with methanol (1:22 M ratios) at 65 °C demonstrated > 98% conversion to biodiesel in 20 min and the ^{1}H NMR spectrum of mutton fat and biodiesel. The quantity of KOH impregnation (5–20 wt.%), the amount of catalyst (1.5–4 wt.%, catalyst/fat), the reaction temperature (45–65 °C), the ratio of fat to methanol (1:11 to 1:22), and the impact of adding oleic acid, water, or palmitic acid (1 wt.%) were the reaction conditions that were optimized.

In order to perform the transesterification of pig fat into biodiesel, an alumina-doped calcium oxide nanocomposite was reported as a heterogeneous catalyst by Cherian's group [82]. Waste egg shells that had been doped with alumina were used to create calcium oxide. The methanol:oil molar ratio, catalyst concentration, reaction duration, and temperature were changed to optimize the generation of biodiesel. The 8% (w/w) (catalyst loading), 9:1 (methanol-to-oil ratio), 65 °C (reaction temperature) at 90 min of reaction period was found to be optimal. They resulted in the highest biodiesel synthesis of 92.5%.

It was reported by Seffati et al. that by employing chicken fat and a $CaO/CuFe_2O_4$ nanocatalyst (shown in Figure 7) during the transesterification process, biodiesel was formed [83]. The impact of several factors, including the methanol-to-oil ratio (4:1–24:1), reaction duration (0.5–6 h), reaction temperature (50–80 °C), and catalyst concentration (0.5–7%), was carefully examined in order to find the biodiesel manufacturing process with the maximum output. The outcome showed that a 15:1 methanol-to-oil ratio, 4 h reaction duration, reaction temperature of 70 °C, and a catalyst concentration of 3%, which was 94.52%, produced the maximum biodiesel yield. The optimally generated biodiesel was then blended with different percentages of gasoline.

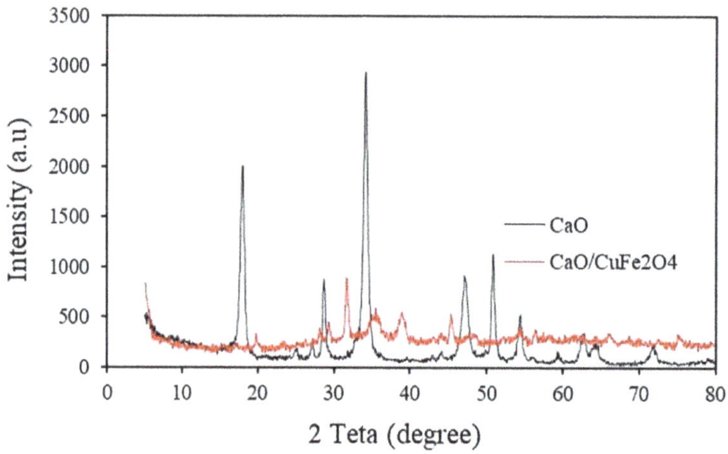

Figure 7. XRD spectrum of the nanocatalyst CaO and $CaO/CuFe_2O_4$. Reprinted with permission from Ref. [83]. 2022, Elsevier.

3.5. Algal Oil

Due to its oleaginous character and quick growth rate, microalgal biomass is a viable feedstock for the production of biofuels. Additionally, since its production does not compete with the area used for growing crops, the conflict between food and fuel is eliminated.

With CaO and 80 wt.% $CaO.Al_2O_3$ as catalysts, Narula et al. investigated the methods to improve the process parameters for the transesterification of algal oil [84]. The process of low-temperature transesterification was chosen. Response surface methodology was used to obtain an 88.89% biodiesel production with a methanol/oil volumetric ratio of (3.2:10) in 125 min at a temperature of 50 °C. It was shown that the volumetric ratio of methanol to oil, catalyst concentration, and reaction duration all significantly impacted yield. Additionally, this concept including a heterogeneous catalyst may be employed in the industry to produce biodiesel from algae oil quickly and cheaply by adjusting the process variables.

In work conducted by Abdala et al., calcium-doped zinc oxide nanocatalysts were produced in a UV shaker and used to examine heterogeneous transesterification of algal oil to biodiesel [85]. The prepared catalyst was exposed to various UV and non-UV light sources, calcium concentrations of 0.01, 0.03, and 0.05 M, and calcination temperatures of 600, 700, and 800 °C. The impacts of the various catalyst preparation parameters were investigated for the transesterification of algal oil, as shown in Figure 8. The catalyst that obtained a 99.18% yield of biodiesel and had a calcium loading of 0.05 M and a calcination temperature of 700 °C was thought to be the most appropriate nanocatalyst. With a 76% yield, the catalyst was utilized three times.

The production of bio-oil from algal biomass and the synthesis and characterization of a noble CaO-based heterogeneous catalyst for the conversion of bio-oil to biodiesel was conducted by Malpani and co-workers [86]. In a Soxhlet system, the bio-oil was extracted from the algal biomass using hexane as the solvent, and then it was transesterified using noble heterogeneous catalysts based on CaO to produce biodiesel. Among all the catalysts prepared, one with a TiO_2:CaO molar ratio of 0.25 and a calcination temperature of 700 °C was shown to be the best suited.

3.6. Biodiesel Production for Biomass-Derived Heterogeneous Catalysts

In a report by Roy and Mohanty, de-oiled microalgae was used as biomass that was subjected to carbonization, followed by sulfonation, to create a new carbon-based solid acid catalyst, as illustrated in Figure 9 [87]. It was found how several catalyst synthesis parameters, such as carbonization reaction temperature, sulfonation time, and concentration of H_2SO_4, affected the catalyst's surface acidity and the conversion of free fatty acids. The FTIR examination and XPS analysis both showed that the de-oiled microalgal biomass-based solid acid catalyst was mostly made up of carboxylic, sulfonic, and phenolic groups. Under ideal transesterification parameters, FAME production of microalgal oil and WCO was 94.23% and 96.25%, respectively. Up to the fourth cycle, the catalyst had a high catalytic activity with FAME conversion > 90%. FT. Yani et al.'s research describes the use of a source of water hyacinth biomass to produce a heterogeneous catalyst to lessen the adverse effects of water hyacinth's extensive use on the environment [88]. Hyacinth biomass that had been crushed up was calcined at 600 degrees to create the catalyst. K_2CO_3 was then added as a co-catalyst using the impregnation technique. The surface topography, particle size, and atom composition of a water hyacinth catalyst with K_2CO_3 impregnation were examined in order to better understand the consequences of the impregnation. It was investigated how well the synthesized catalyst handled the transesterification of palm oil into biodiesel. Methanol and oil had a molar ratio of 12:1, and 15% of the catalyst was loaded. The procedure used 15% of the hyacinth-based catalyst impregnated with 10% K_2CO_3 and produced the best yield (97.57%).

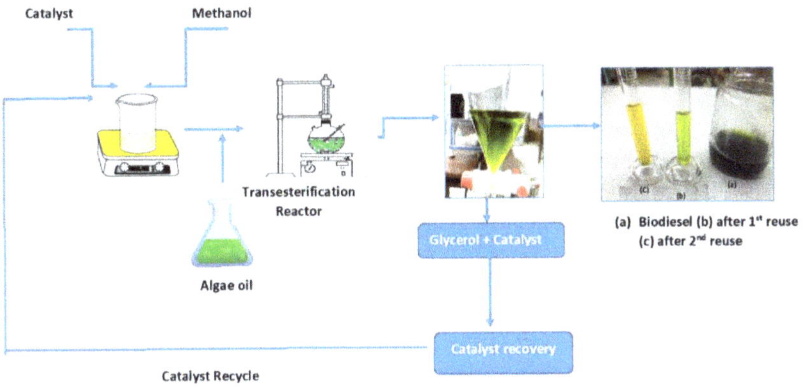

Figure 8. Transesterification of algal oil to biodiesel. Reprinted with permission from Ref. [85]. 2020, American Chemical Society.

To lower the cost of producing biodiesel, a highly effective, low-cost catalyst that can transesterify oil at ambient settings is needed. Tarigan et al. reported that a passion fruit peel could be used to create a new heterogeneous catalyst [89]. Using a factorial design, the catalytic activity of calcined waste passion fruit peel (WPFP), which mainly contains potassium in the forms of chloride and carbonate, has been assessed. This included examining the interactions between the molar ratio of oil to methanol, the weight of the catalyst, and the reaction time with respect to three different reaction conditions: 65, 45, and room temperature. Given that the leaching investigation revealed a drop of 22% in catalyst weight, specific amounts of catalyst must be added in order to reuse the catalyst. The prospect of cost is made possible by calcined WPFP's capacity to catalyze transesterification at room temperature.

Boz et al. used Al_2O_3 nanoparticles coated with KF that were calcined and utilized as heterogeneous catalysts to undergo transesterification of vegetable oil with methanol to produce biodiesel [90]. The study's variables included the ratio of KF to γ-Al_2O_3, calcination temperature, the molar ratio of methanol to oil, temperature and duration of the transesterification process, and catalyst concentration. Under the catalyst preparation and transesterification conditions of 15 wt.% KF loading, 773 °C calcination temperature, eight hours of reaction time at 338 °C, use of 3 wt.% catalysts, and a 15:1 molar ratio of methanol to oil, a methyl ester yield of 97.7 ± 2.14% was produced. The acquired high basicity of the catalyst surface (1.68 mmol/g) and the high surface-to-volume ratio of the γ-Al_2O_3 nanoparticles were thought to be responsible for the unusually high transformation of vegetable oil to biodiesel. Table 1 shows the various other biomass-derived heterogeneous catalysts used for diesel production.

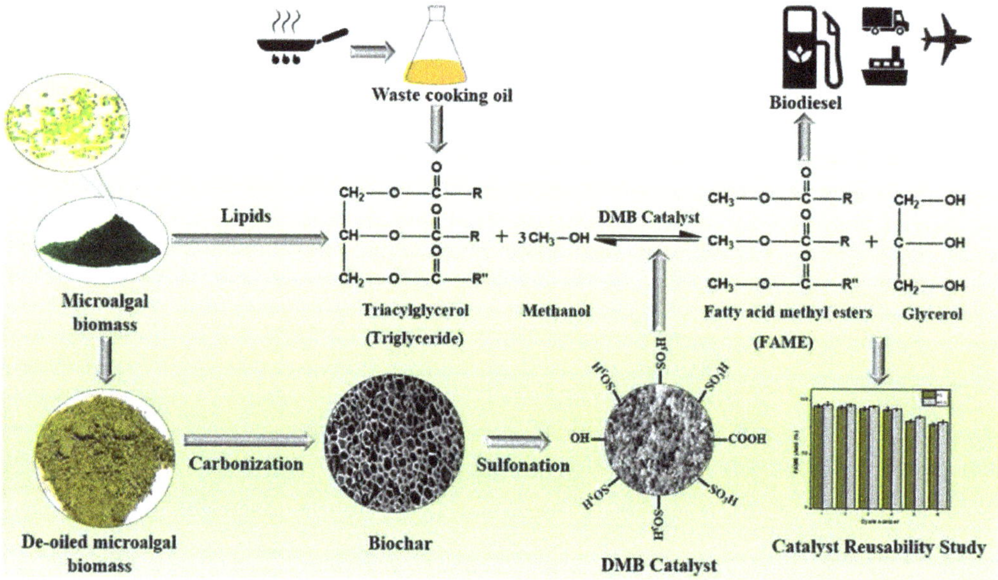

Figure 9. Pictorial representation of biodiesel synthesis using de-oiled microalgal biomass by carbonization, followed by sulfonation, to synthesize a new carbon-based solid acid catalyst. Reprinted with permission from Ref. [87]. 2021, Elsevier.

Table 1. Biomass-derived heterogeneous catalysts for biodiesel production.

Biomass	Feedstock	Yield	Reference
D-glucose	Oleic acid and stearic acid	44 µmol/min	[91]
Starch	Waste cooking oil	95%	[92]
Cellulose	Waste cooking oil	88%	[92]
Sucrose	Waste cooking oil	80%	[92]
Lime stone	Soybean oil	93%	[93]
Lime mud	Peanut oil	94%	[94]
Red mud	Soybean oil	94%	[95]
Dolomite rock	Palm kernel oil	99%	[96]
Chicken eggshell	Palm olein oil	94%	[97]
Snail shell	Palm olein oil	93.20%	[97]
Mussel shell	Chinese tallow oil	90%	[98]
Quail eggshell	Palm oil	98.00%	[99]
Pig bones	Palm oil	>90.00%	[100]
Sheep bones	Canola oil	95.15%	[101]
Turkey bones	mustard oil	91.22%	[102]

4. Optimization of Reaction Parameters for Biodiesel Production Using Heterogeneous Catalysts

One normally needs a large number of results to completely cover the input area in order to optimize the reaction condition. For instance, the methanol-to-oil ratio, temperature, catalyst loading, reaction time, and stirring rate with a fixed type of catalyst and oil feedstock are factors that determine the outcome of biodiesel output. Further discussion of these parameters' effects follows.

The goal of Mohamed et al.'s research was to create a new catalyst that is as effective as the one that was used to make biodiesel from WCO [103]. Fast pyrolysis of rice straw was used to create a heterogeneous solid acid catalyst, which was then subjected to a sulfonation procedure. Concentrated sulfuric acid was used to produce it. The reaction temperature, time, catalyst concentration, temperature, and methanol:oil molar ratio were all investigated as they affected the transesterification process. Biodiesel had a maximum bulk yield of 90.37%, FAME content was approximately 97.71 wt.%, the raw material conversion efficiency was 90.38 wt.%, and FFA% conversion was 91.1% under the ideal conditions of 10 wt.% catalysts and methanol:oil molar ratio of 20 for 6 h at 70 °C.

Mares' group examined the catalytic activity of acai seed ash for the methyl transesterification of soybean oil to produce biodiesel [104]. To find the ideal conditions for catalyst formation, açaí seeds were calcined at various temperatures (500–900 °C) and times (2–5 h). The ideal conditions were 100 °C, an 18:1 alcohol:oil molar ratio, 12.0% (*w*/*w*) catalyst concentration, and a 1 h reaction time, which produced biodiesel with 98.5 ± 0.21% ester content. The catalyst characterization revealed that the high metal oxide content and carbonates with basic surface sites, which make them extremely effective for biodiesel production, gave them their catalytic activity. Following the regeneration procedure, the yield was calculated to be more than 80.0%. The catalyst has several benefits, including being derived from commonly accessible biomass waste, perhaps inexpensive, and being simple to synthesize, making it a green and effective catalyst for the generation of biodiesel.

As a heterogeneous catalyst, waste chicken eggshell-derived CaO particles were synthesized by Attar's group [105]. In order to produce biodiesel from CWO, ultrasonic technology was used along with CaO as a catalyst. Response surface methodology (RSM) based on central composite design (CCD) was used to examine and optimize the effects of reaction parameters on biodiesel yield and specific energy consumption (SEC), such as catalyst loading (6–12 *w*/*w*%), methanol-to-oil ratio (6:1–12:1), ultrasonic power (150–300 W), and reaction time (20–40 min). Except for the impacts of the molar ratio on SEC, the results indicated substantial effects of the factors on biodiesel yield and SEC. Additionally, the ideal parameters for the reaction were a catalyst loading of 6.04 wt.%, an oil-to-methanol molar ratio of 8.33, an ultrasonic power of 299.66, and a reaction duration of 39.84 min, which produced a biodiesel yield of 98.62% and an SEC of 5.01 kJ g^{-1}.

In the use of industrial catalytic processes, catalyst deactivation—the gradual loss of catalytic activity and/or selectivity—is a topic of major and continuing concern. The annual cost to the industry of replacing catalysts and shutting down processes is in the billions. The design and operation of a large-scale catalytic process face significant problems in preventing catalyst degradation. Catalyst deterioration can happen in a variety of ways. One of the dozen impurities present in the feed, for instance, could poison a catalyst. Carbon or coke formed by the cracking/condensation reactions of hydrocarbon reactants, intermediates, and/or products could also block the catalyst's surface, pores, and voids. According to published research, there are three main reasons for the deactivation of CaO-based catalysts for transesterification: leaching of active sites, surface poisoning, and/or pore-clogging [106]. In order to prevent ambient CO_2 interaction with the catalyst right after removal from the reaction mixture, Endalew et al. coated the CaO-based catalyst with glycerol [107]. The collapse of the layered structure is the final factor in the deactivation of CaO-based catalysts. Due to its huge accessible surface area for the catalyst to react on, mesoporous catalysts have attracted a lot of attention for the manufacture of biodiesel [17]. However, the mesopores of CaO-based catalysts can disintegrate at moderate-to-high temperatures.

5. Summary and Future Perspective

One of the key requirements for the development of biodiesel is the need for a more affordable fuel. Traditional catalysis, which uses a homogenous basic catalyst, produces a significant amount of effluent and is susceptible to purifying issues. Therefore, the final result of this method cannot be guaranteed to be pure biodiesel. In this review, we have explained various types of catalysts, with a special focus on the heterogeneous catalyst. Various synthesis procedures in heterogeneous catalysts were also described in detail. The types of heterogeneous catalysts used in biodiesel were also given more attention. The parameters affecting biodiesel production were carefully analyzed for better understanding. We have found that the catalyst's performance is influenced by the temperature and duration of the calcination. Due to its fundamental character, the heterogeneous catalyst made from biomass materials promotes sustained development and yields excellent results in the conversion of triglycerides. More organic compounds need to be examined despite the investigation of a variety of inexpensive biomass-derived catalysts that have been provided.

Researchers concentrated on the manufacture of FAME from various renewable sources as an efficient method in this setting. The manufacturing of biodiesel has been proposed using a number of techniques. Transesterification is seen to be the best option out of all the procedures now in use. Heterogeneous catalysts have many benefits over homogeneous catalysts, including the ease with which continuous reactors may be realized, the generation of cleaner glycerol, and the lack of both the need to neutralize the alkaline catalyst and the requirement to replenish the spent catalyst. These benefits have made it possible for heterogeneous catalysts to provide another potent route for the generation of FAME. The kind of oil, the molar ratio of alcohol to oil, the temperature, and the type of reactor are the primary factors that affect the solid catalyst's reactivity. Therefore, choosing these factors at their best is an important first step. Since only the external-surface active species of the porous solid support are engaged in the reaction, and considering that these catalysts may sometimes be recovered, heterogeneous catalysts are viewed as somewhat promising.

Processing non-edible and waste oil or fat with a high FFA concentration can be conducted using non-catalytic and heterogeneous acid-based-catalyzed biodiesel synthesis processes. Temperature optimization is essential in attaining an acceptable biodiesel yield among the investigated reaction parameters. The mass transfer restriction among the alcohol and oil phases is significantly lessened when co-solvent is used in heterogeneously catalyzed three-phase systems, which increases biodiesel yield while lowering the values of other optimized parameters such as reaction time, reaction temperature, catalyst concentration, and methanol. Algal lipids have far less comprehensive compositional data available than vegetable oils. Few of the more than 40,000 known algae species have FA compositional characteristics. The tremendous compositional heterogeneity between several algae species is evident, nevertheless. Compared to ordinary vegetable oils, certain species exhibit substantially greater degrees of unsaturation.

The current investigation results showed that the basicity and acidity of the catalyst's characteristics are crucial for the generation of biodiesel. Numerous literature studies contend that transesterification directly correlates with the basicity of the catalyst. Similarly to this, the catalyst's acidity determines its esterification activity. With an increase in catalyst acidity, the esterification activity rises. Aside from the fact that these biomass-derived materials are relatively scarce in nature for commercialization prospects, the high energy requisite for catalyst modification to increase its activity, the rapid decline of catalyst efficiency due to the interaction of basic sites with ambient CO_2 and water, as well as the possibility of an active site drop caused by reaction with FFA and leaching of CaO into the polar solvent, invite further research into these resources.

The study's primary goal is to enhance already employed biodiesel production techniques as well as create new methods for producing biodiesel that may satisfy industrial demands. It is challenging to improve upon presently used processes and introduce new ones in order to enhance biodiesel production, particularly in the sector. In addition to the reusability of the catalyst, efficient reactor design, and exploitation of alternative resources, several strategies are investigated to reduce the cost implications. The use of the biomass-derived heterogeneous catalyst for biodiesel synthesis appears to be a viable option since it eliminates the laborious issues associated with homogeneous processes. Exploring trash or biomass as a source of catalysts might lower the price of commercially available solid catalysts and open up new uses for the waste. While the quantitative (yield) influence of catalysts has been extensively covered in the literature review, the relevance of the qualitative function of catalysts is unclear. Because of this, researchers currently have a limited grasp of catalyst behavior while working under dynamic reaction settings. Government assistance in the form of tax exemptions or reductions, as well as the establishment of regulations that promote the use of recycled materials from waste, is required to develop waste into catalysts. In other words, if encouraged, waste-derived catalysts might boost the economy and create jobs. More research and development of catalysts produced from biomass are required to enhance the catalytic performance for the manufacture of biodiesel and other chemical processes.

Author Contributions: Investigation and writing—original draft, S.C.K.; writing—review and editing, S.P.; conceptualization, investigation, and writing—original draft, R.A.; investigation and visualization, A.K.S.; investigation and visualization, M.A.; validation and visualization, S.S.; project administration and supervision, Y.R.L. All authors equally contributed to this work. All authors have read and agreed to the published version of the manuscript.

Funding: This work was supported by the National Research Foundation of Korea (NRF) grant funded by the Korean government MSIT (2021R1A2B5B02002436).

Data Availability Statement: No data availability.

Acknowledgments: The authors thank the National Research Foundation of Korea (NRF) for providing financial support.

Conflicts of Interest: The authors declare no conflict of interest.

References

1. Alshorifi, F.T.; Alswat, A.A.; Salama, R.S. Gold-selenide quantum dots supported onto cesium ferrite nanocomposites for the efficient degradation of rhodamine B. *Heliyon* **2022**, *8*, e09652. [CrossRef] [PubMed]
2. Amjith, L.; Bavanish, B. A review on biomass and wind as renewable energy for sustainable environment. *Chemosphere* **2022**, *293*, 133579. [CrossRef] [PubMed]
3. Demirbas, A. Progress and recent trends in biodiesel fuels. *Energy Convers. Manag.* **2009**, *50*, 14–34.
4. Ashokkumar, V.; Venkatkarthick, R.; Jayashree, S.; Chuetor, S.; Dharmaraj, S.; Kumar, G.; Chen, W.-H.; Ngamcharussrivichai, C. Recent advances in lignocellulosic biomass for biofuels and value-added bioproducts-A critical review. *Bioresour. Technol.* **2022**, *344*, 126195. [CrossRef] [PubMed]
5. Babu, S.; Rathore, S.S.; Singh, R.; Kumar, S.; Singh, V.K.; Yadav, S.; Yadav, V.; Raj, R.; Yadav, D.; Shekhawat, K. Exploring agricultural waste biomass for energy, food and feed production and pollution mitigation: A review. *Bioresour. Technol.* **2022**, *360*, 127566. [CrossRef]
6. Kominko, H.; Gorazda, K.; Wzorek, Z. Effect of sewage sludge-based fertilizers on biomass growth and heavy metal accumulation in plants. *J. Environ. Manag.* **2022**, *305*, 114417. [CrossRef]
7. Agathokleous, E.; Guo, J.; Peñuelas, J. Low doses of toxicants can enhance algae potential as biodiesel and biomass feedstocks. *Renew. Sustain. Energy Rev.* **2022**, *168*, 112858. [CrossRef]
8. Aggarwal, N.K.; Kumar, N.; Mittal, M. Potential of Weed Biomass for Bioethanol Production. In *Bioethanol Production*; Springer: Berlin/Heidelberg, Germany, 2022; pp. 65–71.
9. Rashedi, A.; Gul, N.; Hussain, M.; Hadi, R.; Khan, N.; Nadeem, S.G.; Khanam, T.; Asyraf, M.; Kumar, V. Life cycle environmental sustainability and cumulative energy assessment of biomass pellets biofuel derived from agroforest residues. *PLoS ONE* **2022**, *17*, e0275005. [CrossRef]
10. Beidaghy Dizaji, H.; Zeng, T.; Lenz, V.; Enke, D. Valorization of Residues from Energy Conversion of Biomass for Advanced and Sustainable Material Applications. *Sustainability* **2022**, *14*, 4939. [CrossRef]

11. Pang, S.; Yip, A.; Zhao, P. *Waste Wood Processing Technologies Review*; Envirolink: Māpua, New Zealand, 2022.
12. Konwar, L.J.; Boro, J.; Deka, D. Review on latest developments in biodiesel production using carbon-based catalysts. *Renew. Sustain. Energy Rev.* **2014**, *29*, 546–564. [CrossRef]
13. Borges, M.; Díaz, L.; Gavín, J.; Brito, A. Estimation of the content of fatty acid methyl esters (FAME) in biodiesel samples from dynamic viscosity measurements. *Fuel Process. Technol.* **2011**, *92*, 597–599. [CrossRef]
14. Lukić, I.; Kesić, Ž.; Skala, D. Kinetics of heterogeneous biodiesel synthesis using supported ZnO as catalyst. *Chem. Eng. Technol.* **2014**, *37*, 1879–1884. [CrossRef]
15. Jokiniemi, T.; Ahokas, J. A review of production and use of first generation biodiesel in agriculture. *Agron. Res.* **2013**, *11*, 239–248.
16. Bhuiya, M.; Rasul, M.; Khan, M.M.K.; Ashwath, N.; Azad, A.K.; Hazrat, M. Second generation biodiesel: Potential alternative to-edible oil-derived biodiesel. *Energy Procedia* **2014**, *61*, 1969–1972. [CrossRef]
17. Sivasamy, A.; Cheah, K.Y.; Fornasiero, P.; Kemausuor, F.; Zinoviev, S.; Miertus, S. Catalytic applications in the production of biodiesel from vegetable oils. *ChemSusChem Chem. Sustain. Energy Mater.* **2009**, *2*, 278–300. [CrossRef] [PubMed]
18. Wu, X.; McLaren, J.; Madl, R.; Wang, D. Biofuels from lignocellulosic biomass. In *Sustainable Biotechnology*; Springer: Berlin/Heidelberg, Germany, 2010; pp. 19–41.
19. Kumar, S.; Singhal, M.K.; Sharma, M.P. Utilization of mixed oils for biodiesel preparation: A review. *Energy Sources Part A Recovery Util. Environ. Eff.* **2021**. [CrossRef]
20. Kumar, H.; Sarma, A.; Kumar, P. A comprehensive review on preparation, characterization, and combustion characteristics of microemulsion based hybrid biofuels. *Renew. Sustain. Energy Rev.* **2020**, *117*, 109498. [CrossRef]
21. Prado, C.M.; Antoniosi Filho, N.R. Production and characterization of the biofuels obtained by thermal cracking and thermal catalytic cracking of vegetable oils. *J. Anal. Appl. Pyrolysis* **2009**, *86*, 338–347. [CrossRef]
22. Jahirul, M.I.; Rasul, M.G.; Chowdhury, A.A.; Ashwath, N. Biofuels production through biomass pyrolysis—A technological review. *Energies* **2012**, *5*, 4952–5001. [CrossRef]
23. Leung, D.Y.; Wu, X.; Leung, M. A review on biodiesel production using catalyzed transesterification. *Appl. Energy* **2010**, *87*, 1083–1095. [CrossRef]
24. Orege, J.I.; Oderinde, O.; Kifle, G.A.; Ibikunle, A.A.; Raheem, S.A.; Ejeromedoghene, O.; Okeke, E.S.; Olukowi, O.M.; Orege, O.B.; Fagbohun, E.O. Recent advances in heterogeneous catalysis for green biodiesel production by transesterification. *Energy Convers. Manag.* **2022**, *258*, 115406. [CrossRef]
25. Teo, S.H.; Islam, A.; Mansir, N.; Shamsuddin, M.R.; Joseph, C.G.; Goto, M.; Taufiq-Yap, Y.H. Sustainable biofuel production approach: Critical methanol green transesterification by efficient and stable heterogeneous catalyst. *Renew. Sustain. Energy Rev.* **2022**, *169*, 112889. [CrossRef]
26. Aslam, M. Transformation of 1-G and 2-G liquid biomass to green fuels using hydroprocessing technology: A promising technology for biorefinery development. *Biomass Bioenergy* **2022**, *163*, 106510. [CrossRef]
27. Kalita, P.; Basumatary, B.; Saikia, P.; Das, B.; Basumatary, S. Biodiesel as renewable biofuel produced via enzyme-based catalyzed transesterification. *Energy Nexus* **2022**, *6*, 100087. [CrossRef]
28. Abdullah, A.Z.; Razali, N.; Mootabadi, H.; Salamatinia, B. Critical technical areas for future improvement in biodiesel technologies. *Environ. Res. Lett.* **2007**, *2*, 034001. [CrossRef]
29. Munyentwali, A.; Li, H.; Yang, Q. Review of advances in bifunctional solid acid/base catalysts for sustainable biodiesel production. *Appl. Catal. A Gen.* **2022**, *633*, 118525. [CrossRef]
30. Okechukwu, O.D.; Joseph, E.; Nonso, U.C.; Kenechi, N.-O. Improving heterogeneous catalysis for biodiesel production process. *Clean. Chem. Eng.* **2022**, *3*, 100038. [CrossRef]
31. Mandari, V.; Devarai, S.K. Biodiesel production using homogeneous, heterogeneous, and enzyme catalysts via transesterification and esterification reactions: A critical review. *BioEnergy Res.* **2021**, *15*, 935–961. [CrossRef] [PubMed]
32. Nayab, R.; Imran, M.; Ramzan, M.; Tariq, M.; Taj, M.B.; Akhtar, M.N.; Iqbal, H.M. Sustainable biodiesel production via catalytic and non-catalytic transesterification of feedstock materials–A review. *Fuel* **2022**, *328*, 125254. [CrossRef]
33. Shende, V.S.; Saptal, V.B.; Bhanage, B.M. Recent advances utilized in the recycling of homogeneous catalysis. *Chem. Rec.* **2019**, *19*, 2022–2043. [CrossRef] [PubMed]
34. Mukhtar, A.; Saqib, S.; Lin, H.; Shah, M.U.H.; Ullah, S.; Younas, M.; Rezakazemi, M.; Ibrahim, M.; Mahmood, A.; Asif, S. Current status and challenges in the heterogeneous catalysis for biodiesel production. *Renew. Sustain. Energy Rev.* **2022**, *157*, 112012. [CrossRef]
35. Ibrahim, A.A.; Salama, R.S.; El-Hakam, S.A.; Khder, A.S.; Ahmed, A.I. Synthesis of 12-tungestophosphoric acid supported on Zr/MCM-41 composite with excellent heterogeneous catalyst and promising adsorbent of methylene blue. *Colloids Surf. A Physicochem. Eng. Asp.* **2021**, *631*, 127753. [CrossRef]
36. Tang, Z.-E.; Lim, S.; Pang, Y.-L.; Ong, H.-C.; Lee, K.-T. Synthesis of biomass as heterogeneous catalyst for application in biodiesel production: State of the art and fundamental review. *Renew. Sustain. Energy Rev.* **2018**, *92*, 235–253. [CrossRef]
37. Bakry, A.M.; Alamier, W.M.; Salama, R.S.; El-Shall, M.S.; Awad, F.S. Remediation of water containing phosphate using ceria nanoparticles decorated partially reduced graphene oxide (CeO_2-PRGO) composite. *Surf. Interfaces* **2022**, *31*, 102006. [CrossRef]
38. Salaheldeen, M.; Mariod, A.A.; Aroua, M.K.; Rahman, S.A.; Soudagar, M.E.M.; Fattah, I.R. Current state and perspectives on transesterification of triglycerides for biodiesel production. *Catalysts* **2021**, *11*, 1121. [CrossRef]

39. Zabeti, M.; Daud, W.M.A.W.; Aroua, M.K. Activity of solid catalysts for biodiesel production: A review. *Fuel Process. Technol.* **2009**, *90*, 770–777. [CrossRef]
40. Khan, H.M.; Iqbal, T.; Yasin, S.; Irfan, M.; Abbas, M.M.; Veza, I.; Soudagar, M.E.M.; Abdelrahman, A.; Kalam, M.A. Heterogeneous Catalyzed Biodiesel Production Using Cosolvent: A Mini Review. *Sustainability* **2022**, *14*, 5062. [CrossRef]
41. Ferreira, R.S.B.; dos Passos, R.M.; Sampaio, K.A.; Batista, E. Heterogeneous catalysts for biodiesel production: A review. *Food Public Health* **2019**, *9*, 125–137.
42. Bajaj, A.; Lohan, P.; Jha, P.N.; Mehrotra, R. Biodiesel production through lipase catalyzed transesterification: An overview. *J. Mol. Catal. B Enzym.* **2010**, *62*, 9–14. [CrossRef]
43. Christopher, L.P.; Kumar, H.; Zambare, V.P. Enzymatic biodiesel: Challenges and opportunities. *Appl. Energy* **2014**, *119*, 497–520. [CrossRef]
44. Majidian, P.; Tabatabaei, M.; Zeinolabedini, M.; Naghshbandi, M.P.; Chisti, Y. Metabolic engineering of microorganisms for biofuel production. *Renew. Sustain. Energy Rev.* **2018**, *82*, 3863–3885. [CrossRef]
45. Saka, S.; Kusdiana, D. Biodiesel fuel from rapeseed oil as prepared in supercritical methanol. *Fuel* **2001**, *80*, 225–231. [CrossRef]
46. Ma, F.; Hanna, M.A. Biodiesel production: A review. *Bioresour. Technol.* **1999**, *70*, 1–15. [CrossRef]
47. Stojković, I.J.; Stamenković, O.S.; Povrenović, D.S.; Veljković, V.B. Purification technologies for crude biodiesel obtained by alkali-catalyzed transesterification. *Renew. Sustain. Energy Rev.* **2014**, *32*, 1–15. [CrossRef]
48. Teo, S.H.; Taufiq-Yap, Y.H.; Rashid, U.; Islam, A. Hydrothermal effect on synthesis, characterization and catalytic properties of calcium methoxide for biodiesel production from crude Jatropha curcas. *RSC Adv.* **2015**, *5*, 4266–4276. [CrossRef]
49. Macario, A.; Giordano, G.; Onida, B.; Cocina, D.; Tagarelli, A.; Giuffrè, A.M. Biodiesel production process by homogeneous/heterogeneous catalytic system using an acid–base catalyst. *Appl. Catal. A Gen.* **2010**, *378*, 160–168. [CrossRef]
50. Soltani, S.; Khanian, N.; Choong, T.S.Y.; Asim, N.; Zhao, Y. Microwave-assisted hydrothermal synthesis of sulfonated TiO_2-GO core–shell solid spheres as heterogeneous esterification mesoporous catalyst for biodiesel production. *Energy Convers. Manag.* **2021**, *238*, 114165. [CrossRef]
51. Abdullah, R.F.; Rashid, U.; Hazmi, B.; Ibrahim, M.L.; Tsubota, T.; Alharthi, F.A. Potential heterogeneous nano-catalyst via integrating hydrothermal carbonization for biodiesel production using waste cooking oil. *Chemosphere* **2022**, *286*, 131913. [CrossRef]
52. Cardoso, R.K.; Silva, G.V.; Alves, B.T.; Freire, V.A.; Alves, J.J.; Barbosa, B.V. Evaluation of the effect of Si/Mo and oil/alcohol ratios in the production of biodiesel from soybean oil. *Arab. J. Chem.* **2022**, *15*, 104074. [CrossRef]
53. Atelge, M. Production of biodiesel and hydrogen by using a double-function heterogeneous catalyst derived from spent coffee grounds and its thermodynamic analysis. *Renew. Energy* **2022**, *198*, 1–15. [CrossRef]
54. Fernández, J.V.; Faria, D.N.; Santoro, M.C.; Mantovaneli, R.; Cipriano, D.F.; Brito, G.M.; Carneiro, M.T.W.; Schettino, M.A.; Gonzalez, J.L.; Freitas, J.C. Use of Unmodified Coffee Husk Biochar and Ashes as Heterogeneous Catalysts in Biodiesel Synthesis. *BioEnergy Res.* **2022**. [CrossRef]
55. Yang, L.; Zhang, A.; Zheng, X. Shrimp shell catalyst for biodiesel production. *Energy Fuels* **2009**, *23*, 3859–3865. [CrossRef]
56. Jayakumar, M.; Gebeyehu, K.B.; Selvakumar, K.V.; Parvathy, S.; Kim, W.; Karmegam, N. Waste Ox bone based heterogeneous catalyst synthesis, characterization, utilization and reaction kinetics of biodiesel generation from Jatropha curcas oil. *Chemosphere* **2022**, *288*, 132534. [CrossRef] [PubMed]
57. Chakraborty, R.; Bepari, S.; Banerjee, A. Application of calcined waste fish (Labeo rohita) scale as low-cost heterogeneous catalyst for biodiesel synthesis. *Bioresour. Technol.* **2011**, *102*, 3610–3618. [CrossRef]
58. Nair, P.; Singh, B.; Upadhyay, S.; Sharma, Y. Synthesis of biodiesel from low FFA waste frying oil using calcium oxide derived from Mereterix mereterix as a heterogeneous catalyst. *J. Clean. Prod.* **2012**, *29*, 82–90. [CrossRef]
59. Li, M.; Zheng, Y.; Chen, Y.; Zhu, X. Biodiesel production from waste cooking oil using a heterogeneous catalyst from pyrolyzed rice husk. *Bioresour. Technol.* **2014**, *154*, 345–348. [CrossRef]
60. Daimary, N.; Eldiehy, K.S.; Boruah, P.; Deka, D.; Bora, U.; Kakati, B.K. Potato peels as a sustainable source for biochar, bio-oil and a green heterogeneous catalyst for biodiesel production. *J. Environ. Chem. Eng.* **2022**, *10*, 107108. [CrossRef]
61. Bhatia, S.K.; Gurav, R.; Choi, T.-R.; Kim, H.J.; Yang, S.-Y.; Song, H.-S.; Park, J.Y.; Park, Y.-L.; Han, Y.-H.; Choi, Y.-K. Conversion of waste cooking oil into biodiesel using heterogenous catalyst derived from cork biochar. *Bioresour. Technol.* **2020**, *302*, 122872. [CrossRef]
62. Hazmi, B.; Rashid, U.; Kawi, S.; Mokhtar, W.N.A.W.; Yaw, T.C.S.; Moser, B.R.; Alsalme, A. Palm fatty acid distillate esterification using synthesized heterogeneous sulfonated carbon catalyst from plastic waste: Characterization, catalytic efficacy and stability, and fuel properties. *Process Saf. Environ. Prot.* **2022**, *162*, 1139–1151. [CrossRef]
63. Mahesh, S.E.; Ramanathan, A.; Begum, K.M.S.; Narayanan, A. Biodiesel production from waste cooking oil using KBr impregnated CaO as catalyst. *Energy Convers. Manag.* **2015**, *91*, 442–450. [CrossRef]
64. Alsharifi, M.; Znad, H.; Hena, S.; Ang, M. Biodiesel production from canola oil using novel Li/TiO_2 as a heterogeneous catalyst prepared via impregnation method. *Renew. Energy* **2017**, *114*, 1077–1089. [CrossRef]
65. Wang, Y.-T.; Gao, D.; Yang, J.; Zeng, Y.-N.; Li, J.-G.; Wang, Y.-J.; Wang, X.-M.; Wang, F.-P.; Yu, Q.; Liu, T.-J. Highly stable heterogeneous catalysts from electric furnace dust for biodiesel production: Optimization, performance and reaction kinetics. *Catal. Today* **2022**, *404*, 78–92. [CrossRef]

66. Shimada, G.B.; Cestari, A. Synthesis of heterogeneous catalysts by the hydrolytic Sol-Gel method for the biodiesel production. *Renew. Energy* **2020**, *156*, 389–394. [CrossRef]
67. Salinas, D.; Sepúlveda, C.; Escalona, N.; Gfierro, J.; Pecchi, G. Sol-gel La_2O_3-ZrO_2 mixed oxide catalysts for biodiesel production. *J. Energy Chem.* **2018**, *27*, 565–572. [CrossRef]
68. Maleki, B.; Talesh, S.S.A. Optimization of ZnO incorporation to αFe_2O_3 nanoparticles as an efficient catalyst for biodiesel production in a sonoreactor: Application on the CI engine. *Renew. Energy* **2022**, *182*, 43–59. [CrossRef]
69. Pham, T.; Nguyen, U.; Imamura, K.; Furuta, M.; Maeda, Y. Transesterification using isopropanol as a co-solvent for the production of green biodiesel fuel. *Int. J. Energy Res.* **2022**, *46*, 4352–4361. [CrossRef]
70. Okitsu, K.; Sadanaga, Y.; Takenaka, N.; Maeda, Y.; Bandow, H. A new co-solvent method for the green production of biodiesel fuel–Optimization and practical application. *Fuel* **2013**, *103*, 742–748.
71. Akkarawatkhoosith, N.; Kaewchada, A.; Jaree, A. Simultaneous development of biodiesel synthesis and fuel quality via continuous supercritical process with reactive co-solvent. *Fuel* **2019**, *237*, 117–125. [CrossRef]
72. Bambase, M.E., Jr.; Almazan, R.A.R.; Demafelis, R.B.; Sobremisana, M.J.; Dizon, L.S.H. Biodiesel production from refined coconut oil using hydroxide-impregnated calcium oxide by cosolvent method. *Renew. Energy* **2021**, *163*, 571–578. [CrossRef]
73. Szkudlarek, Ł.; Chałupka, K.; Maniukiewicz, W.; Albińska, J.; Szynkowska-Jóźwik, M.I.; Mierczyński, P. The Influence of Si/Al Ratio on the Physicochemical and Catalytic Properties of MgO/ZSM-5 Catalyst in Transesterification Reaction of Rapeseed Oil. *Catalysts* **2021**, *11*, 1260. [CrossRef]
74. Islam, M.G.U.; Jan, M.T.; Farooq, M.; Naeem, A.; Khan, I.W.; Khattak, H.U. Biodiesel production from wild olive oil using TPA decorated Cr-Al acid heterogeneous catalyst. *Chem. Eng. Res. Des.* **2022**, *178*, 540–549. [CrossRef]
75. Pratika, R.A.; Wijaya, K.; Trisunaryanti, W. Hydrothermal treatment of SO_4/TiO_2 and TiO_2/CaO as heterogeneous catalysts for the conversion of Jatropha oil into biodiesel. *J. Environ. Chem. Eng.* **2021**, *9*, 106547. [CrossRef]
76. Pandiangan, K.D.; Simanjuntak, W.; Jamarun, N.; Arief, S. The use of MgO/SiO_2 as catalyst for transesterification of rubber seed oil with different alcohols. *J. Phys. Conf. Ser.* **2021**, *1751*, 012100. [CrossRef]
77. Khan, S.G.; Hassan, M.; Anwar, M.; Khan, U.M.; Zhao, C. Mussel shell based CaO nano-catalyst doped with praseodymium to enhance biodiesel production from castor oil. *Fuel* **2022**, *330*, 125480. [CrossRef]
78. Rezania, S.; Korrani, Z.S.; Gabris, M.A.; Cho, J.; Yadav, K.K.; Cabral-Pinto, M.M.; Alam, J.; Ahamed, M.; Nodeh, H.R. Lanthanum phosphate foam as novel heterogeneous nanocatalyst for biodiesel production from waste cooking oil. *Renew. Energy* **2021**, *176*, 228–236. [CrossRef]
79. Mohadesi, M.; Aghel, B.; Gouran, A.; Razmehgir, M.H. Transesterification of waste cooking oil using Clay/CaO as a solid base catalyst. *Energy* **2022**, *242*, 122536. [CrossRef]
80. Kataria, J.; Mohapatra, S.; Kundu, K. Biodiesel production from waste cooking oil using heterogeneous catalysts and its operational characteristics on variable compression ratio CI engine. *J. Energy Inst.* **2019**, *92*, 275–287. [CrossRef]
81. Mutreja, V.; Singh, S.; Ali, A. Biodiesel from mutton fat using KOH impregnated MgO as heterogeneous catalysts. *Renew. Energy* **2011**, *36*, 2253–2258. [CrossRef]
82. Cherian, E.; Yazhini, D.; Victor, M.; Baskar, G. Production of biodiesel from pork fat using alumina-doped calcium oxide nanocomposite as heterogeneous catalyst. *Energy Sources Part A Recovery Util. Environ. Eff.* **2021**, *43*, 1386–1395. [CrossRef]
83. Seffati, K.; Honarvar, B.; Esmaeili, H.; Esfandiari, N. Enhanced biodiesel production from chicken fat using $CaO/CuFe_2O_4$ nanocatalyst and its combination with diesel to improve fuel properties. *Fuel* **2019**, *235*, 1238–1244. [CrossRef]
84. Narula, V.; Khan, M.F.; Negi, A.; Kalra, S.; Thakur, A.; Jain, S. Low temperature optimization of biodiesel production from algal oil using CaO and CaO/Al_2O_3 as catalyst by the application of response surface methodology. *Energy* **2017**, *140*, 879–884. [CrossRef]
85. Abdala, E.; Nur, O.; Mustafa, M.A. Efficient biodiesel production from algae oil using Ca-doped ZnO nanocatalyst. *Ind. Eng. Chem. Res.* **2020**, *59*, 19235–19243. [CrossRef]
86. Malpani, M.; Varma, A.; Mondal, P. Production of bio-oil from algal biomass and its upgradation to biodiesel using CaO-based heterogeneous catalysts. *Int. J. Green Energy* **2016**, *13*, 969–976. [CrossRef]
87. Roy, M.; Mohanty, K. Valorization of de-oiled microalgal biomass as a carbon-based heterogeneous catalyst for a sustainable biodiesel production. *Bioresour. Technol.* **2021**, *337*, 125424. [CrossRef]
88. Yani, F.; Ulhaqi, R.; Pratiwi, W.; Pontas, K.; Husin, H. Utilization of water hyacinth-based biomass as a potential heterogeneous catalyst for biodiesel production. *J. Phys. Conf. Ser.* **2019**, *1402*, 055005. [CrossRef]
89. Tarigan, J.B.; Singh, K.; Sinuraya, J.S.; Supeno, M.; Sembiring, H.; Tarigan, K.; Rambe, S.M.; Karo-Karo, J.A.; Sitepu, E.K. Waste Passion Fruit Peel as a Heterogeneous Catalyst for Room-Temperature Biodiesel Production. *ACS Omega* **2022**, *7*, 7885–7892. [CrossRef]
90. Boz, N.; Degirmenbasi, N.; Kalyon, D.M. Conversion of biomass to fuel: Transesterification of vegetable oil to biodiesel using KF loaded nano-γ-Al_2O_3 as catalyst. *Appl. Catal. B Environ.* **2009**, *89*, 590–596. [CrossRef]
91. Takagaki, A.; Toda, M.; Okamura, M.; Kondo, J.N.; Hayashi, S.; Domen, K.; Hara, M. Esterification of higher fatty acids by a novel strong solid acid. *Catal. Today* **2006**, *116*, 157–161. [CrossRef]
92. Deshmane, C.A.; Wright, M.W.; Lachgar, A.; Rohlfing, M.; Liu, Z.; Le, J.; Hanson, B.E. A comparative study of solid carbon acid catalysts for the esterification of free fatty acids for biodiesel production. Evidence for the leaching of colloidal carbon. *Bioresour. Technol.* **2013**, *147*, 597–604. [CrossRef]

93. Kouzu, M.; Kasuno, T.; Tajika, M.; Sugimoto, Y.; Yamanaka, S.; Hidaka, J. Calcium oxide as a solid base catalyst for transesterification of soybean oil and its application to biodiesel production. *Fuel* **2008**, *87*, 2798–2806. [CrossRef]
94. Li, H.; Niu, S.; Lu, C.; Liu, M.; Huo, M. Use of lime mud from paper mill as a heterogeneous catalyst for transesterification. *Sci. China Technol. Sci.* **2014**, *57*, 438–444. [CrossRef]
95. Ngamcharussrivichai, C.; Nunthasanti, P.; Tanachai, S.; Bunyakiat, K. Biodiesel production through transesterification over natural calciums. *Fuel Process. Technol.* **2010**, *91*, 1409–1415. [CrossRef]
96. Ngamcharussrivichai, C.; Wiwatnimit, W.; Wangnoi, S. Modified dolomites as catalysts for palm kernel oil transesterification. *J. Mol. Catal. A Chem.* **2007**, *276*, 24–33. [CrossRef]
97. Viriya-Empikul, N.; Krasae, P.; Nualpaeng, W.; Yoosuk, B.; Faungnawakij, K. Biodiesel production over Ca-based solid catalysts derived from industrial wastes. *Fuel* **2012**, *92*, 239–244. [CrossRef]
98. Hu, S.; Wang, Y.; Han, H. Utilization of waste freshwater mussel shell as an economic catalyst for biodiesel production. *Biomass Bioenergy* **2011**, *35*, 3627–3635. [CrossRef]
99. Cho, Y.B.; Seo, G. High activity of acid-treated quail eggshell catalysts in the transesterification of palm oil with methanol. *Bioresour. Technol.* **2010**, *101*, 8515–8519. [CrossRef]
100. Chen, G.; Shan, R.; Shi, J.; Liu, C.; Yan, B. Biodiesel production from palm oil using active and stable K doped hydroxyapatite catalysts. *Energy Convers. Manag.* **2015**, *98*, 463–469. [CrossRef]
101. Ghanei, R.; Khalili Dermani, R.; Salehi, Y.; Mohammadi, M. Waste animal bone as support for CaO impregnation in catalytic biodiesel production from vegetable oil. *Waste Biomass Valorization* **2016**, *7*, 527–532. [CrossRef]
102. Chakraborty, R.; Das, S.; Bhattacharjee, S.K. Optimization of biodiesel production from Indian mustard oil by biological tri-calcium phosphate catalyst derived from turkey bone ash. *Clean Technol. Environ. Policy* **2015**, *17*, 455–463. [CrossRef]
103. Mohamed, R.; Kadry, G.; Abdel-Samad, H.; Awad, M. High operative heterogeneous catalyst in biodiesel production from waste cooking oil. *Egypt. J. Pet.* **2020**, *29*, 59–65. [CrossRef]
104. Mares, E.K.L.; Goncalves, M.A.; da Luz, P.T.S.; da Rocha Filho, G.N.; Zamian, J.R.; da Conceição, L.R.V. Acai seed ash as a novel basic heterogeneous catalyst for biodiesel synthesis: Optimization of the biodiesel production process. *Fuel* **2021**, *299*, 120887. [CrossRef]
105. Attari, A.; Abbaszadeh-Mayvan, A.; Taghizadeh-Alisaraie, A. Process optimization of ultrasonic-assisted biodiesel production from waste cooking oil using waste chicken eggshell-derived CaO as a green heterogeneous catalyst. *Biomass Bioenergy* **2022**, *158*, 106357. [CrossRef]
106. Oueda, N.; Bonzi-Coulibaly, Y.L.; Ouédraogo, I.W. Deactivation processes, regeneration conditions and reusability performance of CaO or MgO based catalysts used for biodiesel production—A review. *Mater. Sci. Appl.* **2016**, *8*, 94–122. [CrossRef]
107. Endalew, A.K.; Kiros, Y.; Zanzi, R. Heterogeneous catalysis for biodiesel production from Jatropha curcas oil (JCO). *Energy* **2011**, *36*, 2693–2700. [CrossRef]

catalysts

MDPI

Review

Recent Developments on CO_2 Hydrogenation Performance over Structured Zeolites: A Review on Properties, Synthesis, and Characterization

Methene Briones Cutad [1], Mohammed J. Al-Marri [1,2,*] and Anand Kumar [1,*]

[1] Department of Chemical Engineering, College of Engineering, Qatar University, Doha P.O. Box 2713, Qatar; mcutad@qu.edu.qa

[2] Gas Processing Center, College of Engineering, Qatar University, Doha P.O. Box 2713, Qatar

* Correspondence: m.almarri@qu.edu.qa (M.J.A.-M.); akumar@qu.edu.qa (A.K.); Tel.: +974-44034146 (M.J.A.-M.)

Abstract: This review focuses on an extensive synopsis of the recent improvements in CO_2 hydrogenation over structured zeolites, including their properties, synthesis methods, and characterization. Key features such as bimodal mesoporous structures, surface oxygen vacancies, and the Si/Al ratio are explored for their roles in enhancing catalytic activity. Additionally, the impact of porosity, thermal stability, and structural integrity on the performance of zeolites, as well as their interactions with electrical and plasma environments, are discussed in detail. The synthesis of structured zeolites is analyzed by comparing the advantages and limitations of bottom-up methods, including hard templating, soft templating, and non-templating approaches, to top-down methods, such as dealumination, desilication, and recrystallization. The review addresses the challenges associated with these synthesis techniques, such as pore-induced diffusion limitations, morphological constraints, and maintaining crystal integrity, highlighting the need for innovative solutions and optimization strategies. Advanced characterization techniques are emphasized as essential for understanding the catalytic mechanisms and dynamic behaviors of zeolites, thereby facilitating further research into their efficient and effective use. The study concludes by underscoring the importance of continued research to refine synthesis and characterization methods, which is crucial for optimizing catalytic activity in CO_2 hydrogenation. This effort is important for achieving selective catalysis and is paramount to the global initiative to reduce carbon emissions and address climate change.

Keywords: carbon dioxide hydrogenation; structured zeolites; zeolites synthesis and characterization

1. Introduction

Global warming is a significant concern as it is exacerbated by the excessive emissions of greenhouse gases, specifically carbon dioxide (CO_2), which contribute to mounting climate-related issues [1]. CO_2 hydrogenation is gradually emerging as a critical technology for carbon neutrality, which provides a route for CO_2 valorization into valuable chemicals and fuels [2,3]. The process offers a two-fold solution; besides addressing the pivotal issue of CO_2 emissions, it also assists in the sustainable production of heavy hydrocarbons [4]. The selective and optimized transformation of CO_2 represents a current challenge considering its thermodynamic stability and inert properties [5].

There are two primary methods for converting CO_2 into hydrocarbons, both of which have also been industrially applied for decades, reflecting their proven scalability and commercial viability. The first method, modified Fischer–Tropsch synthesis (modified-FTS), combines the reverse water-gas shift (RWGS) reaction and the Fischer–Tropsch reaction [6]. This process not only synergistically utilizes CO_2 but also transforms it into various hydrocarbons, as evidenced by the long-standing operation of SASOL plants in South Africa and the ORYX GTL plant in Qatar, which all utilize FTS technology [7,8]. The second method is the methanol-mediated route, where CO_2 is first converted into methanol and then transformed into hydrocarbons through the methanol-to-hydrocarbon (MTH) process. This pathway is exemplified by the operation of the world's largest methanol plant in Iran, which uses CO_2 as a feedstock [9]. Given the thermal hydrogenation foundation of both technologies, unless otherwise stated, this review will therefore primarily focus on the thermal hydrogenation of CO_2. This approach has been comprehensively explored over structured zeolites in terms of mechanisms, kinetics, catalyst development, and industrial improvements and applications, as reviewed extensively by Ye et al. [10], Ojelade and Zaman [11], and Wang et al. [12], who have detailed the underlying processes and implementations of both the modified-FTS and MTH routes. Structured zeolites, with their unequaled porous structures and acid-base properties, are used as catalysts in a whole range of chemical processes, among them CO_2 hydrogenation. Figure 1 shows the classes of well-known zeolites (e.g., natural [13], and synthetic [14–16]). Both zeolite types have been widely applied not just in catalysis [17], but even in other industries such as water and wastewater treatment, agriculture, biomedicine, laundry detergents, and construction [18–22].

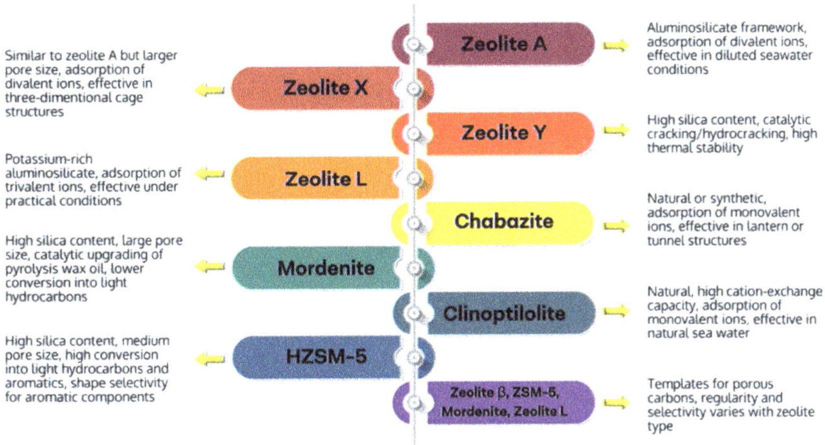

Figure 1. Types of zeolites and their synthesis, application, and performance.

For CO_2 hydrogenation, zeolite application has received considerable attention over the last decade. Figure 2 shows the distribution of several articles displayed on Google Scholar when the phrase "CO_2 hydrogenation over zeolites" is used to search articles. The trend indicates unprecedented growth in the application of zeolites for CO_2 conversion reactions, particularly over the last decade, where an exponential trend has been seen. The recent signs of progress have mainly concentrated on the improvement of the zeolites through careful control of their compositions, porous characteristics, and coupling of thermal and mechanical properties that render promotion of reaction selectivity and enhance the production of longer-chained hydrocarbons such as aromatics, liquid fuels, alcohols, and heavy olefins [5,23]. It should be noted that zeolites by themselves have been widely studied for various applications as aforementioned; nonetheless, the recent developments in the field of CO_2 catalysis and the growing interest of researchers in the application of zeolites for CO_2 conversion create a demand for a review covering the growth in the field, primarily over the last decade, which is the motivation behind this article. This study intends to summarize the recent advances in the area of CO_2 hydrogenation over structured zeolites by offering a synoptic view of the different influential parameters, synthesis techniques, characterization requirements, and catalytic functions. The review critically evaluates the current literature that should bring out new ideas, findings of key innovations, and significant breakthroughs.

It will embrace the rudiments of the process, the specific traits of structured zeolites, and their role in augmenting catalytic activity. In addition, there will be reflections on the challenges and prospects in this field; hence, one can get ideas on structured zeolites and the way they may be used to reduce the global CO_2 problem.

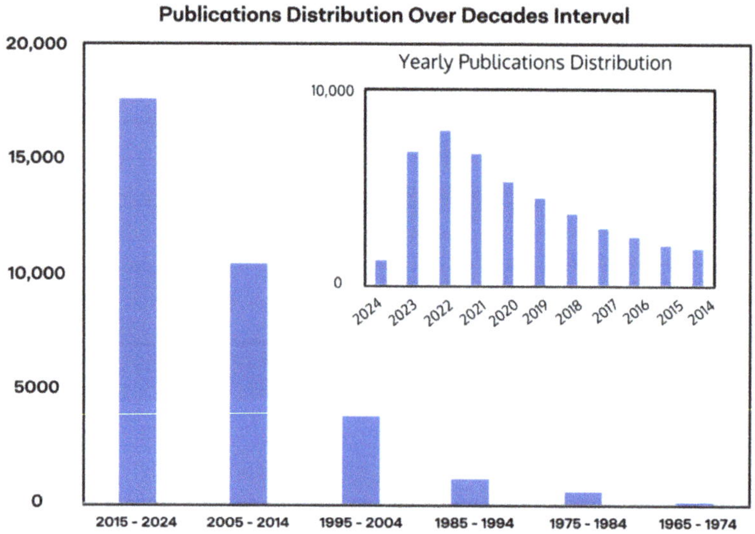

Figure 2. Number of publications displayed by Google Scholar search using the phrase "CO_2 Hydrogenation over zeolites".

2. Thermal CO_2 Hydrogenation over Other Methods: A Brief Overview

In addition to the well-established thermal routes for CO_2 hydrogenation, such as FTS and MTH mentioned in the introduction section, various alternative methods have been explored to expand the range of technologies for CO_2 reduction. Table 1 shows the alternatives, which include electrochemical, photocatalytic, biochemical, and chemo-enzymatic methods, each characterized by distinct mechanisms and specific applications. While each technology has its own merits and limitations, its implementation in industries largely depends on the type of industry, the purity level of the product, and the scale of processing.

Table 1. State-of-the-art overview of CO_2 hydrogenation technologies: assessing advantages and limitations.

Method	Advantages	Disadvantages
Thermal [10–12]	Proven scalability, high product yields, extensively used in industrial settings	Can be energy-intensive; less environmentally friendly without renewable energy integration; catalyst stability issues
Electrochemical [24,25]	Facilitates CO_2 reduction using electricity, can be integrated with renewable electricity sources; scalable; stable long-term	Often a broad product distribution at high conversions; high energy demand; economically challenging
Photocatalytic [26,27]	Utilizes sunlight, environmentally friendly; no need for external energy inputs; economically feasible	Lower efficiency and stability, which hinders scalability; lower yields
Biochemical and Chemo-enzymatic [28–30]	High selectivity; operates under mild conditions; efficient for producing bulk chemicals	Slow reaction rates; complex maintenance of biocatalytic activity; high temperatures can inhibit enzyme activity

However, despite the diversity of available technologies, the foundation of hydrocarbon processing industrial plants still predominantly relies on thermal conversion methods. This is not only evident in traditional sectors like chemical synthesis via the FTS and MTH routes but is also a staple in the oil and gas industries. Processes such as Gas to Liquid (GTL) and Natural Gas Reforming (NGR) for hydrogen production extensively employ thermal methods due to their robustness and scalability [31,32]. These processes are integral to the production workflows in these industries, underlining the predominance of thermal techniques.

The thermal CO_2 hydrogenation process appears more energy-efficient than the solar route. For example, the thermal method benefits from a novel membrane reactor that enhances methanol yield and requires a smaller reactor, which exhibits superior energy efficiency at 70.3% [33]. In contrast, a solar CO_2 hydrogenation process achieves an energy efficiency of 15.5%, with a 26% improvement over such previous methods possible under optimal conditions [34]. Despite the advancements in solar hydrogenation, the thermal approach is suggested to be superior in terms of energy efficiency. This sharp contrast highlights the larger energy savings characteristic of thermal CO_2 hydrogenation. In addition, Zeng et al. [35] highlighted that plasma power (a form of thermal hydrogenation) is more efficient than heating power (a form of solar hydrogenation), even at lower temperatures, showing the highest conversions in the presence of plasma.

Unlike thermal processes, electrochemical processes are energy-intensive, and the lifespan of catalytic materials is low for such processes [36]. Hence, the application of electrochemical conversion in industries remains limited, even today, due to practical challenges in optimizing key parameters such as Faradaic efficiency (FE), Energetic Efficiency (EE), and Current Density (CD), which are crucial for the success of this approach [37,38]. To overcome these obstacles, catalysts must possess rapid electron transmission capabilities, superior electron transport ability, and robust electron-transporting features [36,39].

The primary distinction between photocatalytic CO_2 reduction and electrochemical CO_2 reduction is in their electron sources. In photocatalytic reduction, electrons are generated by exposing semiconductors to light, whereas in electrochemical reduction, electrons are sourced through the application of an electrical current. However, since photocatalytic methods depend on light absorption, they often suffer from limited efficiency, poor selectivity, and insufficient product variety [40]. In addition, such processes are intricate, encompassing various mechanisms like electron and proton transfer as well as the formation and breaking of chemical bonds, which remain poorly comprehended [37]. Furthermore, the process of photocatalytic conversion necessitates substantial capital investment and energy, especially when using semiconductors with wide band gaps [41]. Substantial challenges in substrate-based photocatalysts include the degradation of polymers, the creation of volatile by-products, and the complexities in separating the catalyst from the final product [41]. Therefore, the catalysts used in such a conversion process need to be optimized and treated, which can additionally increase the overall expenses [42].

Biochemical and chemo-enzymatic methods, though innovative and sustainable, face challenges due to the high cost of cofactors, and complex biological systems and demand a need for efficient cofactor regeneration for large-scale applications [43,44]. One of the challenges that the biochemical conversion process faces is the maintenance of an adequate nutrient supply, the collection of end products, and the sustainability of the microbial cultures. It has been suggested that genetic modification of these organisms could overcome such problems [43]. However, this approach can fail in some cases, as the insertion of many gene copies can accidentally inactivate some genes—a situation called genetic knockout [41]. Hence, pretreatment is required, and productivity is lower when compared to that of the thermo-chemical process. Generally, enzymes, often costly and characterized by limited stability, activity, and reusability, face constraints in their industrial usage. Technologies like enzyme modification and immobilization require further advancement to decrease costs and boost enzyme performance and recyclability, thereby enhancing economic viability [45]. Additionally, the slow reaction rate significantly hinders the industrial application of enzymatic CO_2 conversion, necessitating the deployment of biocatalysts to address kinetic issues and achieve a faster, more effective conversion process.

Therefore, considering the extensive application, proven scalability, and integration within the hydrocarbon industrial processes, the thermal route remains the preferred method for CO_2 hydrogenation due to its robustness and higher reaction rates, especially when efficiency and selectivity are paramount. This assertion is corroborated by the above case studies, which clearly illustrated the limitations inherent in alternative methods and the comparative strengths of the thermal approach. Drawing upon these insights, Table 2 encapsulates the array of advantages that thermal methods hold over their electrochemical, photocatalytic, biochemical, and chemo-enzymatic counterparts.

Table 2. Efficacy of thermal conversion against electrochemical, photocatalytic, biochemical, and chemo-enzymatic methods in CO_2 hydrogenation.

Hydrogenation Feature	Thermal Conversion [33–35]	Electrochemical Conversion [36–39]	Photocatalytic Conversion [37,40–42]	Biochemical and Chemo-Enzymatic Conversion [41,43–45]
Energy Efficiency	High, but varies based on the used catalytic materials	Energy-intensive with usually low catalytic material lifespan	Limited by light absorption efficiency	High cofactor costs impact overall efficiency
CO_2 Conversion Rate	High at moderate to high temperatures	Can be high based on applied potential and choice of catalyst	Hampered by poor selectivity and product variety	Less productive, slow reaction rates
Catalyst Durability	Robust; choice of reducible supports enhances durability	Often short lifespan due to rigorous operational conditions	Challenges with catalyst degradation and by-product formation	Enzyme stability and recyclability need enhancement
Capital and Operational Costs	Relatively cost-effective due to higher efficiencies, and easy integration with reforming industries	Relatively high due to the need for optimizing efficiency and current density. Still requires technological maturity for CO_2 conv.	Significant capital investment and energy required, especially for wide bandgap semiconductors	Costly due to the high price of enzymes and cofactors
Technological Maturity	Advanced	Remains limited in industry due to practical challenges	Processes are complex and mechanisms are not fully understood	Requires advancements in enzyme modification and immobilization
Reaction Mechanisms	Complex but widely studied on a variety of catalysts	Complex with limited understanding. Requires intricate tools to monitor reaction intermediates near electrodes in liquid medium	Involves intricate mechanisms that are poorly understood	Constrained by the complexity of biological systems
Selectivity	High and can be tailored for desired outcomes	Low; produces a mix of products. High conversion is often reported with broad product distribution	Often poor due to multiple possible reactions	Can be very selective. Dependent on the specificity of biological pathways
Scalability	Highly scalable and suitable for industrial scale-up	Faces practical challenges in scaling. Requires further technological development.	Limited scalability due to technical and efficiency constraints	Challenging due to the need to maintain microbial cultures and nutrient delivery
Downstream Processing	Simplified due to few by-products	Could require separation of multiple by-products	Complexities in separating the catalyst from the product increase expenses	Demands efficient cofactor regeneration and faces genetic knockout issues
Potential for improvement	Relatively well-established with room for incremental improvements based on catalyst and reactor design	Requires catalysts with fast electron transfer and robust transport features	Needs optimization of photocatalysts and treatment processes	Genetic engineering is suggested but leads to challenges like genetic knockout

Recently, the literature documents the generation of liquid fuels, aromatics, and olefin compounds via the thermal catalytic hydrogenation process applied to CO_2 [12,46,47]. In addition, studies indicate that the inclusion of structured zeolites in thermo-catalytic hydrogenation processes has the potential to enhance the selectivity of hydrocarbon products [48,49]. For example, by integrating a Fe-Zn-Zr catalyst with H-ZSM-5, a notable selectivity for isoalkane products, reaching up to 91.9%, can be achieved [49]. Hence, utilizing the thermal CO_2 hydrogenation approach for zeolites as a catalytic medium brings forth a spectrum of advantages, underscored by their structural features that dictate the pathways of the hydrogenation reaction (more in Section 3.1). In addition, zeolites are highly effective in thermal CO_2 hydrogenation due to their extensive surface area and porosity, which provide numerous active sites, enhancing reaction rates and energy efficiency (more in Section 3.2). Their selective pore structure also plays a crucial role by ensuring molecular sieving, allowing only specific hydrocarbon production, thus increasing selectivity and reducing the need for further separation processes. This makes zeolites exceptionally suitable for controlled and efficient chemical processes. The acidity of zeolites is adjustable, which can be tailored to enhance catalytic activity for thermal CO_2 hydrogenation (more in Section 3.3).

Furthermore, zeolites' thermal stability is also pivotal, as they withstand the necessary high temperatures for thermal hydrogenation, preserving catalytic activity (more in Section 3.4). Additionally, zeolites' ability to be regenerated through thermal treatments allows for their reuse, ensuring economic viability. The ability to regenerate zeolites thermally without significant loss of structural integrity or catalytic activity is a crucial advantage over other catalysts that might degrade or require complex regeneration processes. This is particularly aligned with the characteristics emphasized by the synthesis methods for zeolites; bottom-up and top-down approaches (more in Section 4). This renderability, along potential cost savings and a reduced environmental impact due to less frequent catalyst replacement and decreased waste, positions thermal CO_2 hydrogenation with zeolites as a method that is both efficient and sustainable compared to the aforementioned hydrogenation approaches. Furthermore, zeolites are already widely used in various industrial applications, including oil refining and petrochemical processes [50]. Thus, their use in CO_2 hydrogenation does not require radically new technologies or setups. This compatibility with existing industrial infrastructure and processes reduces the barriers to implementation and capital investment, making it a more feasible option compared to other methods that might require specialized equipment or conditions.

3. Influential Parameters

Zeolites, with their tunable structural and functional properties, are key catalysts in CO_2 hydrogenation processes [51,52]. These structured zeolites offer specific pore structures, a balanced density of acid/basic sites, and the capacity to support and stabilize active metal sites, making them essential for controlled catalytic activity and improved product selectivity. This section explores complex parameters influencing zeolite performance in CO_2 hydrogenation, such as their composition, bimodal mesoporosity, oxygen vacancies, Si/Al ratio, thermal stability, structural integrity, and electrical and plasma interactions. Understanding these factors is crucial not only for defining their role in developing effective and sustainable processes for converting CO_2 into valuable chemicals but also for optimizing zeolites for their potential in industrial applications. Selected examples are presented further to highlight this key aspect of structured zeolites in CO_2 conversion.

3.1. Interplay of Composition and Preparation

Structured zeolites, typically modified with metal-based catalysts, present a multifaceted platform where the interplay between the zeolite topology, metal characteristics, and preparation methods determines the efficiency and outcome of the hydrogenation process. Metal-based catalysts are instrumental in defining product selectivity. These catalysts, when integrated with zeolites, leverage the unique hybrid nature of active sites within the zeolite framework. This configuration supports specific reaction intermediates that dictate the pathways of the hydrogenation reaction, thereby influencing the selectivity toward different hydrocarbons [5]. Wei et al.'s [53] investigation found that the topology of various zeolites, including H-Y and H-BEA, influenced the distribution of hydrocarbons during fuel production, although CO_2 conversion rates remained consistent. Zeolites with ten-membered rings, such as HZSM-5, were notably efficient in enhancing the selectivity towards larger hydrocarbon molecules in the gasoline range (C_5–C_{11}), with HZSM-5 demonstrating a preference for heavier fractions. Acidity levels were critical, with medium acidity in HZSM-5 (160) favoring the desired reaction, whereas stronger acidity prompted excessive cracking of larger hydrocarbons. Popova et al. [54] developed both mono- and bimetallic catalysts with Ni and Ru on ZSM-5 zeolites, noting the impact of nickel (Ni) content and the introduction of ruthenium (Ru) on their properties. Characterization techniques (XRD, TPR-TGA, TEM, XPS) revealed a fine dispersion of Ni and ruthenium oxide (e.g., RuO_2, RuO_4) across the zeolite's surface and pores (more in Section 5). Tests in CO_2 hydrogenation showed that the 10Ni5RuZSM-5 variant exhibited excellent methane production efficiency, achieving 100% conversion with complete selectivity at 400 °C, and displayed promising stability for practical use. Furthermore, the composition of zeolites, particularly when doped with metals such as lithium (Li), can significantly alter catalyst reactivity. For instance, Kitamura et al. [55] investigated CO_2 hydrogenation using Rh zeolites enhanced with Li (see Figure 3a,b). They found that increasing Li levels shifted the primary product from methane to CO and ethanol, with a notable increase in ethanol production when CO was present. FTIR and TPR analyses suggested that Li additives fostered new reaction pathways on the catalyst surface, leading to improved CO_2 adsorption and stabilization of adsorbed CO species. Similarly, incorporating molecules like Na^+, N, and Ge into the ZSM-5 zeolite framework was demonstrated to lower the energy barriers for CO_2 hydrogenation, thereby enhancing the overall catalytic efficacy [56].

Preparative methods also significantly contribute to tuning the characteristics of zeolites for the best performance in CO_2 hydrogenation. The synthesis method of zeolite, such as the choice of template, calcination conditions, and metal loading approaches, can also significantly affect the structural and chemical properties of the catalyst. Liu et al. [57] investigated how alkali metal promotion in Co-zeolite catalysts affects product selectivity, comparing IWI with the IE approach (see Figure 3c,d). Their findings revealed distinct differences: IWI led to a higher yield of olefins and C_{5+} hydrocarbons, attributed to an elevated Si/Al ratio (see Figure 3e). In contrast, the IE method's selectivity remained largely unaffected by changes in the Si/Al ratio, likely due to alkaline poisoning (see Figure 3f). Hence, the zeolite framework topology determines the behavior of CO_2 and hydrogen (H_2) in the active sites. Specific zeolite structures with specific synthesis procedures can selectively adsorb these reactants and promote their transformation into the desired products under the appropriate conditions. For instance, Ramirez et al. [58] explored the significant role of zeolite topology in the direct conversion of CO_2 into valuable hydrocarbons. They reported that catalysts using different zeolite frameworks, specifically MOR and ZSM-5, exhibit markedly different behaviors in their ability to convert CO_2 to olefins and aromatics, respectively. The study also highlighted that the unique structural characteristics of MOR favor olefin production, while the properties of ZSM-5 enhance aromatic synthesis under the same reaction conditions. This variation is attributed to how these zeolite structures interact with potassium superoxide-doped iron oxide catalysts, influencing the formation of surface formate species and the activity of carbocation intermediates (see Figure 4).

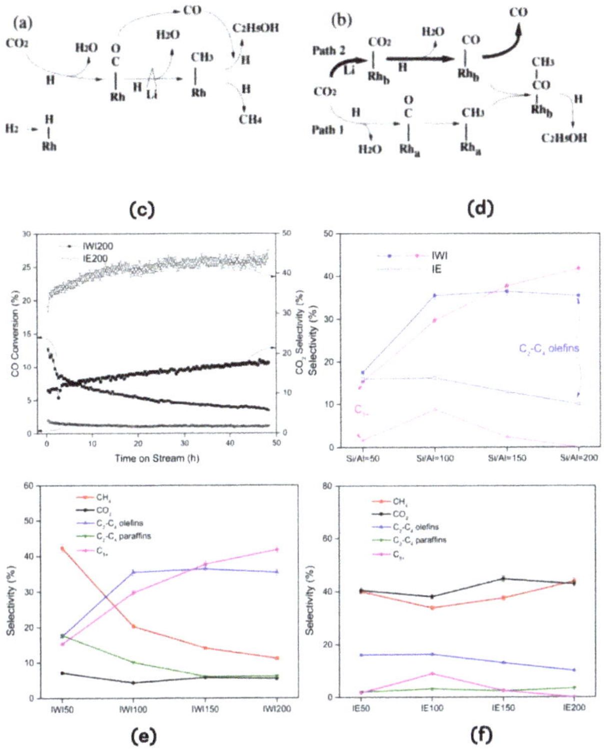

Figure 3. (**a**,**b**) Visual representation of the chemical interactions and mechanisms on a Li/RhY catalyst consisted of 5 wt% Rh and a Li/Rh ratio of 10 [55]. (**c**,**d**) Analysis of the impact of preparative techniques, specifically incipient wetness impregnation (IWI) and ion exchange (IE), on the performance of Co-zeolite catalysts [57], focusing on (**e**) product selectivity for IWI catalysts at different Si/Al ratios and (**f**) product selectivity for IE catalysts at varying Si/Al ratios.

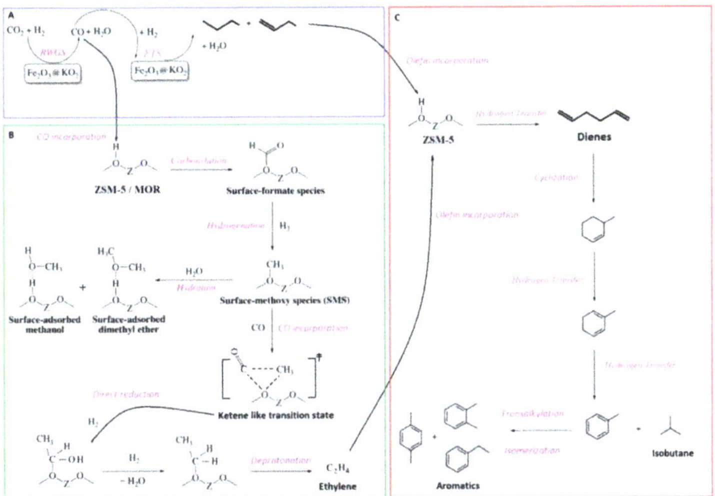

Figure 4. Suggested CO_2 conversion pathways to produce light olefins and aromatics [58]. (**A**) pathway for utilizing only $Fe_2O_3@KO_2$/zeolite catalyst (**B**) CO integration pathway on MOR and ZSM-5

(**C**) Aromatics generation pathway on ZSM-5. ‡ indicates the formation of a short-lived intermediate in the reaction pathway. Specifically, it points to a "ketene-like transition state" during CO incorporation into a surface-adsorbed dimethyl ether.

3.2. Bimodal Mesoporous Structure and Surface Oxygen Vacancies

Bimodal mesoporous zeolites, with their dual-sized pores, optimize CO_2 hydrogenation by boosting mass transfer and catalytic site availability. The adjustment of the zeolite's pore architecture can be accomplished by developing a hierarchical pore system. This system ensures that the zeolite not only retains its intrinsic micropores but also incorporates supplementary meso- or macropores, which may be located either within or between the crystals [59,60]. In this way, the mesopores enhance the ingress of reactants like CO_2 and H_2, while surface oxygen vacancies act as active sites or alter adjacent metal sites to better activate CO_2 and promote essential reduction steps in the hydrogenation process. Song et al. [61] explored the influence of zeolite morphology and pore structure, focusing on aromatics production using a Cu-Fe_2O_3-HZSM-5 catalyst. The study involved commercial zeolites and those synthesized via various methods such as dry-gel, hydrothermal, and phase-transfer techniques. Their research found a preference for mesoporous structures in enhancing aromatic production, particularly noting that the HZSM-5 produced by the phase-transfer method showed an improved selectivity for aromatics from 56.61% to 61.94%, attributed to its rich 5–12 nm pores, which facilitate the formation of aromatics like BTX. In the same investigation, they also proposed an "H recycling mechanism" for CO_2 hydrogenation to aromatics, suggesting that hydrogen species from the zeolite can move to the oxide interface and interact with adsorbed CO_2 at oxygen vacancies. Similarly, Tian et al. [62] developed M(Z, Ca, In)-UIO-66/palygorskite-HZSM-5 zeolite tandem catalysts for converting CO_2 directly into aromatics, employing Diffuse Reflectance Infrared Fourier Transform Spectroscopy (DRIFTS) and Density Functional Theory (DFT) to investigate the reaction mechanisms. The standout, 6%Zn-UIO-662/Z5 at 550 °C, achieved the highest specific surface area of 437.50 m^2/g, optimal acidity, and numerous oxygen vacancies, leading to an impressive CO_2 conversion rate and aromatics selectivity of 88.8%.

The adsorption properties and pore size distributions of Zn-UIO-66/Z5 tandem catalysts across varying calcination temperatures and Zn contents are depicted in Figure 5. The catalysts exhibited type IV isotherms with H4-type hysteresis at higher pressures (P/P0 = 0.7–1.0), indicating the presence of both micropores and mesopores, the latter attributed to the agglomeration state of acidified palygorskite nanofibers. Adjusting crystallization times during Z5 zeolite preparation allowed for controlled interstitial pore formation between nanofiber crystals, thus improving reactant and product diffusion. Notably, the addition of Zn enhanced the adsorption capacities, with optimal results at 6% Zn inclusion. However, excessive calcination temperatures can compromise structural integrity and reduce adsorption capacities. The integration of Zn and UIO-66 into Z5 zeolite and the presence of acidified palygorskite were key to enhancing pore structure and surface area, which in turn improved the catalyst's performance in CO_2 hydrogenation to aromatics. The pore size distribution is predominantly centered around 10–20 nm, contributing to efficient catalysis. Therefore, the mesoporous structure within zeolites enhances the movement of olefin intermediates towards acidic sites. This improved diffusion pathway leads to increased production yields of aromatic compounds. Furthermore, in the work by Yan et al. [63], they discovered that Ni catalysts supported on zeolites with lower Si/Al ratios (such as 4A, Ni-5A, and Ni-13X) possess large Ni nanoparticles that engage weakly with the support and exhibit lower activation temperatures, leading to notable CO_2 methanation activity, with conversions between 70.4–70.9% and a high methane selectivity of 92.4–96.4%. In contrast, zeolites with higher Si/Al ratios and mesoporous structures (such as Ni-ZSM-5 and Ni-BEA), while demonstrating smaller Ni particles and stronger metal-support interaction, resulted in lower CO_2 conversion rates (71.2–73.4%) and methane selectivity, indicating another crucial role of mesopores in catalytic performance. When selectivity decreases in conjunction with the mesoporous structure of zeolites, it suggests that while the mesopores may enhance the accessibility of reactants to the catalytic sites by improving diffusion, they may also allow for a wider range of reactions to occur, which can lead to a broader array of products. In other words, the larger mesopores are not as selective in facilitating only the desired reaction pathways; instead, they might enable side reactions, leading to an assortment of products rather than selectively yielding the target molecule. For CO_2 hydrogenation specifically, if a mesoporous zeolite results in decreased selectivity, it means that while more CO_2 might be converted (enhanced activity), the proportion of the desired product (such as methane in the case of methanation in Yan et al.'s study [63]) is lower because other products are also formed. This could be because mesopores allow not only the desired reactants to access the active sites more easily but also potentially allow unwanted reactants to participate in the reactions, or they might not effectively stabilize the transition states leading to the desired products. Overall, while bimodal mesoporosity and surface oxygen vacancies can improve catalytic activity, their influence on selectivity is a delicate balance that requires careful optimization to favor the desired CO_2 hydrogenation pathways.

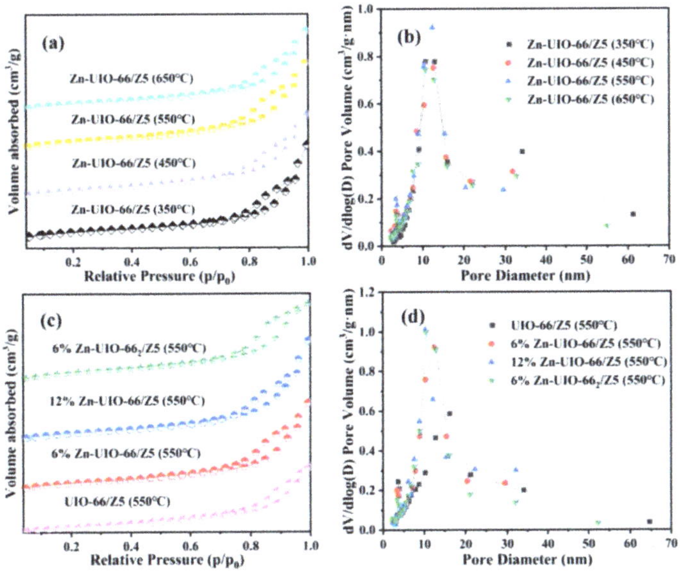

Figure 5. Bimodal mesoporous performance [62]: (**a**) N_2 adsorption–desorption isotherms, and (**b**) pore size distribution for Zn-UIO-66/Z5 at varying calcination temperatures. (**c**) N_2 adsorption–desorption isotherms, and (**d**) pore size distribution for Zn-UIO-66/Z5 with different zinc concentrations.

3.3. Si/Al Ratio

Zeolites with minimal acidity tend to promote hydrogen transfer reactions, often leading to the generation of longer-chain hydrocarbons. In contrast, an excessive number of active sites may be conducive to the formation of saturated hydrocarbons, eventually leading to coking [11]. Thus, the Si/Al ratio must be adjusted with high precision, because this ratio is a direct measure of acidity, to maximize the number of Brønsted acid sites for the best performance. Da Costa-Serra et al. [64] found that Ni-based catalysts mounted on delaminated ITQ-2 zeolite, when presenting higher Si/Al ratios, demonstrated enhanced catalytic activity attributed to increased hydrophobicity. The catalytic outcomes for the ITO-2 and ZSM-5 frameworks varied based on their aluminum content, with high aluminum leading to consistent catalytic behaviors across different zeolite types. Notably, a reduction in aluminum content in the support contributes significantly to catalytic efficiency, as seen in delaminated ITO-2, identified by XRD (see Figure 6a) to have smaller NiO crystallite sizes, resulting in higher CO_2 conversion (50% improvement) and CH_4 selectivity (98%). However, the size distribution of Ni^0 particles in the 5Ni/ITQ-2(∞) sample remains tightly controlled and closely matches the pre-reaction size, measuring 2.8 nm after the reaction compared to 2.4 nm beforehand (see Figure 6b).

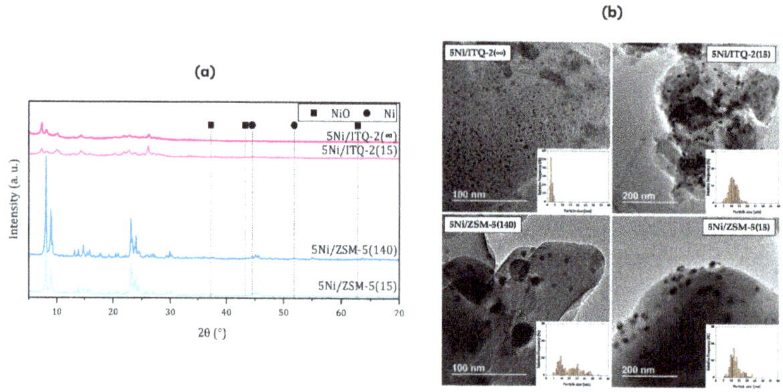

Figure 6. Catalytic Activity of Ni catalysts [64]: (**a**) XRD profiles when mounted on ITQ-2 and ZSM-5. (**b**) HR-TEM image of particle size distribution of Ni^0 in the 5 wt.% Ni catalyst supported on ITQ-2 and ZSM-5 with varying Si/Al ratios following the methanation reaction.

The study by the same authors revealed that higher aluminum content supports (lower Si/Al ratio) yielded similar performances in catalysts due to enhanced Brønsted acidity and hydrophobicity, which are beneficial for reaction progress. Additionally, a high Si/Al ratio improves the dispersion and stability of Ni nanoparticles, influencing their initial dispersion and subsequent thermal redistribution. This relationship highlights the crucial role of aluminum content and Si/Al ratios in determining the properties and effectiveness of these catalysts, which were also previously reported to be significant for methanation processes [65]. Delving deeper, Tada et al. [66] developed multifunctional catalysts, ZnO/ZrO_2, by blending mordenite (MOR) zeolites with varying Si/Al ratios (9, 14, and 104) with ZnO/ZrO_2 and assessed their impact on CO_2 hydrogenation efficiency using a high-pressure flow reactor. Figure 7 illustrates the space-time yield of various hydrocarbons for three different catalysts, namely MOR009, MOR014, and MOR104, over a span of time.

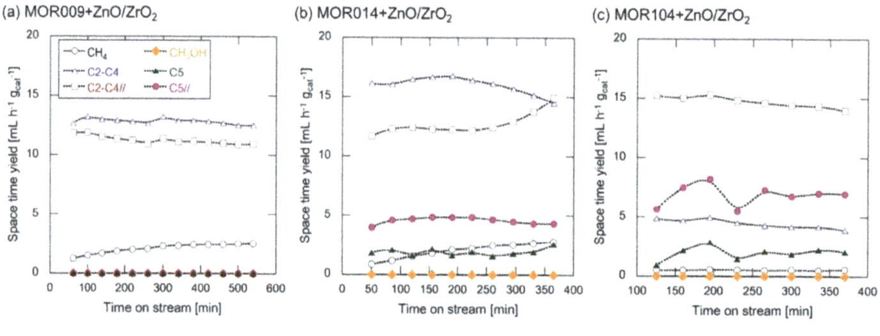

Figure 7. Performance of Zn catalysts supported by MOR zeolites with varied Si/Al ratios and space-time yield of methanol [66]: (**a**) MOR009, (**b**) MOR014, (**c**) MOR104.

For MOR009 (see Figure 7a), the production levels of methane and C_2–C_4 olefins are fairly consistent, though the yield of methanol shows slight fluctuations. Notably, the production of C_5 olefins and paraffins is almost non-existent, which indicates a limited ability to generate higher carbon number hydrocarbons. In contrast, MOR014 in Figure 7b demonstrates a growing yield of C_5 olefins that peaks midway through the assessment period and then modestly declines, while the output of C_5 paraffins remains stable. This behavior suggests a heightened capacity for creating heavier olefins compared to MOR009. Lastly, the MOR104 catalyst in Figure 7c maintains a stable output of C_5 olefins throughout the period, with only minor fluctuations observed. The production rates for C_2–C_4 paraffins are consistent, although C_5 paraffins exhibit some variability. Overall, MOR104 displays a robust capability for producing C_5 olefins, highlighting its effectiveness for sustained light olefin production. For CO_2 aromatization, Cui et al. [67] assessed the impact of zeolite acidity by varying the Si/Al ratio using a Na-modified spinel oxide ZnFeOx coupled with H-ZSM-5 zeolites. They observed that an increased Si/Al ratio led to a decrease in aromatic selectivity due to the reduced density of Brønsted acid sites, which are pivotal for aromatization. Consequently, a higher density of Brønsted acid sites results in greater selectivity for aromatics [68].

However, greater CO_2 adsorption does not necessarily imply greater hydrogenation activity, which is the main problem with typical zeolite catalysts. The hydrogenation process is not solely dependent upon the amount of CO_2 adsorbed. Instead, it is also dependent upon the availability and reactivity of H_2 within the catalyst system and the catalyst's ability to help the reaction between CO_2 and H_2 to form hydrocarbons. A zeolite with a higher CO_2 adsorption capacity, in consequence of its higher basicity (linked to a lower Si/Al ratio), may be able to retain CO_2 more tightly, thus hindering the reaction with H_2, if the CO_2 is not easily released to react with H_2. The equilibrium between the strength of the adsorption and the reactivity is fragile, whereby strong adsorption could interfere with the catalytic process, while weak adsorption might not capture CO_2 efficiently enough for the reaction. Therefore, the ideal scenario for enhanced hydrogenation activity is not just high CO_2 adsorption but also an optimal level of adsorption that allows for effective interaction with H_2, leading to the formation of the desired hydrocarbon products.

3.4. Porosity, Thermal Stability, and Structural Integrity

The porosity, thermal stability, and structural integrity of structured zeolites are also pivotal for optimizing CO_2 hydrogenation processes. These factors are integral to the zeolite's ability to adsorb reactants, withstand high temperatures, and maintain its structure under reaction conditions, which, in turn, affects catalytic performance, product selectivity, and potentially its commercial applicability. In a pioneering study by Xiang et al. [69], an AFX-type SAPO-56 zeolite was innovatively augmented with copper (Cu) nanoparticles, crafting a multifaceted catalyst aimed at converting CO_2 to methanol. This novel catalyst achieved a CO_2 to methanol conversion rate of 16.4% and showcased an impressive 80.9% selectivity for methanol production. In this study, the strategic dispersion of Cu nanoparticles across the extensive pore network of the SAPO-56 zeolite markedly boosted H_2 adsorption and its subsequent dissociation, facilitating widespread H_2 atom availability. These atoms interacted with the zeolite's intrinsic Lewis acid sites, effectively catalyzing methanol synthesis. The catalyst's hierarchically structured porosity was pivotal in steering product selectivity while ensuring exceptional stability, highlighting its potential for industrial deployment in CO_2 reduction initiatives. Similarly, Cui et al. [70] made significant strides in the catalytic conversion of CO_2 to methanol by creating a zeolite-encapsulated Cu/ZnOx@Na-ZSM-5 catalyst with a core–shell structure derived from bimetallic CuZn-HKUST-1 nanoparticles via hydrothermal synthesis. This design also effectively prevented Cu sintering and enhanced the catalyst's thermal stability. The zeolite framework not only contributes to a higher methanol yield by ensuring a synergistic effect and intimate interaction of the encapsulated nanoparticles but also upholds the structural integrity of the catalyst, promoting sustained activity over prolonged durations. These innovations demonstrate the critical role of porosity and the architectural solidity of zeolites in fostering stable and efficient CO_2 hydrogenation processes.

3.5. Electrical and Plasma Interactions

The control of electrical and plasma properties in catalyst design is also essential for achieving high CO_2 conversion rates and selectivity to the desired products operating under various operational conditions. Figure 8 illustrates a setup for plasma-assisted CO_2 hydrogenation using a dielectric barrier discharge (DBD) reactor [71]. Feliz et. al. [72] observed that zeolite structures affect the ionic conductivity and dielectric constant of the catalyst in plasma-assisted reactions. Lower ionic conductivity facilitates a better CO_2 conversion (34%) because of decreased plasma density and CO_2 dissociation at ambient pressure and temperatures under 100 °C. Owing to zeolites' unique structures, they have been recognized as prospective catalyst supports for DBD plasma systems intended for CO_2-to-methane conversion. Attention has been especially given to Ni-based catalysts within these plasma systems [73,74]. For example, the utilization of Ni/beta zeolite catalysts has been associated with a greater selectivity for methane production, with up to 95% conversion for either CO or CO_2 [75]. It was noted that Ni particles could be redistributed under DBD plasma exposure, which might improve the catalytic process. In addition, this treatment was recommended to promote the desorption of the adsorbed molecules from the catalyst, whereby these adsorbed molecules have much weaker bonds compared to their gaseous counterparts. Reactive species that were produced by plasma were believed to play a role in removing the adsorbed CO molecules, which is a key step for achieving higher methane yield, while the accumulation of water on the catalyst's active sites, emerging as a reaction byproduct, poses a significant limitation for thermal CO_2 methanation by obstructing access to these sites, in line with the insights provided by Sabatier's principle [76,77].

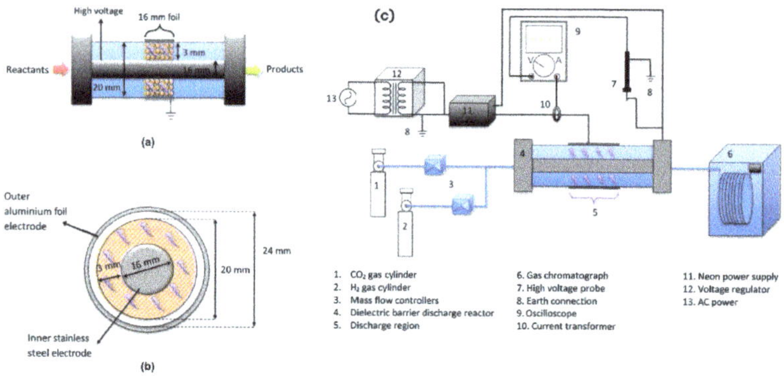

Figure 8. Plasma-assisted catalytic conversion of CO_2 to ethane using a DBD reactor [71]: (**a**) reactor set-up, (**b**) cross-sectional view of the reactor, (**c**) process flow diagram of the overall set-up.

When discussing electrical and plasma interactions in such catalytic systems, it is useful to look at the wider meaning of ionic conductivity and its impact on catalytic processes. The work of Henckel et al. [78] is a relevant example of this, as they used Electrochemical Impedance Spectroscopy (EIS) to investigate the interaction between the catalyst and ionomer, with a particular interest in the transport of hydroxide ions in the electrochemical CO to ethylene conversion. The results confirmed that fine-tuning ionic conductivity can substantially promote the Faradaic efficiency of ethylene and the cell voltage. Although the study did not directly concern zeolite materials and hydrogenation of CO_2, the principles that can be derived from their work are of great interest to CO_2 hydrogenation over structured zeolites. The principle of modulating ionic behavior to maximize catalyst performance is a general concept that is observed in diverse catalyst systems, including zeolites for CO_2 hydrogenation. As it happens in electrochemical systems where both ionic transport and electrical interactions are crucial for the reaction outcomes, in zeolite catalysts, these characteristics can affect the diffusion of reactants and the distribution of active sites and subsequently determine the efficiency and selectivity of the hydrogenation reactions.

4. Synthesis Methods

Structured zeolites can be synthesized in many ways by utilizing precursors in soluble form and reducing them, or by using existing solid compounds and modifying their structures. These methods can be broadly grouped into two categories: bottom-up and top-down approaches [79], as shown in Figure 9a.

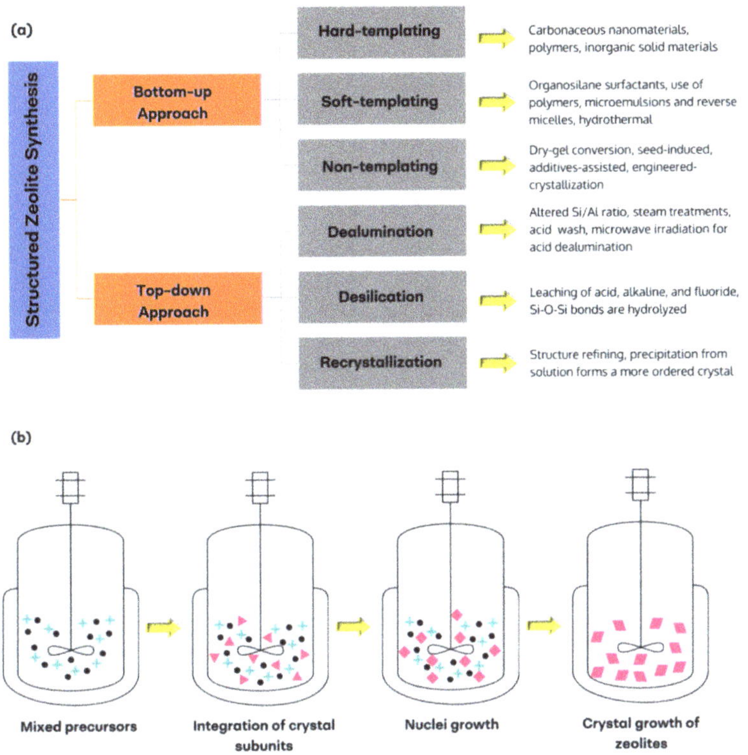

Figure 9. Pathways for zeolite synthesis. (**a**) Conventional routes and their preparation principles. (**b**) Schematic formation of new crystalline phases via hydrothermal synthesis.

4.1. Bottom-Up Approach

The bottom-up approach focuses on creating mesopores within the zeolite structure using a variety of templating methods. Hard-templating involves using rigid carbonaceous materials like carbon black, carbon nanotubes/nanofibers, or 3DOM carbon templates to shape the forming zeolite structure [80–82]. In a study by Wu et al. [83], Ni nanoparticles were encapsulated within microporous graphene-like carbon derived from a NaY zeolite template. The resulting Ni-loaded catalyst demonstrated efficient CO_2 hydrogenation to methane, with 50% conversion and 96% methane selectivity at high space-time yields, showcasing remarkable long-term stability. Subsequently, soft-templating is less rigid and facilitates the crystallization of zeolites by forming interactions with zeolite precursors or frameworks through a range of covalent and electrostatic forces, including van der Waals interactions, hydrogen bonds, and ionic attractions [84]. It utilizes quaternary ammonium surfactants such as cetyltrimethylammonium bromide (CTAB) or cationic polyelectrolytes such as polydiallyldimethylammonium chloride (PDDA-Cl) and polydiallyldimethylammonium chloride (PDADMAC) [85–87]. These create microporous zeolite crystallization, while the organosilane's hydrophobic alkyl chains direct mesopore formation [88]. However, the conventional hydrothermal synthesis method (see Figure 9b), is notable for its substantial byproduct generation, including waste materials and gases such as CO_2 and N_2, and its prolonged crystallization process typically leads to reduced production efficiency [89]. Therefore, alternative 'green' synthesis routes have been sustainably employed to overcome such concerns. For example, Tian et al. [90] described an innovative approach for crafting SAPO-34 zeolites with layered porosity using only rice husk as the silicon source. The resulting bio-SAPO-34 exhibited high efficiency in converting CO_2 to olefins, with an impressive C_2–C_4 olefin selectivity of 94.5%. The study found that CH_3O^* species were pivotal intermediates, with DRIFTS analysis showing their formation on the ZnZrOx catalyst surface and subsequent transfer to the acid sites of bio-SAPO-34 for effective carbon–carbon coupling. In addition, and interestingly, Din et al. [91] performed the green synthesis of bimetallic catalysts supported on zeolites, whereby it involved the deposition of Cu and Co onto the zeolite rather than the synthesis of the zeolite framework itself, which did not require extensive washing of precipitates. The process included dissolving the metal acetates in water, adding zeolite to the solution, adjusting the pH with sodium hydroxide, and maintaining the mixture at a specified temperature while stirring. The final product was then dried, resulting in the formation of bimetallic Cu-Co/zeolite catalysts. The ideal balance of Cu:Co at a 1:1 ratio was determined to be the most effective among the various metal combinations tested, leading to the highest observed rate of methanol synthesis.

Furthermore, non-templating methods rely on alternative strategies like one-pot synthesis, dry-gel conversion, or seed-induced crystallization without using traditional templates [54,89]. For instance, Wei et al. [53] developed a Na–Fe_3O_4/HZSM-5 catalyst using a one-pot synthesis technique. This catalyst demonstrated the ability to transform CO_2 into hydrocarbons within the gasoline range (C_5–C_{11}), achieving a selectivity of 78% among hydrocarbons and limiting methane to just 4%, with a CO_2 conversion rate of 22% under conditions pertinent to industry. They built the catalyst at the molecular level and relied on chemical reactions between iron salts and NaOH in solution to spontaneously form the catalyst without a structural template.

The bottom-up approach to structured zeolite synthesis is not without difficulties. A principal issue is the intrinsic fragility of hierarchical zeolite structures, which could result in loose intermolecular interactions, thereby affecting mechanical stability and charge transport [92]. This can be an especially serious issue when zeolites are used under the high-temperature and high-pressure conditions characteristic of CO_2 hydrogenation processes. One possible approach is to optimize the strength and performance of the materials by linking the molecules through strong covalent bonding that involves the reaction directly on the surfaces [92]. Furthermore, the bottom-up approach sometimes struggles with the controlled physical and chemical transformations within liquid droplets, a process that is integral to the synthesis of zeolites [93]. An alternative method, encompassing sol-gel and aerosol-assisted approaches, is proposed due to its enhanced focus on controlling the physical and chemical properties during the synthesis of nanomaterials, particularly resultant zeolites [94,95]. Another challenge in the bottom-up synthesis of zeolites is managing intercrystallite diffusion, which can impede reactant and product movement within the crystalline lattice [96]. To address this, an optimal strategy involves enhancing porosity through methods like dealumination, desilication, and recrystallization. These techniques, which now fall under the top-down approach, modify existing zeolite structures to improve molecular accessibility and diffusion pathways.

4.2. Top-Down Approach

The top-down approach involves modifying pre-existing zeolite structures through post-synthesis treatments to enhance their properties, rather than constructing new materials from the ground up. This method focuses on altering existing materials to develop structures with novel characteristics. Top-down approaches play a vital role in the development of hierarchical zeolite structures, which boast improved catalytic properties due to increased surface area and pore volume, as well as the surface properties of the zeolites. Dealumination is a process where aluminum atoms are removed from the zeolite framework through methods such as acid leaching or steam treatment, impacting the Si/Al ratio [97–99]. Its counterpart, desilication, involves the removal of silica atoms, typically by treating them with a base solution [100]. This procedure enhances mesoporosity within the zeolite structure [101]. Recrystallization typically involves dissolving the zeolite in a suitable solvent and then causing the material to precipitate out of the solution, forming a more ordered crystalline structure [102,103]. The parameters of the process, such as temperature, concentration, and choice of solvent, can significantly influence the outcome [89]. For example, Oishi et al. [48] employed a multi-top-down approach to modify MOR zeolite, initially transforming it into a slurry via milling and subsequently recrystallizing it under heat in an autoclave, resulting in nanosized MOR particles. The processed MOR underwent ion exchange with NH_4Cl, fine-tuning its ionic makeup. They then performed dealumination, using different sulfuric acid concentrations to adjust the Si/Al ratio, thus tailoring the zeolite's acidity and enhancing its catalytic abilities. Lastly, defect healing with a fluoride solution was conducted on the dealuminated MOR to mend structural imperfections, optimizing the material. Following the modifications, the enhanced MOR sample exhibited improved catalysis in CO_2 hydrogenation to lower olefins, notably doubling the olefin-to-paraffin ratio from 0.69 to 1.4, thus signifying a more selective and efficient reaction. Furthermore, Liu et al. [57] demonstrated alternative top-down approaches such as incipient wetness impregnation (IWI) and ion exchange (IE) for a series of Co/ZSM-5 catalysts to facilitate CO_2 hydrogenation to value-added chemicals via the modified-FTS pathway. They found that the selectivity for CO_2 hydrogenation on catalysts prepared by IWI strongly depends on the Si/Al ratio. In contrast, such a ratio has little impact on catalysts produced via IE. Enhancing the Si/Al ratio in impregnated catalysts usually improves olefin production, which is associated with acidic features and cobalt reducibility. However, in the same study, altering the Si/Al ratio in ion-exchanged catalysts did not influence the acid properties of K-ZSM-5 for catalytic performance. This lack of effect can be attributed to the introduction of alkali metals, which increased the number of basic sites and reduced the crystallinity. Therefore, it is essential to avoid excessive modification of atoms, as this can lead to the destruction of the crystal structure, thus underscoring the necessity for controlled and balanced treatment processes [50].

The top-down methodology for synthesizing zeolites also encapsulates a set of distinctive challenges that command innovative solutions to further enhance their catalytic performance. Firstly, inadequate characterization methods used for zeolites synthesized by the top-down approach prove to be one of the serious limitations whereby it is crucial to disclose the specific T-sites where the metal centers are situated, as these locations dictate the geometry of the active centers within the zeolites and, consequently, define their reactivity [104,105]. To address this challenge, Dapsens et al. [106] emphasized the utilization of advanced X-ray Absorption Spectroscopy methods, such as the integration of Extended X-ray Absorption Fine Structure (EXAFS) and X-ray Absorption Near Edge Structure (XANES), in overcoming analytical limitations. These techniques were pivotal in the studies conducted by Liu et al. [57] and Ding et al. [107], which focused on the characterization of Co/ZSM-5 and Cu@NaBeta catalysts, respectively. Their investigations demonstrated that the synergistic interaction between metal nanoparticles and the zeolitic framework plays a critical role in forming high-performance catalysts for CO_2 hydrogenation to olefins and ethanol accordingly. Next, the structural integrity of catalysts can be compromised by water generated during reactions. Specifically, during a CO_2 methanation reaction, zeolites are vulnerable to dealumination, a process in which aluminum atoms are stripped from the framework. This occurs under high operating temperatures and pressures in the equipment, particularly in the presence of water molecules, leading to partial degradation or impairment of the zeolite structure [108]. Such conditions can also result in the clogging of the zeolite pores as the released aluminum accumulates within the structure, further compromising its effectiveness. To mitigate these concerns, Bacariza et al. [109] suggested that commercial USY zeolites, typically used in fluid catalytic cracking (FCC) industrial processes and subjected to ultra-stabilization treatments including controlled steaming for dealumination, could serve as effective supports for CO_2 methanation catalysts. The same authors also pointed out that employing USY zeolite as a support demonstrated promising results, with no structural damage observed in the samples after both conventional and deactivation tests, even when exposed to water during the reaction. In addition, Verboekend et al. noted desilication issues like limited Si/Al ratio control, over-extraction of silicon, structural damage, ineffective pore agents, and uneven mesopores affecting zeolite performance [110]. Optimizing zeolite desilication involves careful control of alkali treatment conditions and the use of pore-directing agents to ensure consistent pore size while maintaining the zeolite's structure and acid sites [111]. Selective treatment is key to achieving ideal mesoporosity without compromising catalytic efficiency. Using tetrapropylammonium hydroxide (TPAOH) for selective desilication, for example, Dai et al. [112] engineered silicalite-1 with precise mesoporosity, achieving nanocube sizes up to 600 nm. This approach also improved CO_2–hydrocarbon conversion, with a marked increase in C_{5+} hydrocarbon selectivity when iron-based catalysts were supported on these tailored zeolites. Similarly, Sharma et al. [113] designed a novel two-bed catalytic system for CO_2 hydrogenation via the MTH pathway, leveraging a desilicated HZSM-5 zeolite to boost hydrocarbon yields, with a notable hydrocarbon selectivity of up to 71.2%. This system, through selective desilication, enhanced the production of longer-chain liquid hydrocarbons, impressively improving C_8–C_{12} selectivity from 29.2% to 42.4%. Finally, recrystallization concerns are not singled out because they can be part of a broader set of synthesis conditions that also include dealumination and desilication. Overall, these conditions collectively impact the structural and compositional modifications of zeolites during the top-down approach.

5. Characterization Techniques

From the preceding sections, it is clear that catalyst characterization techniques are indispensable tools for gaining a comprehensive understanding of the catalytic mechanisms, performance, and intricate behavior of zeolites in catalysis, such as CO_2 hydrogenation processes. These advanced techniques in Table 3 help elucidate the structural and chemical properties of catalysts, which are critical for optimizing their activity and stability under tailored reaction conditions. In this section, we will delve deeper into the specific methods and their applications in detailing the structural adaptations and chemical dynamics of zeolites, which facilitate their efficacy in CO_2 hydrogenation. We will explore how these characterization methods contribute to a better understanding of the interaction between catalysts and reactants and how this knowledge can lead to the development of more efficient catalytic systems. For the subsequent discussions, we will only highlight specific characterization methods from each category, providing detailed illustrations of how these techniques are applied to enhance our understanding of zeolite catalysis in CO_2 hydrogenation.

Table 3. Advanced characterization techniques for structured zeolites in catalytic reactions like CO_2 hydrogenation.

Category	Method	Application
Molecular and Chemical Structure Analysis [5,114,115]	Fourier-transform infrared (FTIR) spectroscopy	Quantify absorption spectra in chemical bonds and functional groups in molecules
	Raman spectroscopy	Postulate information about molecular vibrations, crystal structures, and phase transitions
	Nuclear Magnetic Resonance (NMR) spectroscopy	Provide detailed information on the framework aluminum distribution and the nature of acid sites
	Small-angle X-ray Scattering (SAXS)	Analyze the structural details of materials at the nanoscale level by detecting inhomogeneities and phase separation within

Table 3. *Cont.*

Category	Method	Application
Crystallographic and Phase Analysis [54,115,116]	X-ray diffraction (XRD)	Assess crystalline structures, crystal phases, and crystal defects
	Powder X-ray Diffraction (PXRD)	Analyze powdered crystalline materials for crystal structure identification
	Selected Area Electron Diffraction (SAED)	Obtain crystallographic information from a sample area
Surface and Elemental Analysis [91,117,118]	X-ray photoelectron spectroscopy (XPS)	Examine the chemistry of the surface, including aspects such as elemental composition, chemical and empirical states, and the electronic state of elements
	X-ray Absorption Spectroscopy (XAS)	Determine local geometric/electronic structural order
	Auger Electron Spectroscopy (AES)	Detect emitted energy of electrons from the catalyst surface
Microscopy and Imaging [115,119]	Scanning Electron Microscopy (SEM) Transmission Electron Microscopy (TEM)	Generate high-resolution images of the surface, internal structure, morphology, and crystallography of nanomaterials
Thermal Analysis [4,5,54,115,120]	Temperature-Programmed Reduction-Thermogravimetric Analysis (TPR-TGA)	TPR: Measure the change in chemical state upon heating TGA: Measure changes in physical and chemical states upon heating
	Temperature-Programmed Desorption (TPD)	Investigate adsorption and desorption behaviors on surface interactions and binding energies
	Temperature-Programmed Oxidation (TPO)	Evaluate oxidation behaviors, particularly in carbonaceous materials, catalyst deactivation investigations
	Temperature-Programmed Reaction (TPRe)	Study reaction kinetics, and catalytic stability under different thermal environments
	Temperature-Programmed Surface Reaction (TPSR)	Focus on surface reactions; mechanisms of surface-mediated reactions
	Temperature-Programmed Reduction/Oxidation (TPR-O)	Explore redox properties for redox reactions
	Temperature Programmed Reduction-Differential Thermogravimetry (TPR-DTG)	Determine the temperatures at which reduction events occur and the quantitative aspects amount of oxygen removed from an oxide
	Temperature-Programmed Ammonia Desorption (TPAD)	Observe ammonia-desorption for acid catalysis

Molecular and chemical structure analysis techniques, such as FTIR and Raman spectroscopy, measure molecular vibrations and identify chemical bonds to reveal the composition of substances. NMR spectroscopy offers a detailed view of atomic-level structure, while SAXS enables the investigation of nanoscale inhomogeneities and phase distribution within materials. In a study by Ramirez et al. [5], NMR spectroscopy provided detailed insights into the selective catalytic roles of various zeolites (MOR, ZSM-22, FER, and ZSM-5) to explore the role of CO at the RWGS reaction phase of the modified-FTS route for CO_2 hydrogenation to alkenes, aromatic compounds, and alkanes. Figure 10a highlights how MOR zeolites are inclined towards producing lighter olefins, whereas ZSM-22 zeolites preferentially lead to heavier olefin production. Additionally, ZSM-5 achieves complete ethylene conversion, while MOR and ZSM-22 exhibit the lowest, with MOR specifically favoring paraffin production when ^{13}CO is absent. However, the introduction of ^{13}CO sees MOR attain the highest conversion rate at around 4%, with ZSM-5 and ZSM-22 at roughly 1% (Figure 10b). Solid-state NMR confirms COs direct incorporation into the zeolite structure (Figure 10c), and a notable carbonyl peak at 177.5 ppm associates with methyl acetate suggests a ketene intermediate in the reaction (Figure 10d). This set of NMR figures paints a comprehensive picture of how each zeolite type uniquely influences the reaction pathway and product distribution.

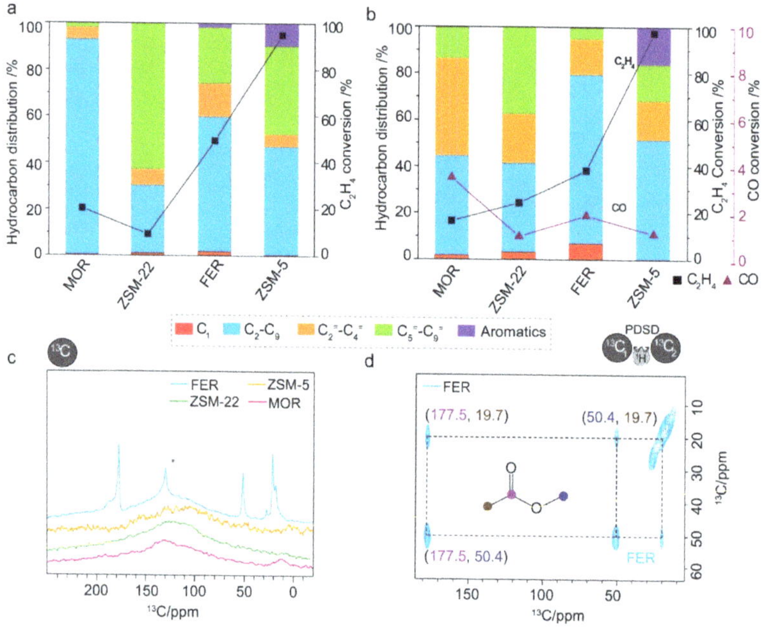

Figure 10. NMR spectroscopic analysis by Ramirez et al. [5] detailed the catalytic activity of various zeolites (e.g., MOR, ZSM-22, FER, ZSM-5), where (**a**) revealed the behavior of these zeolites with ethylene and (**b**) showed the dynamics with enriched ^{13}CO under hydrogen-rich conditions. (**c**) The isotopic enrichment of ^{13}CO correlated with its consumption during the reaction, and (**d**) captured the identification of methyl acetate in FER zeolites, suggesting the involvement of ketene intermediates.

Crystallographic and phase analysis methods encompass techniques like XRD, which examines the arrangement of atoms within crystals and identifies phases and defects. PXRD is tailored to analyze powdered samples for structural identification, while SAED provides crystallographic information from targeted areas within a specimen. On the other hand, microscopy and imaging techniques like SEM and TEM are pivotal for capturing high-resolution images that reveal surface features, internal structures, morphologies, and the crystallography of zeolites. Krachuamram et al. [115] conducted a study to fine-tune the synthesis of NaA/NaX zeolite to maximize surface area and pore volume, finding the best conditions to be a 5 day aging and a 9 h crystallization process in the presence of CTAB and heptane. When used as a support for Ni impregnation at various loadings, the zeolite synthesized under these conditions showed the highest catalytic activity for CO_2 methanation at a 15% Ni level, attributed to the optimal dispersion of Ni on the zeolite surface. Their XRD analyses indicated a notable reduction in the crystallinity of the zeolite, with peak intensities for NaX and NaA decreasing upon Ni impregnation (Figure 11a), and NiO reflections becoming more pronounced. The specific surface area of the zeolite increased significantly from 166.5 m^2/g to 1411 m^2/g with Ni addition (Figure 12b), suggesting a dispersion of Ni that did not disrupt the zeolite's structure, as confirmed by SEM images (Figure 12a). Despite these structural changes, the catalytic efficiency remained primarily reliant on gas adsorption capacities rather than on the surface area, with the presence of Ni-enhancing adsorption sites for H_2 and CO_2 (Figure 11b).

Surface and elemental analysis for zeolites typically employ techniques such as XPS, which analyzes surface chemistry, including elemental composition and states; XAS, for assessing the local structure and electronic state order; and AES, which detects the energies of electrons emitted from surfaces, providing details on the surface's elemental composition. Conversely, under varying thermal conditions, TPR-TGA allows for the study of reduction processes and zeolite stability. TPD can help determine the strength and nature of adsorption sites by measuring the gases desorbed as the temperature increases. Similarly, TPR-O and TPR-DTG provide insight into the oxidative and reductive characteristics of zeolites, while TPSR and TPAD are employed to examine surface interactions and acidity, respectively.

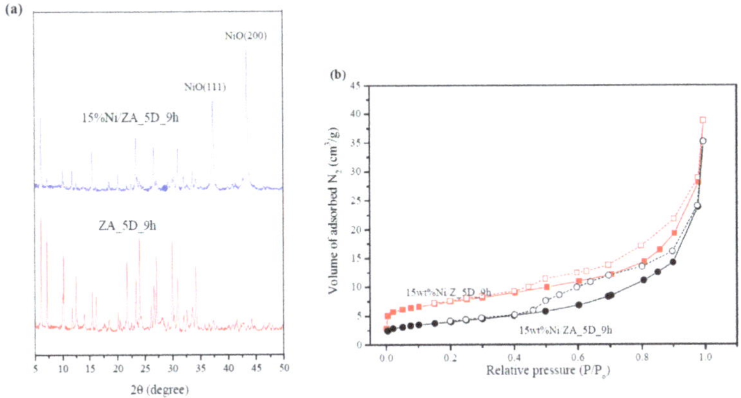

Figure 11. Krachuamram et al. [115] demonstrated the (**a**) XRD profiles of the zeolite framework and the zeolite after 15% nickel has been added. (**b**) Shows the N_2 adsorption–desorption curves for the zeolite treated with an organic phase (Z_5D_9h) compared to the zeolite not treated with an organic phase.

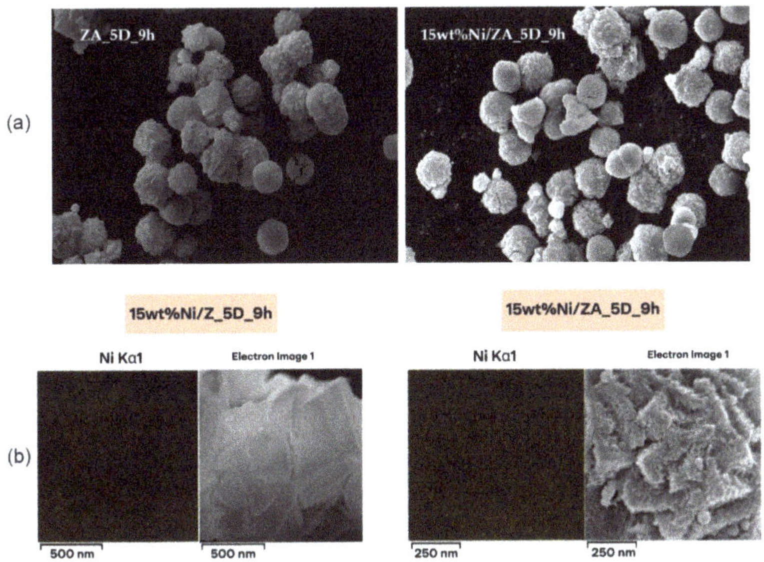

Figure 12. Krachuamram et al. [115] presented (**a**) SEM visuals of the zeolite before and after nickel impregnation and (**b**) SEM images and elemental mapping of the zeolite supports with and without the addition of the organic phase (Z_5D_9h).

Popova et al. [54] developed micro-mesoporous ZSM-5 catalysts modified with Ni and Ru through wet impregnation, examining how Ni content and the addition of Ru influenced the catalysts' properties by utilizing TPR-DTG to analyze, and they observed distinct reduction peaks. For monometallic samples with 10%Ru, a significant reduction peak occurred at 190 °C, while samples with 5%Ni and 10%Ni exhibited two peaks spanning temperatures of 195–205 °C and 325–360 °C, respectively, suggesting Ni's influence on the reduction process and dispersion on the zeolite (Figure 13a). The XPS analysis post-reduction offered insights into the surface chemistry, indicating a prevalence of RuO_2 for the monometallic samples (e.g., 5RuZSM-5) and a more balanced presence of reduced and oxidized Ru states in the bimetallic samples, such as 10Ni5RuZSM-5 (Figure 13c). These findings aligned with their TPR-DTG results, revealing that bimetallic samples with higher Ni content have more Ni accessible on the surface, which enhanced catalytic activity due to a greater number of active sites, whereby 10Ni5RuZSM-5 exhibits a single, wider peak reaching its apex at 268 °C (Figure 13b). As a result, the 10Ni5RuZSM-5 catalyst showcased high efficiency, achieving 100% selectivity for methane production from CO_2 at 400 °C. This performance, coupled with its stability and the potential for reuse, underscores the catalyst's applicability in practical settings.

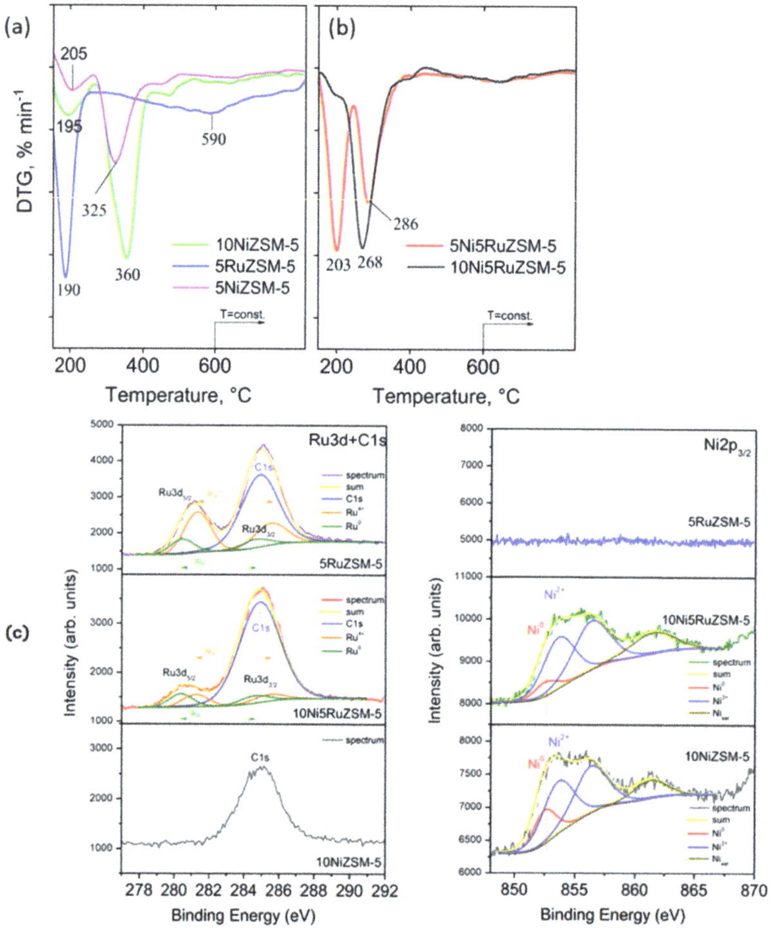

Figure 13. Popova et al. [54] performed TPR-DTG analysis of ZSM-5 zeolites modified with Ni and Ru in (a) single-metal and (b) dual-metal configurations, and (c) XPS profiles for Ru (left) and Ni (right) on the 10NiZSM-5, 5RuZSM-5, and 10Ni5RuZSM-5 zeolites post-reduction.

In common practice, a combination of these characterization techniques in Table 3 often provides a more complete picture. For instance, SEM might be paired with XRD to correlate surface morphology with crystalline structure, or TPR-TGA might be used alongside XPS to evaluate how thermal treatments affect surface chemistry. The choice depends on whether the study aims to optimize the catalyst's structural properties, understand the chemical processes occurring during the reaction, or improve the catalyst's stability and reusability. A multi-technique approach often yields the most comprehensive understanding of the catalytic system.

6. Conclusions

CO_2 hydrogenation over structured zeolites has emerged as a significant technological approach for carbon neutrality, leveraging the conversion of CO_2 into valuable hydrocarbons. Two main methods, the modified-FTS and the MTH route, effectively channel CO_2 into a variety of hydrocarbons, with zeolites playing a pivotal role due to their unique catalytic properties. The thermal hydrogenation method stands out over electrochemical, photocatalytic, biochemical, and chemo-enzymatic approaches for its high efficiency, selectivity, and proven scalability. This positions zeolites as an optimal catalyst for CO_2 hydrogenation, capable of undergoing regeneration and withstanding the high temperatures required for these processes.

Subsequently, the efficacy of structured zeolites in CO_2 hydrogenation depends on the synergy between their composition, preparation, and unique structural properties. Careful calibration of the Si/Al ratio is critical, as it influences acidity and CO_2 adsorption, which are key factors for selectivity and conversion efficiency. Bimodal mesoporous zeolites improve catalytic activity due to enhanced mass transfer and accessible sites but require careful optimization to maintain product specificity. The integration of metals such as Cu and Ni within zeolite frameworks, coupled with plasma treatments, has been shown to boost performance, balancing strong adsorption with necessary reactivity. Attention to electrical and plasma interactions within the catalyst design is therefore paramount, ensuring high conversion rates and desired selectivity in CO_2 hydrogenation processes.

In addition, optimizing zeolite synthesis through bottom-up and top-down methods is critical for enhancing CO_2 hydrogenation. Bottom-up strategies, such as various templating techniques, improve mass transfer by overcoming the diffusion limitations inherent in microporous zeolites and by embedding metals like Cu and Co to augment hydrocarbon production. To combat issues like stability under extreme conditions, these methods are evolving to reinforce molecular bonds and manage intracrystalline diffusion. Top-down methods, including dealumination and recrystallization, further refine zeolite structures, bolstering mesoporosity and catalytic selectivity without altering the crystal integrity. Advanced spectroscopy (e.g., EXAFS and XANES) plays a key role in this fine-tuning, guiding the precise distribution of metal nanoparticles and ensuring optimal catalytic functionality. Furthermore, CO_2 hydrogenation over zeolites benefits from a suite of characterization techniques. FTIR and NMR reveal molecular intricacies and active sites critical for the catalytic process, while phase analyses like XRD detail crystalline structures that influence functionality. Elemental analyses like XPS provide surface analysis crucial for CO_2 interaction, and microscopic imaging like SEM gives insight into morphology and metal distribution, affecting activity and selectivity. Thermal methods, like TPR-TGA gauge stability and reducibility, are essential for catalyst durability.

Looking forward, the trajectory of CO_2 hydrogenation using structured zeolites is set towards transformative progress. Further development of synthesis approaches (bottom-up and top-down), together with sophisticated characterization methods as listed in Table 3, is fundamental for realizing the whole potential of zeolite as an efficient catalyst. With the world moving towards greener and more sustainable options, structured zeolites play a more and more significant role in holistically converting CO_2 into high-value-added chemical products. Achieving optimal selectivity is a delicate balance that is centered on numerous factors, including the accessibility of active sites, metal particle dynamics, support structure interactions, and overall porosity, each shaped by synthesis and characterization methods. Future research will focus on overcoming current limitations, as remarked in this review, and innovating to enhance catalyst efficiency and selectivity, thereby contributing to the wider goal of reducing the carbon footprint and addressing climate change.

Funding: This research was funded by the Qatar National Research Fund (a member of Qatar Foundation), grant number NPRP14S–0302–210011.

Acknowledgments: This publication was made possible by an NPRP grant (NPRP14S–0302–210011) from the Qatar National Research Fund (a member of Qatar Foundation). The statements made herein are solely the responsibility of the author(s). The authors would also like to gratefully acknowledge the support of Graduate Teaching/Research Assistantship (GTRA) from Qatar University in conducting this research.

Conflicts of Interest: The authors declare no conflicts of interests.

References

1. Rafiee, A.; Khalilpour, K.R.; Milani, D.; Panahi, M. Trends in CO_2 conversion and utilization: A review from process systems perspective. *J. Environ. Chem. Eng.* **2018**, *6*, 5771–5794. [CrossRef]
2. North, M.; Styring, P. Perspectives and visions on CO_2capture and utilisation. *Faraday Discuss.* **2015**, *183*, 489–502. [CrossRef] [PubMed]
3. Huang, C.H.; Tan, C.S. A Review: CO_2 Utilization. *Aerosol Air Qual. Res.* **2014**, *14*, 480–499. [CrossRef]
4. Nezam, I.; Zhou, W.; Gusmão, G.S.; Realff, M.J.; Wang, Y.; Medford, A.J.; Jones, C.W. Direct aromatization of CO_2 via combined CO_2 hydrogenation and zeolite-based acid catalysis. *J. CO2 Util.* **2021**, *45*, 101405. [CrossRef]
5. Ramirez, A.; Gong, X.; Caglayan, M.; Nastase, S.-A.F.; Abou-Hamad, E.; Gevers, L.; Cavallo, L.; Chowdhury, A.D.; Gascon, J. Selectivity descriptors for the direct hydrogenation of CO_2 to hydrocarbons during zeolite-mediated bifunctional catalysis. *Nat. Commun.* **2021**, *12*, 5914. [CrossRef] [PubMed]
6. Murciano, R.; Serra, J.M.; Martínez, A. Direct hydrogenation of CO_2 to aromatics via Fischer-Tropsch route over tandem K-Fe/Al_2O_3+H-ZSM-5 catalysts: Influence of zeolite properties. *Catal. Today* **2023**, *427*, 114404. [CrossRef]
7. Dry, M.E.; Hoogendoorn, J.C. Technology of the Fischer-Tropsch Process. *Catal. Rev.* **1981**, *23*, 265–278. [CrossRef]
8. Ibrahim, I.; Harrigan, F. Qatar's economy: Past, present and future. *QScience Connect.* **2012**, *2012*, 9. [CrossRef]
9. Holm-Larsen, H. CO_2 reforming for large scale methanol plants—An actual case. *Stud. Surf. Sci. Catal* **2001**, *136*, 441–446. [CrossRef]
10. Ye, R.-P.; Ding, J.; Gong, W.; Argyle, M.D.; Zhong, Q.; Wang, Y.; Russell, C.K.; Xu, Z.; Russell, A.G.; Li, Q.; et al. CO_2 hydrogenation to high-value products via heterogeneous catalysis. *Nat. Commun.* **2019**, *10*, 5698. [CrossRef] [PubMed]
11. Ojelade, O.A.; Zaman, S.F. A review on CO_2 hydrogenation to lower olefins: Understanding the structure-property relationships in heterogeneous catalytic systems. *J. CO2 Util.* **2021**, *47*, 101506. [CrossRef]
12. Wang, D.; Xie, Z.; Porosoff, M.D.; Chen, J.G. Recent advances in carbon dioxide hydrogenation to produce olefins and aromatics. *Chem* **2021**, *7*, 2277–2311. [CrossRef]
13. Wang, C.; Li, J.; Sun, X.; Wang, L.; Sun, X. Evaluation of zeolites synthesized from fly ash as potential adsorbents for wastewater containing heavy metals. *J. Environ. Sci.* **2009**, *21*, 127–136. [CrossRef] [PubMed]
14. Melaningtyas, G.S.A.; Krisnandi, Y.K.; Ekananda, R. Synthesis and characterization of NaY zeolite from Bayat natural zeolite: Effect of pH on synthesis. *IOP Conf. Ser. Mater. Sci. Eng.* **2019**, *496*, 012042. [CrossRef]
15. Dimitar, G.; Bogdanov, B.; Angelova, K.; Markovska, I.; Hristov, Y. Synthetic Zeolites—Structure, Clasification, Current Trends in Zeolite Synthesis. International Science Conference. Available online: https://www.researchgate.net/profile/Irena-Markovska-2/publication/322211658_SYNTHETIC_ZEOLITES_-_STRUCTURE_CLASIFICATION_CURRENT_TRENDS_IN_ZEOLITE_SYNTHESIS_REVIEW/links/5a990a4f0f7e9ba4297769ff/SYNTHETIC-ZEOLITES-STRUCTURE-CLASIFICATION-CURRENT-TRENDS-IN-ZEOLITE-SYNTHESIS-REVIEW.pdf (accessed on 12 May 2024).
16. Ruiz, A.Z.; Ban, T.; Calzaferri, G.; Brühwiler, D. Synthesis of Zeolite L. Tuning Size and Morphology. *Monatshefte Fuer Chemie/Chem. Mon.* **2005**, *136*, 77–89. [CrossRef]
17. Sun, Q.; Wang, N.; Yu, J. Advances in Catalytic Applications of Zeolite-Supported Metal Catalysts. *Adv. Mater.* **2021**, *33*, 2104442. [CrossRef]
18. Tran, Y.T.; Lee, J.; Kumar, P.; Kim, K.-H.; Lee, S.S. Natural zeolite and its application in concrete composite production. *Compos. Part B Eng.* **2019**, *165*, 354–364. [CrossRef]
19. Cataldo, E.; Salvi, L.; Paoli, F.; Fucile, M.; Masciandaro, G.; Manzi, D.; Masini, C.M.; Mattii, G.B. Application of Zeolites in Agriculture and Other Potential Uses: A Review. *Agronomy* **2021**, *11*, 1547. [CrossRef]
20. Bacakova, L.; Vandrovcova, M.; Kopova, I.; Jirka, I. Applications of zeolites in biotechnology and medicine—A review. *Biomater. Sci.* **2018**, *6*, 974–989. [CrossRef] [PubMed]
21. Kalló, D. Applications of Natural Zeolites in Water and Wastewater Treatment. *Rev. Miner. Geochem.* **2001**, *45*, 519–550. [CrossRef]
22. Fruijtier-Pölloth, C. The safety of synthetic zeolites used in detergents. *Arch. Toxicol.* **2008**, *83*, 23–35. [CrossRef] [PubMed]
23. Wittcoff, H.A.; Reuben, B.G.; Plotkin, J.S. *Industrial Organic Chemicals*; John Wiley & Sons: Hoboken, NJ, USA, 2012.

24. Overa, S.; Ko, B.H.; Zhao, Y.; Jiao, F. Electrochemical Approaches for CO_2 Conversion to Chemicals: A Journey toward Practical Applications. *Accounts Chem. Res.* **2022**, *55*, 638–648. [CrossRef] [PubMed]
25. Perry, S.C.; Leung, P.-K.; Wang, L.; de León, C.P. Developments on carbon dioxide reduction: Their promise, achievements, and challenges. *Curr. Opin. Electrochem.* **2020**, *20*, 88–98. [CrossRef]
26. Nguyen, T.P.; Nguyen, D.L.T.; Nguyen, V.-H.; Le, T.-H.; Vo, D.-V.N.; Trinh, Q.T.; Bae, S.-R.; Chae, S.Y.; Kim, S.Y.; Van Le, Q. Recent Advances in TiO_2-Based Photocatalysts for Reduction of CO_2 to Fuels. *Nanomaterials* **2020**, *10*, 337. [CrossRef] [PubMed]
27. Shehzad, N.; Tahir, M.; Johari, K.; Murugesan, T.; Hussain, M. A critical review on TiO_2 based photocatalytic CO_2 reduction system: Strategies to improve efficiency. *J. CO2 Util.* **2018**, *26*, 98–122. [CrossRef]
28. Bhatia, S.K.; Bhatia, R.K.; Jeon, J.-M.; Kumar, G.; Yang, Y.-H. Carbon dioxide capture and bioenergy production using biological system—A review. *Renew. Sustain. Energy Rev.* **2019**, *110*, 143–158. [CrossRef]
29. Thakur, I.S.; Kumar, M.; Varjani, S.J.; Wu, Y.; Gnansounou, E.; Ravindran, S. Sequestration and utilization of carbon dioxide by chemical and biological methods for biofuels and biomaterials by chemoautotrophs: Opportunities and challenges. *Bioresour. Technol.* **2018**, *256*, 478–490. [CrossRef] [PubMed]
30. Sharma, T.; Sharma, S.; Kamyab, H.; Kumar, A. Energizing the CO_2 utilization by chemo-enzymatic approaches and potentiality of carbonic anhydrases: A review. *J. Clean. Prod.* **2020**, *247*, 119138. [CrossRef]
31. Ribun, V.; Boichenko, S.; Kale, U. Advances in gas-to-liquid technology for environmentally friendly fuel synthesis: Analytical review of world achievements. *Energy Rep.* **2023**, *9*, 5500–5508. [CrossRef]
32. Boretti, A.; Banik, B.K. Advances in Hydrogen Production from Natural Gas Reforming. *Adv. Energy Sustain. Res.* **2021**, *2*, 2100097. [CrossRef]
33. Atsonios, K.; Panopoulos, K.D.; Kakaras, E. Thermocatalytic CO_2 hydrogenation for methanol and ethanol production: Process improvements. *Int. J. Hydrogen Energy* **2016**, *41*, 792–806. [CrossRef]
34. Do, T.N.; Kim, J. Process development and techno-economic evaluation of methanol production by direct CO_2 hydrogenation using solar-thermal energy. *J. CO2 Util.* **2019**, *33*, 461–472. [CrossRef]
35. Zeng, Y.; Chen, G.; Liu, B.; Zhang, H.; Tu, X. Unraveling Temperature-Dependent Plasma-Catalyzed CO_2 Hydrogenation. *Ind. Eng. Chem. Res.* **2023**, *62*, 19629–19637. [CrossRef] [PubMed]
36. Chatzilias, C.; Martino, E.; Katsaounis, A.; Vayenas, C.G. Electrochemical promotion of CO_2 hydrogenation in a monolithic electrochemically promoted reactor (MEPR). *Appl. Catal. B Environ.* **2021**, *284*, 119695. [CrossRef]
37. Zhu, Q. Developments on CO_2-utilization technologies. *Clean Energy* **2019**, *3*, 85–100. [CrossRef]
38. Jhong, H.; Brushett, F.R.; Kenis, P.J.A. The Effects of Catalyst Layer Deposition Methodology on Electrode Performance. *Adv. Energy Mater.* **2013**, *3*, 589–599. [CrossRef]
39. Kalaitzidou, I.; Makri, M.; Theleritis, D.; Katsaounis, A.; Vayenas, C. Comparative study of the electrochemical promotion of CO_2 hydrogenation on Ru using Na^+, K^+, H^+ and O^{2-} conducting solid electrolytes. *Surf. Sci.* **2016**, *646*, 194–203. [CrossRef]
40. Lo, A.-Y.; Taghipour, F. Review and prospects of microporous zeolite catalysts for CO_2 photoreduction. *Appl. Mater. Today* **2021**, *23*, 101042. [CrossRef]
41. Saravanan, A.; Kumar, P.S.; Vo, D.-V.N.; Jeevanantham, S.; Bhuvaneswari, V.; Narayanan, V.A.; Yaashikaa, P.R.; Swetha, S.; Reshma, B. A comprehensive review on different approaches for CO_2 utilization and conversion pathways. *Chem. Eng. Sci.* **2021**, *236*, 116515. [CrossRef]
42. Kumar, B.; Llorente, M.; Froehlich, J.; Dang, T.; Sathrum, A.; Kubiak, C.P. Photochemical and Photoelectrochemical Reduction of CO_2. *Annu. Rev. Phys. Chem.* **2012**, *63*, 541–569. [CrossRef] [PubMed]
43. Yaashikaa, P.R.; Kumar, P.S.; Varjani, S.J.; Saravanan, A. A review on photochemical, biochemical and electrochemical transformation of CO_2 into value-added products. *J. CO2 Util.* **2019**, *33*, 131–147. [CrossRef]
44. Zhang, Z.; Zhang, X.; Ji, X. Developing and Regenerating Cofactors for Sustainable Enzymatic CO_2 Conversion. *Processes* **2022**, *10*, 230. [CrossRef]
45. Long, N.V.D.; Lee, J.; Koo, K.-K.; Luis, P.; Lee, M. Recent Progress and Novel Applications in Enzymatic Conversion of Carbon Dioxide. *Energies* **2017**, *10*, 473. [CrossRef]
46. Gao, P.; Zhang, L.; Li, S.; Zhou, Z.; Sun, Y. Novel Heterogeneous Catalysts for CO_2 Hydrogenation to Liquid Fuels. *ACS Central Sci.* **2020**, *6*, 1657–1670. [CrossRef] [PubMed]
47. Ojelade, O.A. CO_2 Hydrogenation to Gasoline and Aromatics: Mechanistic and Predictive Insights from DFT, DRIFTS and Machine Learning. *Chempluschem* **2023**, *88*, e202300301. [CrossRef] [PubMed]
48. Oishi, R.; Li, D.; Okazaki, M.; Kinoshita, H.; Ochiai, N.; Yamauchi, N.; Kobayashi, Y.; Wakihara, T.; Okubo, T.; Tada, S.; et al. Precise tuning of the properties of MOR-type zeolite nanoparticles to improve lower olefins selectivity in composite catalysts for CO_2 hydrogenation. *J. CO2 Util.* **2023**, *72*, 102491. [CrossRef]
49. Wang, X.; Yang, G.; Zhang, J.; Song, F.; Wu, Y.; Zhang, T.; Zhang, Q.; Tsubaki, N.; Tan, Y. Macroscopic assembly style of catalysts significantly determining their efficiency for converting CO_2 to gasoline. *Catal. Sci. Technol.* **2019**, *9*, 5401–5412. [CrossRef]
50. Azhari, N.J.; Nurdini, N.; Mardiana, S.; Ilmi, T.; Fajar, A.T.; Makertihartha, I.; Subagjo; Kadja, G.T. Zeolite-based catalyst for direct conversion of CO_2 to C_{2+} hydrocarbon: A review. *J. CO2 Util.* **2022**, *59*, 101969. [CrossRef]
51. Hölderich, W.; Hesse, M.; Näumann, F. Zeolites: Catalysts for Organic Syntheses. *Angew. Chem. Int. Ed.* **1988**, *27*, 226–246. [CrossRef]
52. Yilmaz, B.; Müller, U. Catalytic Applications of Zeolites in Chemical Industry. *Top. Catal.* **2009**, *52*, 888–895. [CrossRef]

53. Wei, J.; Ge, Q.; Yao, R.; Wen, Z.; Fang, C.; Guo, L.; Xu, H.; Sun, J. Directly converting CO_2 into a gasoline fuel. *Nat. Commun.* **2017**, *8*, 15174, Erratum in **2017**, *8*, 16170. [CrossRef] [PubMed]
54. Popova, M.; Oykova, M.; Dimitrov, M.; Karashanova, D.; Kovacheva, D.; Atanasova, G.; Szegedi, Á. CO_2 Hydrogenation to Renewable Methane on Ni/Ru Modified ZSM-5 Zeolites: The Role of the Preparation Procedure. *Catalysts* **2022**, *12*, 1648. [CrossRef]
55. Bando, K.K.; Soga, K.; Kunimori, K.; Arakawa, H. Effect of Li additive on CO_2 hydrogenation reactivity of zeolite supported Rh catalysts. *Appl. Catal. A Gen.* **1998**, *175*, 67–81. [CrossRef]
56. Chan, B.; Radom, L. Design of Effective Zeolite Catalysts for the Complete Hydrogenation of CO_2. *J. Am. Chem. Soc.* **2006**, *128*, 5322–5323. [CrossRef] [PubMed]
57. Liu, R.; Leshchev, D.; Stavitski, E.; Juneau, M.; Agwara, J.N.; Porosoff, M.D. Selective hydrogenation of CO_2 and CO over potassium promoted Co/ZSM-5. *Appl. Catal. B Environ.* **2021**, *284*, 119787. [CrossRef]
58. Ramirez, A.; Chowdhury, A.D.; Dokania, A.; Cnudde, P.; Caglayan, M.; Yarulina, I.; Abou-Hamad, E.; Gevers, L.E.; Ould-Chikh, S.; De Wispelaere, K.; et al. Effect of Zeolite Topology and Reactor Configuration on the Direct Conversion of CO_2 to Light Olefins and Aromatics. *ACS Catal.* **2019**, *9*, 6320–6334. [CrossRef]
59. Pérez-Ramírez, J.; Christensen, C.H.; Egeblad, K.; Christensen, C.H.; Groen, J.C. Hierarchical zeolites: Enhanced utilisation of microporous crystals in catalysis by advances in materials design. *Chem. Soc. Rev.* **2008**, *37*, 2530–2542. [CrossRef] [PubMed]
60. Maghfirah, A.; Ilmi, M.; Fajar, A.; Kadja, G. A review on the green synthesis of hierarchically porous zeolite. *Mater. Today Chem.* **2020**, *17*, 100348. [CrossRef]
61. Song, G.; Li, M.; Yan, P.; Nawaz, M.A.; Liu, D. High Conversion to Aromatics via CO_2-FT over a CO-Reduced Cu-Fe_2O_3 Catalyst Integrated with HZSM-5. *ACS Catal.* **2020**, *10*, 11268–11279. [CrossRef]
62. Tian, H.; Gao, P.; Yang, X.; Jiao, C.; Zha, F.; Chang, Y.; Chen, H. Tandem composite of M (Zn, Ga, In)-UIO-66/(HZSM-5)-palygorskite for hydrogenation of carbon dioxide to aromatics. *Chem. Eng. J.* **2023**, *466*, 143267. [CrossRef]
63. Yan, P.; Peng, H.; Wu, X.; Rabiee, H.; Weng, Y.; Konarova, M.; Vogrin, J.; Rozhkovskaya, A.; Zhu, Z. Impact of varied zeolite materials on nickel catalysts in CO_2 methanation. *J. Catal.* **2024**, *432*, 115439. [CrossRef]
64. da Costa-Serra, J.F.; Cerdá-Moreno, C.; Chica, A. Zeolite-Supported Ni Catalysts for CO_2 Methanation: Effect of Zeolite Structure and Si/Al Ratio. *Appl. Sci.* **2020**, *10*, 5131. [CrossRef]
65. Bacariza, M.C.; Graca, I.; Lopes, J.M.; Henriques, C. Enhanced activity of CO_2 hydrogenation to CH_4 over Ni based zeolites through the optimization of the Si/Al ratio. *Microporous Mesoporous Mater.* **2018**, *267*, 9–19. [CrossRef]
66. Tada, S.; Li, D.; Okazaki, M.; Kinoshita, H.; Nishijima, M.; Yamauchi, N.; Kobayashi, Y.; Iyoki, K. Influence of Si/Al ratio of MOR type zeolites for bifunctional catalysts specific to the one-pass synthesis of lower olefins via CO_2 hydrogenation. *Catal. Today* **2023**, *411–412*, 113828. [CrossRef]
67. Cui, X.; Gao, P.; Li, S.; Yang, C.; Liu, Z.; Wang, H.; Zhong, L.; Sun, Y. Selective Production of Aromatics Directly from Carbon Dioxide Hydrogenation. *ACS Catal.* **2019**, *9*, 3866–3876. [CrossRef]
68. Wei, J.; Yao, R.; Ge, Q.; Xu, D.; Fang, C.; Zhang, J.; Xu, H.; Sun, J. Precisely regulating Brønsted acid sites to promote the synthesis of light aromatics via CO_2 hydrogenation. *Appl. Catal. B Environ.* **2021**, *283*, 119648. [CrossRef]
69. Xiang, M.; Shi, Z.; Zhang, X.; Gao, Z.; Guo, J.; Wu, Z.; Ma, S.; Bai, J.; Zhang, W.; Deng, Y.; et al. Facile synthesis of hierarchical SAPO-56 zeolite as a highly efficient catalyst for CO_2 hydrogenation to methanol. *Fuel* **2024**, *361*, 130663. [CrossRef]
70. Cui, W.-G.; Li, Y.-T.; Yu, L.; Zhang, H.; Hu, T.-L. Zeolite-Encapsulated Ultrasmall Cu/ZnO_x Nanoparticles for the Hydrogenation of CO_2 to Methanol. *ACS Appl. Mater. Interfaces* **2021**, *13*, 18693–18703. [CrossRef] [PubMed]
71. Ashford, B.; Poh, C.-K.; Ostrikov, K.; Chen, L.; Tu, X. Plasma-catalytic CO_2 hydrogenation to ethane in a dielectric barrier discharge reactor. *J. CO2 Util.* **2022**, *57*, 101882. [CrossRef]
72. Feliz, M.Q.; Polaert, I.; Ledoux, A.; Fernandez, C.; Azzolina-Jury, F. Influence of ionic conductivity and dielectric constant of the catalyst on DBD plasma-assisted CO_2 hydrogenation into methanol. *J. Phys. D Appl. Phys.* **2021**, *54*, 334003. [CrossRef]
73. Bacariza, M.; Biset-Peiró, M.; Graça, I.; Guilera, J.; Morante, J.; Lopes, J.; Andreu, T.; Henriques, C. DBD plasma-assisted CO_2 methanation using zeolite-based catalysts: Structure composition-reactivity approach and effect of Ce as promoter. *J. CO2 Util.* **2018**, *26*, 202–211. [CrossRef]
74. Chen, H.; Mu, Y.; Shao, Y.; Chansai, S.; Xu, S.; Stere, C.E.; Xiang, H.; Zhang, R.; Jiao, Y.; Hardacre, C.; et al. Coupling non-thermal plasma with Ni catalysts supported on BETA zeolite for catalytic CO_2 methanation. *Catal. Sci. Technol.* **2019**, *9*, 4135–4145. [CrossRef]
75. Jwa, E.; Lee, S.; Lee, H.; Mok, Y. Plasma-assisted catalytic methanation of CO and CO_2 over Ni–zeolite catalysts. *Fuel Process. Technol.* **2013**, *108*, 89–93. [CrossRef]
76. Da Costa, P.; Hasrack, G.; Bonnety, J.; Henriques, C. Ni-based catalysts for plasma-assisted CO_2 methanation. *Curr. Opin. Green Sustain. Chem.* **2021**, *32*, 100540. [CrossRef]
77. Medford, A.J.; Vojvodic, A.; Hummelshøj, J.S.; Voss, J.; Abild-Pedersen, F.; Studt, F.; Bligaard, T.; Nilsson, A.; Nørskov, J.K. From the Sabatier principle to a predictive theory of transition-metal heterogeneous catalysis. *J. Catal.* **2015**, *328*, 36–42. [CrossRef]
78. Henckel, D.; Saha, P.; Intia, F.; Taylor, A.K.; Baez-Cotto, C.; Hu, L.; Schellekens, M.; Simonson, H.; Miller, E.M.; Verma, S.; et al. Elucidation of Critical Catalyst Layer Phenomena toward High Production Rates for the Electrochemical Conversion of CO to Ethylene. *ACS Appl. Mater. Interfaces* **2024**, *16*, 3243–3252. [CrossRef] [PubMed]

79. Mitchell, S.; Pinar, A.B.; Kenvin, J.; Crivelli, P.; Kärger, J.; Pérez-Ramírez, J. Structural analysis of hierarchically organized zeolites. *Nat. Commun.* **2015**, *6*, 8633. [CrossRef] [PubMed]
80. Jacobsen, C.J.H.; Madsen, C.; Houzvicka, J.; Schmidt, I.; Carlsson, A. Mesoporous Zeolite Single Crystals. *J. Am. Chem. Soc.* **2000**, *122*, 7116–7117. [CrossRef]
81. Schmidt, I.; Boisen, A.; Gustavsson, E.; Ståhl, K.; Pehrson, S.; Dahl, S.; Carlsson, A.; Jacobsen, C.J.H. Carbon Nanotube Templated Growth of Mesoporous Zeolite Single Crystals. *Chem. Mater.* **2001**, *13*, 4416–4418. [CrossRef]
82. Chen, H.; Wydra, J.; Zhang, X.; Lee, P.-S.; Wang, Z.; Fan, W.; Tsapatsis, M. Hydrothermal Synthesis of Zeolites with Three-Dimensionally Ordered Mesoporous-Imprinted Structure. *J. Am. Chem. Soc.* **2011**, *133*, 12390–12393. [CrossRef] [PubMed]
83. Wu, J.; Jin, Z.; Wang, B.; Han, Y.; Xu, Y.; Liang, Z.; Wang, Z. Nickel Nanoparticles Encapsulated in Microporous Graphenelike Carbon (Ni@MGC) as Catalysts for CO_2 Methanation. *Ind. Eng. Chem. Res.* **2019**, *58*, 20536–20542. [CrossRef]
84. Pal, N.; Bhaumik, A. Soft templating strategies for the synthesis of mesoporous materials: Inorganic, organic–inorganic hybrid and purely organic solids. *Adv. Colloid Interface Sci.* **2013**, *189–190*, 21–41. [CrossRef] [PubMed]
85. Möller, K.; Yilmaz, B.; Müller, U.; Bein, T. Hierarchical Zeolite Beta via Nanoparticle Assembly with a Cationic Polymer. *Chem. Mater.* **2011**, *23*, 4301–4310. [CrossRef]
86. Kadja, G.T.; Suprianti, T.R.; Ilmi, M.M.; Khalil, M.; Mukti, R.R.; Subagjo. Sequential mechanochemical and recrystallization methods for synthesizing hierarchically porous ZSM-5 zeolites. *Microporous Mesoporous Mater.* **2020**, *308*, 110550. [CrossRef]
87. Xiao, F.; Wang, L.; Yin, C.; Lin, K.; Di, Y.; Li, J.; Xu, R.; Su, D.S.; Schlögl, R.; Yokoi, T.; et al. Catalytic Properties of Hierarchical Mesoporous Zeolites Templated with a Mixture of Small Organic Ammonium Salts and Mesoscale Cationic Polymers. *Angew. Chem.* **2006**, *118*, 3162–3165. [CrossRef]
88. Choi, M.; Cho, H.S.; Srivastava, R.; Venkatesan, C.; Choi, D.-H.; Ryoo, R. Amphiphilic organosilane-directed synthesis of crystalline zeolite with tunable mesoporosity. *Nat. Mater.* **2006**, *5*, 718–723. [CrossRef]
89. Pan, T.; Wu, Z.; Yip, A.C.K. Advances in the Green Synthesis of Microporous and Hierarchical Zeolites: A Short Review. *Catalysts* **2019**, *9*, 274. [CrossRef]
90. Tian, P.; Zhan, G.; Tian, J.; Tan, K.B.; Guo, M.; Han, Y.; Fu, T.; Huang, J.; Li, Q. Direct CO_2 hydrogenation to light olefins over $ZnZrO_x$ mixed with hierarchically hollow SAPO-34 with rice husk as green silicon source and template. *Appl. Catal. B Environ.* **2022**, *315*, 121572. [CrossRef]
91. Din, I.U.; Din, I.U.; Alotaibi, M.A.; Alotaibi, M.A.; Alharthi, A.I.; Alharthi, A.I.; Al-Shalwi, M.N.; Al-Shalwi, M.N.; Alshehri, F.; Alshehri, F. Green synthesis approach for preparing zeolite based Co-Cu bimetallic catalysts for low temperature CO_2 hydrogenation to methanol. *Fuel* **2022**, *330*, 125643. [CrossRef]
92. Gourdon, A. On-Surface Covalent Coupling in Ultrahigh Vacuum. *Angew. Chem. Int. Ed.* **2008**, *47*, 6950–6953. [CrossRef] [PubMed]
93. Debecker, D.P.; Le Bras, S.; Boissière, C.; Chaumonnot, A.; Sanchez, C. Aerosol processing: A wind of innovation in the field of advanced heterogeneous catalysts. *Chem. Soc. Rev.* **2018**, *47*, 4112–4155. [CrossRef] [PubMed]
94. Thilakarajan, P. Synthesis and Evaluation of Nanocatalysts for CO_2 Hydrogenation to Methanol: A Comprehensive Review. *Int. J. High Sch. Res.* **2024**, *6*, 118–126.
95. Bahari, N.A.; Isahak, W.N.R.W.; Ba-Abbad, M.M. The Effectiveness of Various Bimetallic on Iron-Zeolite Catalyst by Carbon Dioxide Hydrogenation. *IOP Conf. Ser. Earth Environ. Sci.* **2019**, *268*, 012110. [CrossRef]
96. Gao, X.; Deng, S.; Kawi, S. Zeolite-based catalytic membrane reactors for thermo-catalytic conversion of CO_2. *iScience* **2022**, *25*, 105343. [CrossRef] [PubMed]
97. Maghfirah, A.; Susanti, Y.; Fajar, A.T.N.; Mukti, R.R.; Kadja, G.T.M. The role of tetraalkylammonium for controlling dealumination of zeolite Y in acid media. *Mater. Res. Express* **2019**, *6*, 094002. [CrossRef]
98. Fajar, A.; Nurdin, F.; Mukti, R.; Subagjo; Rasrendra, C.; Kadja, G. Synergistic effect of dealumination and ceria impregnation to the catalytic properties of MOR zeolite. *Mater. Today Chem.* **2020**, *17*, 100313. [CrossRef]
99. Le Hua, Z.; Zhou, J.; Shi, J.L. Recent advances in hierarchically structured zeolites: Synthesis and material performances. *Chem. Commun.* **2011**, *47*, 10536–10547. [CrossRef]
100. Verboekend, D.; Mitchell, S.; Milina, M.; Groen, J.C.; Pérez-Ramírez, J. Full Compositional Flexibility in the Preparation of Mesoporous MFI Zeolites by Desilication. *J. Phys. Chem. C* **2011**, *115*, 14193–14203. [CrossRef]
101. Wardani, M.K.; Kadja, G.T.M.; Fajar, A.T.N.; Subagjo, S.; Makertihartha, I.G.B.N.; Gunawan, M.L.; Suendo, V.; Mukti, R.R. Highly crystalline mesoporous SSZ-13 zeolite obtained via controlled post-synthetic treatment. *RSC Adv.* **2019**, *9*, 77–86. [CrossRef] [PubMed]
102. Wakihara, T.; Tatami, J. Top-down Tuning of Nanosized Zeolites by Bead-milling and Recrystallization. *J. Jpn. Pet. Inst.* **2013**, *56*, 206–213. [CrossRef]
103. Kurniawan, T.; Muraza, O.; Hakeem, A.S.; Al-Amer, A.M. Mechanochemical Route and Recrystallization Strategy To Fabricate Mordenite Nanoparticles from Natural Zeolites. *Cryst. Growth Des.* **2017**, *17*, 3313–3320. [CrossRef]
104. Airi, A.; Signorile, M.; Bonino, F.; Quagliotto, P.; Bordiga, S.; Martens, J.A.; Crocellà, V. Insights on a Hierarchical MFI Zeolite: A Combined Spectroscopic and Catalytic Approach for Exploring the Multilevel Porous System Down to the Active Sites. *ACS Appl. Mater. Interfaces* **2021**, *13*, 49114–49127. [CrossRef]
105. Jia, X.; Khan, W.; Wu, Z.; Choi, J.; Yip, A.C. Modern synthesis strategies for hierarchical zeolites: Bottom-up versus top-down strategies. *Adv. Powder Technol.* **2019**, *30*, 467–484. [CrossRef]

106. Dapsens, P.Y.; Mondelli, C.; Pérez-Ramírez, J. Design of Lewis-acid centres in zeolitic matrices for the conversion of renewables. *Chem. Soc. Rev.* **2015**, *44*, 7025–7043. [CrossRef] [PubMed]
107. Ding, L.; Shi, T.; Gu, J.; Cui, Y.; Zhang, Z.; Yang, C.; Chen, T.; Lin, M.; Wang, P.; Xue, N.; et al. CO_2 Hydrogenation to Ethanol over Cu@Na-Beta. *Chem* **2020**, *6*, 2673–2689. [CrossRef]
108. Bacariza, M.C.; Graça, I.; Westermann, A.; Ribeiro, M.F.; Lopes, J.M.; Henriques, C. CO_2 Hydrogenation Over Ni-Based Zeolites: Effect of Catalysts Preparation and Pre-reduction Conditions on Methanation Performance. *Top. Catal.* **2016**, *59*, 314–325. [CrossRef]
109. Bacariza, M.C.; Graça, I.; Lopes, J.M.; Henriques, C. Tuning Zeolite Properties towards CO_2 Methanation: An Overview. *ChemCatChem* **2019**, *11*, 2388–2400. [CrossRef]
110. Verboekend, D.; Pérez-Ramírez, J. Design of hierarchical zeolite catalysts by desilication. *Catal. Sci. Technol.* **2011**, *1*, 879–890. [CrossRef]
111. Gackowski, M.; Tarach, K.; Kuterasiński, Ł.; Podobiński, J.; Sulikowski, B.; Datka, J. Spectroscopic IR and NMR studies of hierarchical zeolites obtained by desilication of zeolite Y: Optimization of the desilication route. *Microporous Mesoporous Mater.* **2019**, *281*, 134–141. [CrossRef]
112. Dai, C.; Zhang, A.; Li, L.; Hou, K.; Ding, F.; Li, J.; Mu, D.; Song, C.; Liu, M.; Guo, X. Synthesis of Hollow Nanocubes and Macroporous Monoliths of Silicalite-1 by Alkaline Treatment. *Chem. Mater.* **2013**, *25*, 4197–4205. [CrossRef]
113. Sharma, P.; Ho, P.H.; Di, W.; Creaser, D.; Olsson, L. Novel catalyst configuration to boost the yield of longer hydrocarbons from methanol-mediated CO_2 hydrogenation. *J. CO2 Util.* **2023**, *74*, 102549. [CrossRef]
114. Gac, W.; Zawadzki, W.; Słowik, G.; Kuśmierz, M.; Dzwigaj, S. The state of BEA zeolite supported nickel catalysts in CO_2 methanation reaction. *Appl. Surf. Sci.* **2021**, *564*, 150421. [CrossRef]
115. Krachuamram, S.; Kidkhunthod, P.; Poo-Arporn, Y.; Kamonsutthipaijit, N.; Chanapattharapol, K.C. On the Optimization of Ni/A and Ni/X Synthesis Procedure toward Active and Selective Catalysts for the Production of CH_4 from CO_2. *Catalysts* **2022**, *12*, 823. [CrossRef]
116. Velisoju, V.K.; Cerrillo, J.L.; Ahmad, R.; Mohamed, H.O.; Attada, Y.; Cheng, Q.; Yao, X.; Zheng, L.; Shekhah, O.; Telalovic, S.; et al. Copper nanoparticles encapsulated in zeolitic imidazolate framework-8 as a stable and selective CO_2 hydrogenation catalyst. *Nat. Commun.* **2024**, *15*, 2045. [CrossRef]
117. Ticali, P.; Salusso, D.; Ahmad, R.; Ahoba-Sam, C.; Ramirez, A.; Shterk, G.; Lomachenko, K.A.; Borfecchia, E.; Morandi, S.; Cavallo, L.; et al. CO_2 hydrogenation to methanol and hydrocarbons over bifunctional Zn-doped ZrO_2/zeolite catalysts. *Catal. Sci. Technol.* **2021**, *11*, 1249–1268. [CrossRef]
118. Wang, X.; Jeong, S.Y.; Jung, H.S.; Shen, D.; Ali, M.; Zafar, F.; Chung, C.-H.; Bae, J.W. Catalytic activity for direct CO_2 hydrogenation to dimethyl ether with different proximity of bifunctional Cu-ZnO-Al_2O_3 and ferrierite. *Appl. Catal. B Environ.* **2023**, *327*, 122456. [CrossRef]
119. Wang, C.; Guan, E.; Wang, L.; Chu, X.; Wu, Z.; Zhang, J.; Yang, Z.; Jiang, Y.; Zhang, L.; Meng, X.; et al. Product Selectivity Controlled by Nanoporous Environments in Zeolite Crystals Enveloping Rhodium Nanoparticle Catalysts for CO_2 Hydrogenation. *J. Am. Chem. Soc.* **2019**, *141*, 8482–8488. [CrossRef] [PubMed]
120. Dang, S.; Gao, P.; Liu, Z.; Chen, X.; Yang, C.; Wang, H.; Zhong, L.; Li, S.; Sun, Y. Role of zirconium in direct CO_2 hydrogenation to lower olefins on oxide/zeolite bifunctional catalysts. *J. Catal.* **2018**, *364*, 382–393. [CrossRef]

catalysts

MDPI

Article

Effects of Site Geometry and Local Composition on Hydrogenation of Surface Carbon to Methane on Ni, Co, and NiCo Catalysts

Sebastian Godoy [1], Prashant Deshlahra [2,*], Francisco Villagra-Soza [1], Alejandro Karelovic [1] and Romel Jimenez [1,*]

1 Carbon and Catalysis Laboratory (CarboCat), Department of Chemical Engineering, Faculty of Engineering, University of Concepción, Concepción 4070409, Chile
2 Department of Chemical and Biological Engineering, Tufts University, Medford, MA 02155, USA
* Correspondence: prashant.deshlahra@tufts.edu (P.D.); romeljimenez@udec.cl (R.J.)

Abstract: Surface carbon deposits deactivate Ni and Co catalysts in reactions involving hydrocarbons and CO_x. Electronic properties, adsorption energies of H, C, and CH_x species, and the energetics of the hydrogenation of surface C atom to methane are studied for (100) and (111) surfaces of monometallic Ni and Co, and bimetallic NiCo. The bimetallic catalyst exhibits a Co→Ni electron donation and a concomitant increase in the magnetization of Co atoms. The CH_x species resulting from sequential hydrogenation are more stable on Co than on Ni atoms of the NiCo surfaces due to more favorable (C-H)–Co agostic interactions. These interactions and differences between Co and Ni sites are more significant for (111) than for (100) bimetallic surfaces. On (111) surfaces, CH is the most stable species, and the first hydrogenation of C atom exhibits the highest barrier, followed by the CH_3 hydrogenation steps. In contrast, on (100) surfaces, surface C atom is the most stable species and CH_2 or $*CH_3$ hydrogenations exhibit the highest barriers. The Gibbs free energy profiles suggest that C removal on (111) surfaces is thermodynamically favorable and exhibits a lower barrier than on the (100) surfaces. Thus, the (100) surfaces, especially Ni(100), are more prone to C poisoning. The NiCo(100) surfaces exhibit weaker binding of C and CH_x species than Ni(100) and Co(100), which improves C poisoning resistance and lowers hydrogenation barriers. These results show that the electronic effects of alloying Ni and Co strongly depend on the local site composition and geometry.

Keywords: alloy; cobalt; coking; DFT; hydrogenation; intermetallic; nickel

1. Introduction

H_2 and small carbon-containing species such as CO, CO_2, and CH_4 are produced and utilized as feedstocks in large industrial processes, making their interconversions important for the economy and environment. Prominent examples of such processes include CO and CO_2 methanation, Fischer–Tropsch synthesis, methane reforming using H_2O (steam reforming) or CO_2 (dry reforming of methane, DRM), decomposition of larger hydrocarbons, and H_2 production. The rational selection and design of catalysts that are active at milder reaction conditions, selective for the desired products, and can be used uninterruptedly for longer periods lead to more efficient and greener processes with lower energy requirements. Noble metals (e.g., Ru, Rh, and Pt) show high activity for methanation and DRM reactions [1,2] while being resistant to deactivation, but their scarcity and high cost limit their industrial applications. More abundant and cheaper Ni-based catalysts exhibit high selectivity for CH_4 in methanation processes, and high activity for both methanation and DRM, but suffer from rapid deactivation due to carbon deposition. Therefore, understanding molecular details and site requirements for C deposition in Ni-based catalysts is important.

Cobalt is also more abundant than noble metals and shows high activity for methanation and DRM. More importantly, Co shows a higher resistance to carbon poisoning. However, it forms carbon deposits (sheets and filaments) in other reactions, such as ethanol dehydrogenation [3], and has been found to deactivate by oxidation at high temperatures used in DRM processes [4,5]. Co forms homogeneous alloys with nickel over wide composition and temperature ranges [6], making it a good candidate for a bimetallic CoNi catalyst. Alloying or doping Ni-based catalysts with transition metals has been shown to affect the resistance to carbon deposition and, in turn, the activity and selectivity for DRM and methanation [2,4,7–9]. This fine-tuning of catalytic (and electronic) properties via alloying has received significant attention in recent theoretical and experimental studies [2,8,10–12] and is a promising area for the development of more active, selective, and stable catalysts for a wide range of reactions.

For example, activity for the decomposition of ethanol on Co-mixed oxides was improved by the formation of active NiCo phases, but began to decrease after 14 h, showing significant deposition of carbonaceous species (carbon nano-filaments and multi-walled nanotubes according to the particle size) [13]. Takanabe et al. [4] experimentally studied Co-Ni/TiO_2 catalysts in DRM conditions, showing that while >80%Ni alloys underwent C formation and >90%Co alloys deactivate by metal oxidation, the intermediate NiCo alloys performed stably with high activity. Adding oxygen to the feed has been reported to considerably decrease coke formation during DRM on Ni-rich NiCo catalysts [14]. The higher oxophilicity (affinity to bind O and OH) of Co surfaces has been proposed to propitiate oxidation and removal of surface carbon deposits, making it more resistant to coke poisoning [8,15]. Tu et al. [8] studied (111) surfaces by DFT and proposed that the higher surface O coverage on Co(111) and CoNi(111) allows an O-assisted dissociative adsorption of CH_4 (not favored on pure Ni surfaces) that enhances methane activation in the DRM conditions. Chen and Yang recently studied NiCo step B_5 sites by DFT, suggesting that higher *C coverage on Ni steps may relate to carbon deposition and higher *O coverage on Co steps may relate to deactivation by oxidation. Ou et al. [2] used DFT to study CH_4 dehydrogenation to surface *C on clean and Co-doped Ni(111) surfaces, finding a negative effect of the Co doping for direct dissociative adsorption of CH_4, which was suggested to be the rate-determining step. They also reported that the incorporation of Co weakened *C adsorption and lowered the barriers for the desired *C + *H → *CH step, which suggested the thermodynamically and kinetically enhanced removal of surface carbon on Co-doped (111) surfaces. The *CH dehydrogenation has been found to be a kinetically relevant barrier, sometimes competing with the dissociative adsorption of CH_4 as the rate-determining step for surface *C production in various DRM-DFT studies of the close-packed Co [1,16], Ni(111) [17,18], and NiCo(111) surfaces [7,9,19], with one barrier or the other being higher depending on the specific system and conditions considered. For example, Liu et al. [19] agree with the results of Ou et al. [2] for higher CH_4 dissociative adsorption barriers on NiCo(111) systems, but the formation of surface *C from *CH has been proposed as the highest overall barrier on Ni(111) [19] and Co(111) [7], also in agreement with Li et al.'s results [9] for Ni alloying with 9-group metals, and closely competing with CH_4 dissociation on NiCo(111) [7].

Notably, these and most other studies use the close-packed (111) (and similar (0001) for Co) surfaces in density functional theory (DFT) modeling, with a few recent exceptions [20]. Although the (111) surfaces show clear differences between Ni and Co, other surfaces that are also present on monometallic Ni [21,22] and Co [23] clusters in smaller but still significant proportions may exhibit very different binding properties that make them worth studying in comparison to the (111) surface. For example, on Ni-based single-atom alloys, the CO dissociation has been shown to be endothermic on (111) surfaces but exothermic on (100) surfaces [11]. The dissociation routes (direct or H-assisted) are favored for Ni surfaces on (100) and step sites [24], which are also more reactive than close-packed facets for both CH_4 steam reforming and the formation of C deposits over certain crystal sizes [25]. The hydrogenation of $*CH_3$ (or the methane dissociative adsorption) has been suggested as the rate-limiting step for CH_4 formation (or DRM) on various *fcc* and *hcp* Co surfaces [26,27] and activation barriers related to the corresponding surface *d*-band centers [26], but with different dominating paths and deactivation methods for Co(111) and Co(211) [26]. Stronger *C binding with lower mobility and considerably lower *CH dehydrogenation barriers are reported for Co [16] and Ni [17] (100) surfaces compared to the close-packed (111) surfaces [16], and the strong square planar adsorption of *C on Ni and Co (100) sites also promotes significant surface restructuration [28].

Thus, the surface geometry significantly affects CH_x reactions on both monometallic Ni and Co surfaces. The most studied close-packed (111) facets tend to be less active for CH_x conversions, but also more resistant to the formation of surface C formation, than the (100) facets. Overall activity, selectivity, and deactivation resistance may come from a combination of these (and other relevant) surfaces. These strong differences between monometallic facets and reported properties of NiCo(111) alloys led us to comparatively study paths for the elimination of surface *C on the monometallic and NiCo alloys of both (111) and (100) surfaces. The results presented here analyze in detail the differences between the reactivity of these surfaces and their links to the surface electronic properties resulting from the from the Ni–Co alloying.

2. Results and Discussion

2.1. Structure and Electronic Properties of Ni, Co, and NiCo

Nickel nanoparticles exhibit an *fcc* crystal structure. Cobalt can form both *fcc* and *hcp* (hexagonal close-packed) phases, depending on the preparation temperature and nanoparticle size [29], which also affects its catalytic properties [23]. For homogeneous NiCo alloys, the *fcc* phase is experimentally observed when the Ni content is greater than 35 wt.% [6] and forms naturally from $NiCoO_2$ mixed oxide after reduction [13]. Only the *fcc* phase is considered in this work to avoid crystallographic effects and focus on alloying effects. The lattice constants fitting the RPBE-derived bulk energies to the Murnaghan equation of state are 3.55 Å for Ni, Co, and NiCo, within the accepted DFT error [20] but slightly larger than experimental values (near 3.52 Å) [6,30–32].

From the bulk model, (111) and (100) surfaces were constructed. Figure 1 shows the unique adsorption sites of the monometallic (a and b, Co as an example) and bimetallic (b and c) slab models. The Ni_2Co or $NiCo_2$ local surface composition of three-fold (111) sites distinguishes Ni-rich or Co-rich sites. The four-fold (100) sites are analogously distinguished by a subsurface Ni or Co atom (h_N and h_C in Figure 1d, respectively).

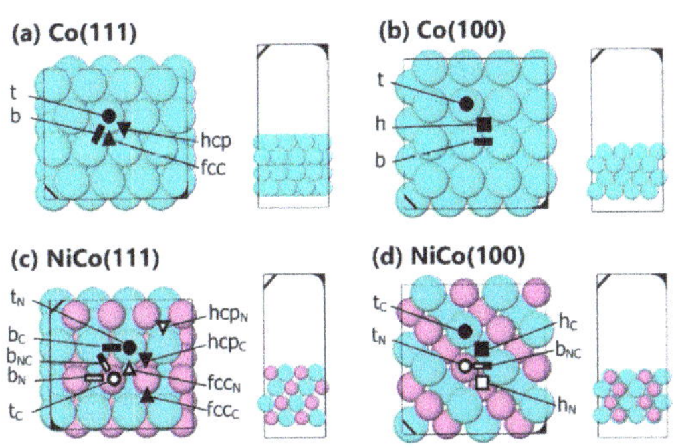

Figure 1. Unique adsorption sites on surface models: (**a**) Co(111), (**b**) Co(100), (**c**) NiCo(111), and (**d**) NiCo(100). Symbols according to site geometry: • t for top, ▲ fcc, ▼ hcp, — b for bridge, and ■ h for four-fold hollow sites, filled symbols for Co or Co-rich sites, open symbols on Ni or Ni-rich sites.

The atomic magnetic moments derived from the DDEC6 method for bulk Co and Ni are 1.66 and 0.65 μ_B, respectively (Table 1), in agreement with Bader partitions (SI.2, Table S2) and previously published values [5,33,34]. Surface atoms in monometallic (111) and (100) surfaces remain essentially neutral (DDEC6 electronic charges ~±0.01 e, Table 1) but exhibit higher magnetic moments than the bulk atoms (1.72 and 1.78 μ_B for Co, 0.67 and 0.70 μ_B for Ni, Table 1). DDEC6 and Bader charges show electron density donation from Co atoms to the more electronegative Ni atoms in bulk and surface models. The magnetic moment of Co@NiCo is higher than monometallic bulk Co and monometallic surface Co, but the magnetic moment of Ni@NiCo for the bulk and surfaces remains essentially the same as monometallic bulk Ni and slightly lower than monometallic surface Ni. The extent of charge donation and magnetic moment change in Co is higher for the less coordinated (100) surfaces than the (111) surface. This increase in magnetic moment of Co upon alloying but slight decrease for Ni is not explained by simple electronic donation and requires consideration of their corresponding density of states, as discussed next.

Table 1. DDEC6 atomic (average) electronic charges and spin magnetic moments.

	Electronic Charges (e)			Spin Magnetic Moment (μ_B)		
	Bulk	**(111)$_{surf.}$**	**(100)$_{surf.}$**	**Bulk**	**(111)$_{surf.}$**	**(100)$_{surf.}$**
Co	0	−0.01	+0.01	1.66	1.72	1.82
Co$_{(@ NiCo)}$	+0.04	+0.04	+0.10	1.72	1.81	1.89
Ni$_{(@ NiCo)}$	−0.04	−0.05	−0.09	0.65	0.64	0.65
Ni	0	−0.01	+0.01	0.65	0.67	0.70

The projected density of states for the *d*-band (*d*-PDOS) of Co and Ni atoms in mono- and inter-metallic bulk, (111), and (100) surfaces are shown in Figure 2.

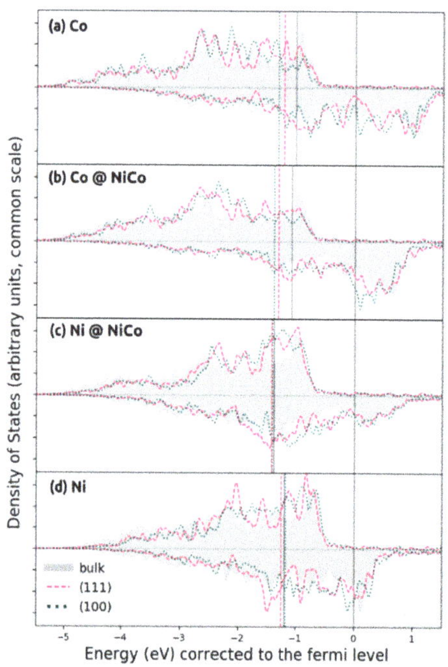

Figure 2. *d*-band projected density of states (PDOS) in bulk, (100), and (111) surfaces for (**a**) Co, (**b**) Co in NiCo, (**c**) Ni in NiCo, and (**d**) Ni. Vertical lines show the corresponding *d*-band centers.

The spin-up components of Co atoms (Figure 2a,b) appear at significantly lower energies than the spin-down component. Spin-up bands are completely filled, while the spin-down bands span across the Fermi level, leading to a considerable fraction of empty states above the Fermi level. The Ni atoms (Figure 2c,d) contain relatively more symmetric spin-up and spin-down components at the low-energy end with the spin-up also completely filled but fewer unoccupied states above the Fermi level for the spin-down component. These differences in *d*-bands are consistent with the higher magnetization of Co than Ni. This asymmetry in band structure suggests that a simple Co→Ni electron donation will make Co band occupancies more asymmetric by withdrawing spin-down electrons near the Fermi level, consistent with the observed increase in Co magnetic moment. A corresponding increase in Ni spin-down electrons will decrease the Ni magnetic moment, qualitatively explaining the trends in Table 1. However, these magnetization changes are also affected by changes in the band structure caused by the interactions of Ni and Co bands in the alloy, as discussed next.

For bulk metals and surfaces, the alloying and the Co→Ni electron donation increase the amount of Co empty states just above the Fermi level, with only a slight reduction of occupied states around −0.5 eV. A similar reduction around that energy is also observed for the Ni atoms in the alloy, but in this case the density of unoccupied states is also reduced around the Fermi level and spread to higher energies, while the occupied states concentrate at lower energies compared to the monometallic Ni (~−1.2 eV).

These changes in band and filled states' distributions can be represented by the Fermi-corrected *d*-PDOS first moment (i.e., the *d*-band center e_d, Table 2 and vertical lines in Figure 2). For monometallic surfaces, the reported e_d values are higher for Ni(100) than Ni(111) [17], but lower for Co(100) than Co(111) [1], which is consistent with the trends in Table 2. Such changes are attributed to the polarization of Co atoms. For Ni and Ni@NiCo atoms, there is little difference in e_d between bulk and both surfaces, consistent with the smaller effect on magnetization in Table 1 compared to Co atoms. More importantly, the overall effect of the Ni–Co interaction is a reduction of the *d*-band centers for both Co and Ni in all bulk, (111), and (100) surfaces. Ni atoms also show a larger downward shift from mono- to bi-metallic even when only occupied levels are considered (see SI.3).

Table 2. Average *d*-band centers (e_d in eV) for bulk, (111), and (100) surfaces.

	Bulk	$(111)_{surf.}$	$(100)_{surf.}$
Co	−1.01	−1.23	−1.32
Co@ NiCo	−1.08	−1.32	−1.38
NiCo (average)	−1.24	−1.38	−1.40
Ni @ NiCo	−1.39	−1.43	−1.41
Ni	−1.21	−1.26	−1.18

The atomic charges, band structures, and band center energies suggest that alloying significantly modifies the electronic properties of both Ni and Co atoms. Next, we discuss the effect of such modification on binding energies of surface species on monometallic and bimetallic surfaces.

2.2. Geometries and Adsorption Energies of C, H, and CH_x Species

Previous DFT studies have shown that C, H, and CH_x (x = 1–3) species preferentially bind to three- and four-fold sites of (111) and (100) Ni and Co surfaces, respectively, with the exception of CH_3 that prefers a bidentate bridge site on (100) surfaces [35,36]. Our results are consistent with those reports, and we further probe in detail the differences between hcp and fcc sites, and differences from the local site composition on the (111) and (100) intermetallic surfaces. Figure 3 shows adsorption energies for the most stable geometries of adsorbed H, C, CH, CH_2, and CH_3 species on surfaces, and the symbol shapes for the different site geometries are ▲ fcc,▼ hcp, — bridge, and ■ hollow (same as Figure 1, open and filled symbols represent Ni-rich and Co-rich sites, respectively) (images, energies, and geometric parameters can be found in SI.7).

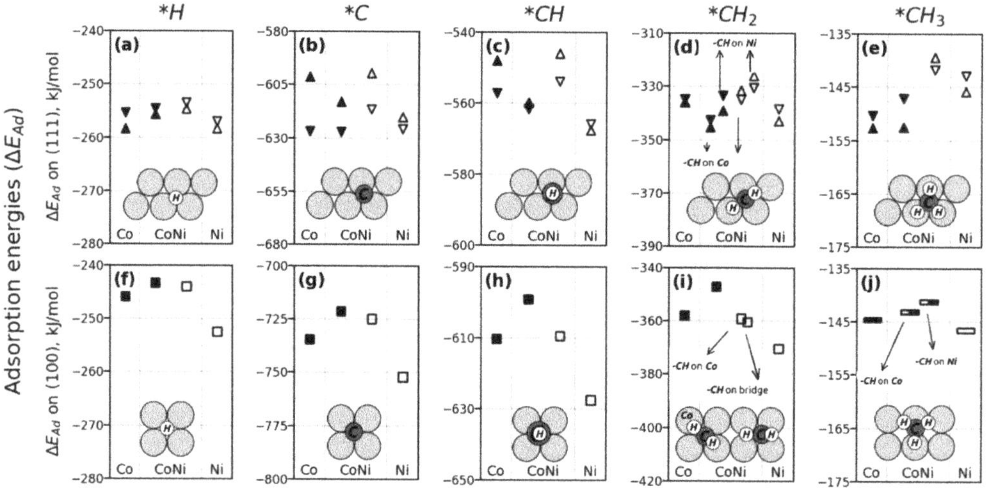

Figure 3. Adsorption energies (ΔE_{ad}) on Co, NiCo, and Ni for H (**a**,**f**), C (**b**,**g**), CH (**c**,**h**), CH_2 (**d**,**i**), and CH_3 (**e**,**j**) on (111) and (100) surfaces, respectively. Sites geometries: ▲ fcc, ▼ hcp, — bridge, and ■ hollow, using filled symbols on Co and Co-rich sites and open symbols on Ni and Ni-rich sites. Top views of geometries are shown as insets.

2.2.1. H and C_n Adsorptions

Figure 3 shows that C and H atoms exhibit similar binding strengths on Ni(111) and Co(111) surfaces (Figure 3a,b), but both bind significantly more strongly on Ni(100) than Co(100) (Figure 3f,g). Moreover, while H binds more strongly on the (111) fcc sites than on the four-fold sites of (100) surfaces, C binds over 100 kJ/mol more strongly on the (100) surface than on the (111) hcp sites. For C on (111) surfaces, Co and the Co-rich intermetallic sites show similar binding energies, close to Ni and significantly stronger than the Ni-rich intermetallic site. For the intermetallic sites, weaker adsorptions are found for H on both surfaces and for C on (100), which is qualitatively consistent with the lower band centers according to the *d*-band model [37]. This anti-synergistic effect is more clear and significantly greater on the (100) surface than on the previously studied (111) surface [9,19], especially for C on NiCo(100), being 9 kJ/mol less stable than on Co(100) and 30 kJ/mol than on Ni(100). Weaker C and H binding also suggests a higher surface mobility that can promote hydrogenation, lessen carbon deposits, and render intermetallic sites less prone to poisoning.

The strong binding of C is consistent with the well-reported deactivation due to the formation of surface C deposits, especially for Ni-based catalysts. Such deactivation is also observed for Co in DRM conditions [27] but not usually reported, for example, under milder methanation conditions. The build-up of a small amount of C was studied through the formation of C-chains from atomic *C (Figure 4; images, energies, and geometric parameters are presented in SI.8), that are reported to be thermodynamically stable and have higher mobility than adatoms and serve as a precursor of star-like formations, rings, and later graphitic overlayers [38].

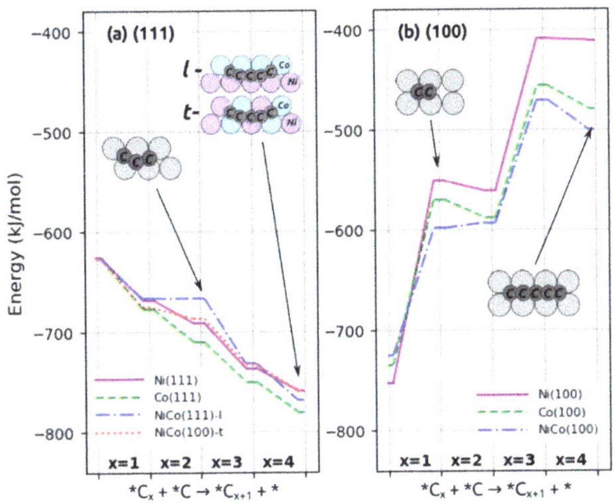

Figure 4. Surface C addition, reaction of electronic energies on (**a**) (111) and (**b**) (100) surfaces.

On (111) surfaces, the C-chains become more stable with increasing chain length (Figure 4a), compared to surface C species. In contrast, on (100) surfaces, the longer chains are less stable, which suggests that the high stability of C adsorbed on the four-fold M_4 sites hinders the formation of C-C bonds. Other reconstruction or deactivation processes [28,39] may be more favorable on (100) surfaces. For the (111) intermetallic surface, two chain directions, one along and the other across the Co–Ni interface lines (Figure 4a, insets l- and t-), exhibit some differences in the energies of C_2 and C_3 chains but give similar energies for longer chains. For (100), C_5 chains are similar or more stable than C_4 chains, and C_3 are more stable than C_2. On both C_2 and C_4 chains, one terminal C ends on a less stable bridge site. Thus, the position of terminal C atoms instead of the direction of chains appears to dominate the formation energy of short linear C_n.

2.2.2. Adsorption of CH_x Species

CH: From Figure 3c, on (111) surfaces, the hcp sites are preferred on monometallic Co and the NiCo alloy, but fcc sites are preferred on Ni(111). The CH species binding is strongest on Ni, followed by Co-rich NiCo sites, Co(111), and lastly, Ni-rich NiCo, for their respective preferred sites. These energy differences span a range of about 10 kJ/mol from the strongest binding Ni(111) and weakest binding Ni-rich NiCo(111). From Figure 3h, all (100) sites bind CH species at least 30 kJ/mol more strongly than the (111) sites of similar composition. Among the (100) sites, the CH binding is the strongest on Ni(100), followed by Co(100) and the Ni-rich NiCo(100) sites (17 kJ/mol weaker). In this case, the energy difference between the strongest binding Ni and weakest binding Co-rich NiCo sites is over 30 kJ/mol. Compared to pure Ni and Co surfaces, both NiCo sites appear less effective to stabilize C-containing surface species.

CH_2: CH_2 species adsorbs with an H atom above the C atom and the second H (with longer C-H bond) in a top or bridge position (Figure 3d,i insets, and SI.7, M-H distances between 1.7 and 1.8 Å). The fcc sites are preferred on Co(111), Ni(111), and the Co-rich NiCo(111) sites. The Co-rich NiCo(111) site adsorbs CH_2 most strongly, followed by Ni(111), Co(111), and 10 kJ/mol more strongly than the Ni-rich NiCo(111) sites. On NiCo(111), $^*CH_2$ orientations with the second H on top of a Co atom are preferred over that on top of Ni atoms by at least 5 kJ/mol, suggesting that (C-H)–Co agostic interactions (i.e., interactions of C-H bonds with metal atoms [18,40]) are stronger than (C-H)–Ni interactions. Adsorption of CH_2 on Ni(100) is about 12 kJ/mol stronger than on Co(100) and the Ni-rich NiCo(100) sites, but 27 kJ/mol stronger than the Co-rich sites on NiCo(100). On the NiCo(100) surfaces, the preference for Ni-rich sites suggests a predominant effect of the more stable Ni–Carbon interaction, as observed for C and CH (Figure 3g,h), rather than the (C-H)–Co agostic interactions that exist in (111) surfaces.

CH_3: On (111) surfaces, CH_3 sits in a three-fold site with its three H atoms on top of surface metal atoms (inset in Figure 3e, and SI.7, M-H distance between 2.0 and 2.1 Å). The fcc sites bind CH_3 more strongly than the hcp sites by about 3 kJ/mol on both monometallic sites and by 5.6 kJ/mol on the NiCo(111) Co-rich sites. The Co and Co-rich (111) fcc sites exhibit the strongest binding, 6.7 kJ/mol stronger than Ni(111) and over 11 kJ/mol stronger than the hcp Ni-rich site on NiCo(111). On (100) surfaces, the CH_3 species binds preferentially through C on a bridge site and one of the three H atoms over a surface metal (inset Figure 3j). The binding on Ni(100) is slightly stronger than on Co(100) (2 kJ/mol) but both monometallic sites bind CH_3 more strongly than the NiCo(100) surface. In contrast to other CH_x species, the Co-rich NiCo(100) sites bind CH_3 more strongly than the Ni-rich NiCo(100), and all (100) surfaces bind CH_3 weaker than the (111) surfaces and the Co-rich NiCo(100) site is preferred to the Ni-rich NiCo(100) site.

In summary, adsorption energies of H, C, and CH_x species exhibit complex trends resulting from electronic and geometric factors. The adsorption energy of CH_x species on (100) and (111) surfaces becomes less negative (weaker binding) for more hydrogenated species. On the (100) surface, all species exhibit similar trends for composition effects on adsorption energy, with bimetallic sites binding species less strongly than monometallic sites, which potentially makes *C adsorption energy a descriptor for CH_x species. Similar trends have been described for CH_x species on other (111) surfaces [41], but no significant trends exist between CH_3 and C adsorption energies for Ni, Co, and NiCo (111) surfaces (see CH_x vs. C relations in SI.9). For the three-fold sites on NiCo, adsorption of C and CH_x species is stronger on the Co-rich sites than on the Ni-rich sites regardless of the distinction between fcc and hcp geometries. In contrast, on the four-fold NiCo(100) sites C, CH, and CH_2 show a preferred adsorption on Ni-rich NiCo sites that is consistent with the stronger adsorption on Ni(100) compared to Co(100). For CH_2 on (111) and CH_3 on both surfaces, the preferred adsorption configuration allows favorable alignment of H atoms with surface metal atoms at short distances (1.7–2.1 Å), suggesting a role of agostic interactions between electron-deficient metal and C-H σ-bonds. The (C-H)–Co agostic interactions are stronger than (C-H)–Ni interactions, which accounts for stronger adsorption on Co(111) than on Ni(111), contrary to other CH_x species. For the (100) surfaces, the Ni(100) is preferred to Co(100) and the Ni-rich sites are preferred on NiCo(100), which suggests that the strong interaction with Ni atoms in four-fold sites is the predominant effect. Weaker adsorption on Ni-rich sites is observed for all species and surfaces compared to the monometallic Ni and it is also weaker than monometallic Co in most cases, which suggest a strong anti-synergy from alloying Ni and Co in the less coordinated surface. From the Co → Ni electronic donation, the slightly more negative Ni atoms (also with fewer states near the Fermi level) form weaker binding sites for CH_x. In turn, the more electronically unsaturated Co atoms offer stronger binding sites and are able to further stabilize C-H σ-bonds through agostic interactions. These Co atoms in the alloy provide more stable M-(C-H) centers that may also function as transition states for hydrogenation/dehydrogenation reactive steps, as discussed later.

2.3. Elementary Steps and Energetics for Hydrogenation of Surface Carbon

The sequential hydrogenation of an adsorbed C atom to methane was studied through the elementary steps in Scheme 1. Figure 5 shows the free energy reaction profile on the (111) and (100) surfaces (265 °C, 25 kPa H_2, and 1 kPa CH_4). Free energies, electronic energies, and the electronic energy profile are shown in the Supplementary Information (SI.10 and SI.11, respectively). The free energy for dissociative H_2 co-adsorption is positive, suggesting absence of a pool of bound H atoms. Therefore, H atoms are added sequentially for each hydrogenation and referenced to gas-phase H_2-free energy. Regardless, H_2 dissociation and migration steps exhibit low barriers [42–45] (~10 kJ/mol) and are considered kinetically irrelevant compared to the hydrogenation steps. The transition states for the H_2 dissociation are therefore not shown in Figure 5 for brevity. Figures 6 and 7 show the geometries of steps R1–R4 on the (111) and (100) surfaces of the alloy (other surfaces are shown in SI.10, geometric parameters in Tables S12 and S13), and the exploration of different reaction paths for the kinetically relevant steps converged to the paths presented here.

(R1)	$^*C + ½\ H_{2(g)}$	→	$^*CH$
(R2)	$^*CH + ½\ H_{2(g)}$	→	$^*CH_2$
(R3)	$^*CH_2 + ½\ H_{2(g)}$	→	$^*CH_3$
(R4)	$^*CH_3 + ½\ H_{2(g)}$	→	$^*CH_{4(phys)}$
(D)	$^*CH_{4(phys)}$	→	$^* + CH_{4(g)}$

Scheme 1. Steps involved in the hydrogenation of *C to CH_4. Surface sites are represented by *.

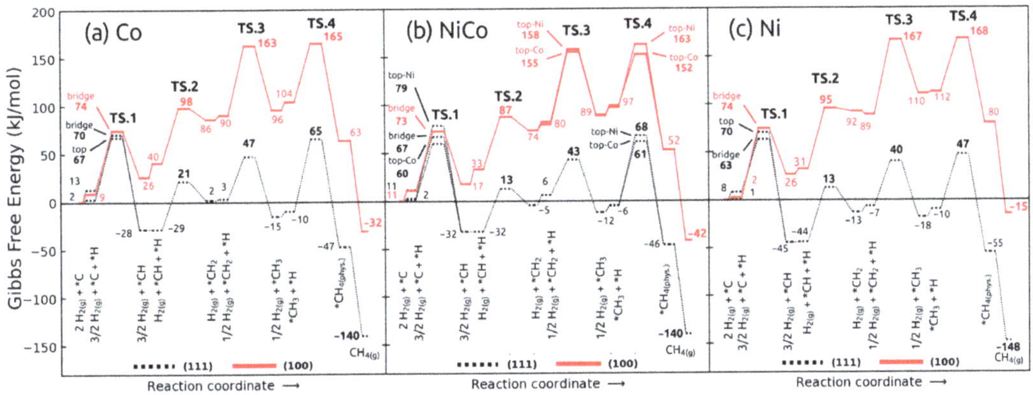

Figure 5. Gibbs free energy profiles for the hydrogenation of *C to CH_4(g) on (**a**) Co, (**b**) NiCo and (**c**) Ni, (111) (dotted lines) and (100) (solid lines) surfaces (265 °C, 25 kPa H_2, 1 kPa CH_4).

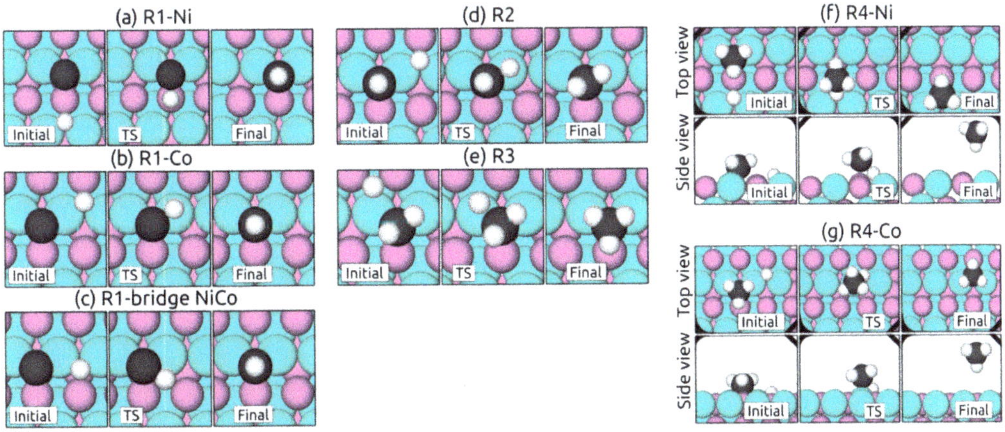

Figure 6. Hydrogenation sequence on the NiCo(111) surface to form (**a–c**) *CH, (**d**) $^*CH_2$, (**e**) $^*CH_3$ and (**f,g**) $^*CH_{4(phys)}$. Atom colors: Cyan—Co, purple—Ni, gray—C, white—H.

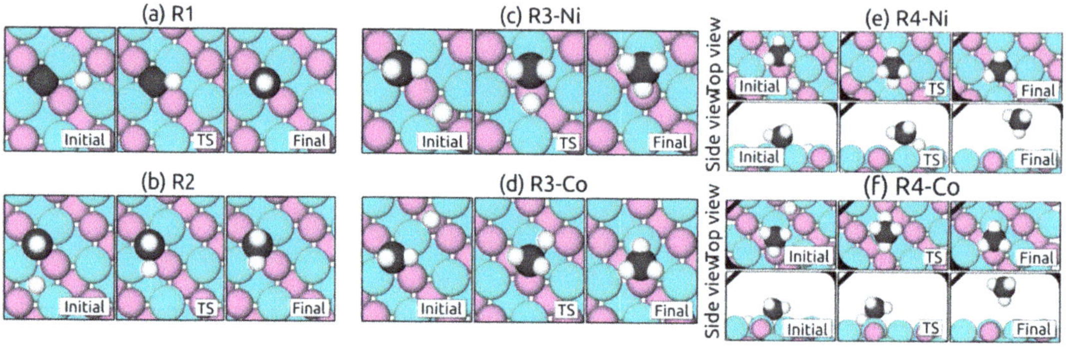

Figure 7. Hydrogenation steps on the NiCo(100) surface to form (**a**) *CH, (**b**) CH_2, (**c**,**d**) $*CH_3$ and (**e**,**f**) $*CH_{4(phys)}$. Atom colors: Cyan—Co, purple—Ni, gray—C, white—H.

2.3.1. Hydrogenation Steps on (111) Surfaces

The sequential hydrogenation of *C species requires H_2 dissociation at an empty site and the migration of a *H species close to a CH_x species. The C-H bond formation transition state for the first H addition (step R1, Scheme 1) can involve an H atom atop a surface metal atom or in the bridge position between two metal atoms (Figure 6a–c). The transition states with H atoms on top of metal atoms are more stable on Co(111) and NiCo(111) surfaces, consistent with the importance of (C-H)–metal agostic interactions observed in Section 2.2.2. In the case of NiCo(111), transitions started with H atop a Co atom are more stable than those atop Ni atoms. Subsequent H additions (steps R2–R4) also involve similar interactions involving H atoms in top positions and favoring Co atoms over Ni in the intermetallic surfaces.

On all (111) surfaces, the first hydrogenation step R1 (Scheme 1) is exoergic (dotted lines; $\Delta G_r < 0$; Figure 5), favoring the formation of *CH from *C. The *CH is the lowest free energy CH_x species on the (111) surface for Co(111), NiCo(111), and Ni(111), making it the most abundant surface species. The transition states for the second (R2; TS.2) and third hydrogenations (R3; TS.3) on all (111) surfaces exhibit lower free energy than the first hydrogenation (TS.1). Thus, interconversion between *CH, $*CH_2$, and $*CH_3$ can occur more frequently than between *C and *CH. The transition state for the hydrogenation of $*CH_3$ to form methane (TS.4) has higher free energy than the hydrogenation of *CH (TS.2) and $*CH_2$ (TS.3) on all (111) surfaces, close to the free energy barrier for the first hydrogenation (TS.1) on Co(111) and NiCo(111), but 16 kJ/mol lower on Ni(111). This suggests that the first (R1) and last (R4) hydrogenations have similar kinetic relevance on Co(111) and NiCo(111) surfaces, but the first hydrogenation limits the rate on Ni(111) at the simulated conditions. Conditions that inhibit the H_2 dissociative adsorption (higher ΔG_r, e.g., for lower H_2 pressure or higher temperature) will push the later hydrogenation barriers (TS.2, TS.3, and TS.4) to higher energies, thus making the last hydrogenation more relevant (see SI.11). At the reaction conditions, the complete hydrogenation of *C towards $CH_{4(g)}$ is thermodynamically favored on all (111) surfaces ($\Delta G_r < -100$ kJ/mol), similar on Co(111) and NiCo(111), and slightly more negative on Ni(111). The free energies of *CH, $CH_{4(g)}$, and the TS.4 transition state on NiCo(111) are similar to Co(111) (e.g., TS.4 ΔG 65, 61, and 47 kJ/mol in Co, NiCo, and Ni; Figure 5), suggesting that the equimolar intermetallic behaves kinetically more similar to the monometallic Co(111) than Ni(111). Ni catalysts are experimentally known to deactivate by C poisoning more than Co catalysts, but the TS.1 barriers are slightly lower on Ni(111) than on Co(111) and the TS.4 barriers for Ni(111) are 14 kJ/mol lower than for Co(111) or NiCo(111), suggesting a more facile elimination of *C through hydrogenation on Ni(111) surfaces. Therefore, even though the combination of *C in $*C_n$ chains is favored on (111) surfaces (Section 2.2.1), the flat (111) surfaces are not the primary reason for deactivation processes by C poisoning.

2.3.2. Hydrogenation Steps on (100) Surfaces

The transition states for sequential hydrogenation of *C involve a bridging position of the first and second added H atoms (R1, R2; Figure 7). The third and the fourth H atoms are near the top of a metal atom in their respective hydrogenation transition states (R3, R4). The free energy of the first hydrogenation transition states (TS.1; Figure 5) on (100) surfaces is within 5 kJ/mol of values on corresponding (111) surfaces. However, on (100) surfaces, the first hydrogenation step R1 is endoergic ($\Delta G_r > 17$ kJ/mol) and all TS.2 free energies are higher than TS.1. This suggests a favored equilibrium towards *C rather than *CH. Moreover, for all (100) surfaces, the transition state energies are in the order TS.1 < TS.2 < TS.3~TS.4. Thus, all $*CH_x$ species favor further dehydrogenation and *C is the most abundant surface species. In particular, the small TS.2 (reverse) barriers compared to $*CH_2$ suggest an equilibrium entirely displaced towards the dehydrogenation and any $*CH_2$ formed quickly dissociates to *CH, especially on the Ni(100) surface. TS.4 presents the highest barrier for methane formation on all Ni(100) and Co(100) surfaces, but within 5 kJ/mol of the $*CH_2$ hydrogenation barrier (TS.3). The significant dehydrogenation barriers for $*CH_3$ (>57 kJ/mol, with TS.3 as the reverse barrier) suggest that $*CH_3$ species may also be relevant for other reaction conditions. For instance, previous studies of the DMR process on (100) monometallic surfaces [16,17] coincide with the dissociative adsorption of $CH_{4(g)}$ as the highest barrier towards *C.

The TS.1 free energies are similar on all (100) surfaces (~74 kJ/mol), but the values for TS.2, TS.3, and TS.4 are lower on NiCo(100) than on monometallic (100) surfaces, suggesting a more facile hydrogenation of *C on NiCo(100) surfaces consistent with the experimentally observed C poisoning resistance of NiCo catalysts [46]. In particular, the higher TS.4 barrier on Ni(100) compared to Co(100) and NiCo(100) suggest a more difficult elimination of *C from the four-fold sites by hydrogenation on Ni(100). These *C do not recombine favorably to C_n chains (Section 2.2.1) but are hard to eliminate by hydrogenation (specially on the Ni(100) surface), and this suggests that *C species may distribute on (100) surfaces, blocking four-fold sites. The accumulation of *C on four-fold sites is related to deactivation near step edges [39] and propitiates surface restructuring of flat surfaces [28] (though such reconstructions are easier on Co than Ni surfaces). These results suggest that (100) surfaces are more relevant for explaining deactivation phenomena and the greater resistance of NiCo bimetallic sites to C poisoning.

The formation of $CH_{4(g)}$ from *C is thermodynamically more favored on NiCo(100) than on Co(100) and significantly less favored on Ni(100). This trend is consistent with the trends in adsorption energy discussed in Section 2.2, with all species (especially *C) binding more strongly on Ni(100) and significantly weaker on the NiCo(100) surface.

2.3.3. Effect of Density of States on Transition State Stabilization

As discussed in Section 2.3.1 and 2.3.2, all transition states on (111) and (100) surfaces involving H atoms on top of a metal atom are stabilized more when that metal atom is Co rather than Ni (Figures 5–7). Similarly, as shown in Section 2.2, CH_x species involving agostic interactions with Co are more stable than that with Ni. To analyze the difference between top-Co and top-Ni TS.4 transition states on the NiCo(111) and NiCo(100) surfaces in more detail, the atom-projected density of states (PDOS) for metal (Ni or Co), C, and H atoms was computed as shown in Figure 8. The interaction of the forming bond with the metal, i.e., M–(C-$H_{(1)}$), is observed between −7.3 and −6.3 eV, while the non-interacting $H_{(3)}$ and $H_{(4)}$ only have a peak in the same position as the C atoms near −5.6 eV. On the (100) surfaces, the positions of the reacting peaks (−6.78 eV for Co, −6.83 eV for Ni; Figure 8b,d) are closer to the Fermi level than on (111) surfaces (−7.16 eV for Co, −7.22 eV for Ni; Figure 8a,c). On both (111) and (100) surfaces, these peaks are slightly closer to the Fermi level for the top-Co transition states, in agreement with reports for (111) bimetallic surfaces [2]. The $H_{(2)}$ atom in TS.4 is near the metal surface in the opposite position to the forming C-$H_{(1)}$ bond. This $H_{(2)}$ atom shows populations overlapping with the M-$H_{(1)}$ populations, and the non-reactive $H_{(3)}$ and $H_{(4)}$ populations, suggesting some interaction of $H_{(2)}$ with the metal that is consistent with the M–(C-H) interactions proposed in Section 2.2.2 and the high stability of the Co over Ni transition states. Therefore, the electronic and binding effects of alloying Ni and Co translate to kinetic effects that differentiate NiCo surfaces from the expected average of monometallic Ni and Co surfaces. Agostic interactions of the Co and Ni atoms (and to a smaller extent on Ni atoms) with the forming C-$H_{(1)}$ bond and for the opposite C-$H_{(2)}$ bond are also observed in the spin density isosurfaces in Figure S6 of the Supplementary Information (SI.13).

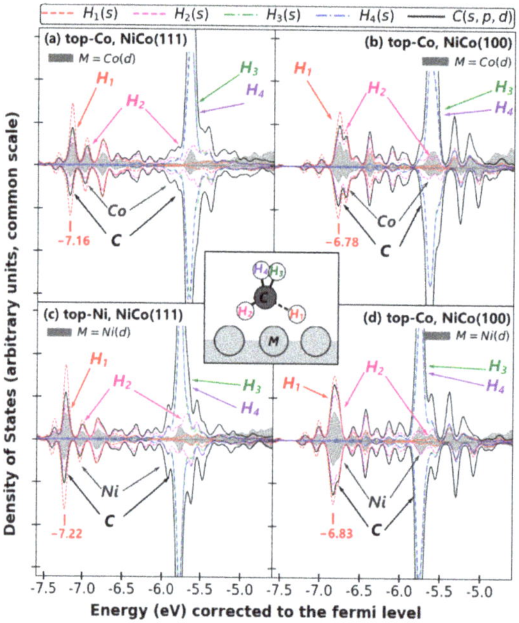

Figure 8. PDOS for the transition state $^*CH_3$ + $^*H \rightarrow CH_{4(phys\cdot)}$ + 2*: (**a**) top-C path on NiCo(111), (**b**) top-Co path on NiCo(100), (**c**) top-Ni on NiCo(111), and (**d**) top-Ni on NiCo(100).

The electronic adsorption energies (Section 2.2.2), the reaction energies to form key intermediates, and transition states' barriers suggest that the intermetallic surfaces behave more similarly to monometallic Co than Ni. Although, it should be noted that the effects of the Ni–Co interaction discussed here are specific to the local coordination degree, a specific reaction path, and are strongly related to the site composition, with Ni atoms in the alloy providing generally less stable adsorption and reaction sites compared to Co atoms in the alloy and monometallic Ni surfaces. The (111) surfaces are usually the only surfaces considered when studying the effects of alloying metals, but for a better description of the differences between bimetallic and monometallic catalysts, the key insights on the Ni–Co interactions derived from this work should be extended to study sites on other kinetically relevant surfaces (e.g., steps and kinks).

3. Methodology

The direct hydrogenation path from surface C to methane on Ni, Co, and NiCo surfaces was studied using density functional theory (DFT), as implemented in the VASP 5.4 code [47–50]. The revised PBE functional (RPBE [51]) was employed in spin-polarized calculations to appropriately deal with nickel and cobalt magnetism. Core electrons were described by the projector-augmented wave method [52,53]. A set of plane waves up to 460 eV was employed and a 0.05 Methfessel-Paxton [54] smearing of orbital populations was used to speed up calculations. All electronic energies were then extrapolated to 0 K. Energy differences under 10^{-6} eV between iterations are required for electronic convergence.

Face-centered cubic (fcc) crystal models were used to fit lattice parameters according to the Murnaghan equation of state [55] (see Section 2.1) and estimate atomic charges and magnetization of bulk monometallic Ni, Co, and equimolar NiCo intermetallic sites (see Supplementary Information, SI.1). One-sided slab models of (111) and (100) surfaces were constructed in 20 Å periodic cells using 4 layers of 16 atoms with o(4 × 4) arrangement and about a 15 Å vacuum between a slab and its periodic image. The resulting supercell sizes were 10.0 Å × 8.7 Å × 22.3 Å for (111) surfaces and 10.0 Å × 10.0 Å × 22.2 Å for (100) surfaces. Then, 12 × 12 × 12 and 4 × 4 × 1 Monkhorst-Pack [56] k-point meshes were employed for bulk and surface models, respectively. The top two layers and any adsorbed species were relaxed until all forces were below 0.02 eV/Å. Atomic charges and magnetism were estimated through Bader partitions [57] and the density-derived electrostatic and chemical (DDEC6) method [58] in the chargemol code [59] (see Supplementary Information, SI.2). Projected densities of states (PDOS) and their *d*-band centers were computed (see SI.3).

The electronic adsorption energy for A is defined as $\Delta E_{Ad} = E_{S \cdot A} - E_S - E_A$, where $E_{S \cdot A}$, E_S, and E_A are, respectively, the electronic energy of the surface with the adsorbed species, clean surface, and free relaxed A species in a 15 Å empty box. Lower E_{Ad} values correspond to a stronger binding to the surface. Vibrational frequencies were computed through diagonalization of the hessian matrix constructed from finite displacements of atomic coordinates assuming parabolic potentials. Only the degrees of vibrational freedom of the free gases (H2, CH4) and adsorbed species were considered. Small frequencies were truncated to 100 cm^{-1} (see SI.4). Real frequencies allow checking for configurational stability and estimating zero-point vibrational energies (ZPVE), entropy, and enthalpy contributions to compute Gibbs free energies (see SI.5). Transition states (TS) were approximated by climbing image-nudged elastic band (CI-NEB) reaction paths [60] and then refined through a dimer relaxation [61] until all forces were below 0.02 eV/Å (see SI.6). A single imaginary frequency and its visualization confirmed that the TS connected initial and final structures. The reaction-free energy ($\Delta G_{r,s}$) of a step is the difference between its final and initial free energies, and activation-free energy barriers (ΔG_a) are the difference between the free energy of the activated geometry and the free energy of the reference initial state (*C + $4H_{2(g)}$ on the different surfaces at conditions described before).

4. Conclusions

For the NiCo alloy, DDEC6 and Bader charge partitions show that Ni atoms draw electronic charge from Co atoms. These charge donations are greater for the less coordinated (100) surfaces and disproportionately affect the spin magnetic moment of surface atoms, with a slight decrease for Ni atoms in the alloy regardless of the coordination degree but a significant increase for Co atoms in the alloy bulk and surfaces. The PDOS for the intermetallic surfaces shows a significant reorganization with more occupied and unoccupied states around the Fermi level for Co atoms in the alloy compared to monometallic Co. In contrast, Ni atoms show a lower population near the Fermi level, with grouping of occupied states below and unoccupied states above the Fermi level. The electronic adsorption energies of gaseous CH_x fragments become less negative (weaker binding) with the hydrogenation degree. C, CH, and CH_2 species bind more strongly on (100) surfaces, while H and CH_3 species bind slightly more strongly on (111) surfaces. On NiCo(111), adsorption of CH_x species is more stable on Co-rich sites and favors orientations that allow (C-H)–Co interactions. C and CH_x species bind more strongly on Ni(100) than on Co(100). On NiCo(100), the Ni-rich sites are more stable than Co-rich sites, though both are significantly less stable than the average of the stability on monometallic Ni(100) and Co(100), suggesting a strong anti-synergy effect of alloying on binding strengths. Unlike the (111) surfaces, the formation of short C_n chains from *C is unfavorable on (100) surfaces due to the particularly strong binding of C on four-fold sites.

The hydrogenation of *C to $CH_{4(g)}$ is similarly favored on all (111) surfaces. The first hydrogenation is the highest barrier on (111) surfaces, followed by the last hydrogenation barrier, and *CH is the most stable surface species. Ni(111) is found to offer lower hydrogenation barriers than Co(111), with the first hydrogenation barrier being close to NiCo(111). Thus, the elimination of *C is more favored on Ni(111) and NiCo(111) than on Co(111). On (100) surfaces, the full hydrogenation towards CH_4 is less favorable on Ni(100) than Co(100) and most favored on NiCo(100), which is attributed to the weakly bonded CH_x species on NiCo(100). The hydrogenation of *CH_2 and *CH_3 are the main reaction barriers. *C instead of *CH is the most stable surface species. These computed free energies suggest that the greater deactivation by C poisoning of Ni catalysts compared to Co catalysts can be explained based on the energetics on the (100) surfaces, rather than (111). These energetics on (100) surfaces also show enhanced resistance to *C formation for the NiCo catalysts.

For the kinetically relevant steps on the NiCo surfaces, lower barriers are observed for transition states involving H on top of surface Co rather than surface Ni atoms, which is consistent with electron-deficient Co atoms on the alloy surfaces better-stabilizing hydrogenation transition states. This also explains the binding strength trends for NiCo(111), though the interaction of C with Ni is still more stable on the four-fold (100) sites, showing that the effects of the bimetallic interaction are strongly affected by surface coordination. Less coordinated sites and surface restructuring may also be involved in deactivation, but (111) and (100) are likely more abundant sites in fresh catalysts. DFT results in this work provide insight at the electronic level of key aspects that differentiate the kinetic behavior of alloyed NiCo from monometallic Ni and Co surfaces of different coordination degrees.

Supplementary Materials: The following supporting information can be downloaded at: https://www.mdpi.com/article/10.3390/catal12111380/s1, Table S1: Optimized lattice parameters. Table S2: Bader and DDEC6 atomic electronic charges. Table S3: Bader and DDEC6 atomic spin magnetic moments. Table S4: Average occupied *d*-band centers (up to the Fermi level) for bulk and surfaces. Figure S1: *d*-band projected density of states (PDOS) in bulk, (100), and (111) surfaces. Table S5: Frequencies from hessian matrices (different steps). Table S6: Translational and rotational thermodynamic corrections. Table S7: Vibrational frequencies, physisorbed CH_4. Figure S2: *C adsorption energies as descriptors of *CH_x adsorption energies. Tables S8 and S9: Gibbs free energy differences for initial geometries (ΔG_i) with *C+4$H_{2(g)}$ as a reference, for reaction steps ($\Delta G_{r,s}$), activation step barriers ($\Delta G_{a,s}$), and reverse step barriers (in parenthesis) on (111) and (100) surfaces. Tables S10 and S11: Electronic energy differences for initial geometries (ΔE_i) with *C+4$H_{2(g)}$ as a reference, for reaction steps ($\Delta E_{r,s}$), activation step barriers ($\Delta E_{a,s}$), and reverse step barriers (in parenthesis) on (111) and (100) surfaces. Scheme S1: Reaction steps for the construction of energy profiles. Figures S3–S5: Alternative electronic and free energy profiles. Tables S12 and S13: Relevant geometric parameters of the hydrogenation steps (with H*) on (111) and (100) surfaces. Figure S6: Spin densities for the activated state of the *CH_3 hydrogenation [6,11,26,30–32,55,62].

Author Contributions: Conceptualization, S.G., P.D. and R.J.; methodology, S.G. and P.D.; software, S.G.; formal analysis, S.G. and P.D.; data curation, S.G. and F.V.-S.; investigation, S.G., F.V.-S., A.K. and R.J.; writing—original draft preparation, S.G.; writing—review and editing, P.D.; supervision, P.D., A.K. and R.J.; funding acquisition, R.J. All authors have read and agreed to the published version of the manuscript.

Funding: Authors acknowledge financial support of the FONDECYT regular project 1170610. S.G. acknowledges financial support of the ANID national scholarship, ANID BECAS/DOCTORADO NACIONAL 2018/21180468. F.V.-S. acknowledges financial support of the ANID national scholarship, ANID BECAS/DOCTORADO NACIONAL 2021/21210586. Powered@NLHPC: This research was partially supported by the supercomputing infrastructure of the NLHPC (ECM-02). Some computations were performed with resources provided by the Kultrun Astronomy Hybrid Cluster via the projects Conicyt Programa de Astronomia Fondo Quimal 2017 QUIMAL170001, Conicyt PIA ACT172033, and Fondecyt Iniciacion 11170268. P.D. acknowledges support from the National Science Foundation (award number 2034911) and computing resources from eXtreme Science and Engineering Discovery environment (project number TG-CTS150005).

Data Availability Statement: The data presented in this study are detailed in the supplementary information, models and further details are available on request from the corresponding author.

Acknowledgments: S.G acknowledges the technical support and recommendations from the supercomputing infrastructure of the NLHPC team and the Kultrun Astronomy Hybrid Cluster team.

Conflicts of Interest: The authors declare no conflict of interest.

References

1. Huang, H.; Yu, Y.; Zhang, M. Mechanistic insight into methane dry reforming over cobalt: A density functional theory study. *Phys. Chem. Chem. Phys.* **2020**, *22*, 27320–27331. [CrossRef] [PubMed]
2. Ou, Z.; Ran, J.; Niu, J.; Zhang, Z.; Deng, T.; He, Z.; Qin, C. Effect of active site and charge transfer on methane dehydrogenation over different Co doped Ni surfaces by density functional theory. *Int. J. Hydrogen Energy* **2020**, *45*, 31849–31862. [CrossRef]
3. Ashok, A.; Kumar, A.; Bhosale, R.; Saad, M.A.S.; AlMomani, F.; Tarlochan, F. Study of ethanol dehydrogenation reaction mechanism for hydrogen production on combustion synthesized cobalt catalyst. *Int. J. Hydrogen Energy* **2017**, *42*, 23464–23473. [CrossRef]
4. Takanabe, K.; Nagaoka, K.; Nariai, K.; Aika, K. Titania-supported cobalt and nickel bimetallic catalysts for carbon dioxide reforming of methane. *J. Catal.* **2005**, *232*, 268–275. [CrossRef]
5. Kocić, S.; Corral Valero, M.; Schweitzer, J.-M.; Raybaud, P. Surface speciation of Co based Fischer-Tropsch catalyst under reaction conditions: Deactivation by coke or by oxidation? *Appl. Catal. A Gen.* **2020**, *590*, 117332.
6. Liu, Y.; Yang, H.; Liu, Y.; Jiang, B.; Ding, J.; Woodward, R. Thermally induced fcc to hcp martensitic transformation in Co–Ni. *Acta Mater.* **2005**, *53*, 3625–3634. [CrossRef]
7. Liu, H.; Wang, B.; Fan, M.; Henson, N.; Zhang, Y.; Towler, B.F.; Harris, H.G. Study on carbon deposition associated with catalytic CH4 reforming by using density functional theory. *Fuel* **2013**, *113*, 712–718. [CrossRef]
8. Tu, W.; Ghoussoub, M.; Singh, C.V.; Chin, Y.-H.C. Consequences of Surface Oxophilicity of Ni, Ni-Co, and Co Clusters on Methane Activation. *J. Am. Chem. Soc.* **2017**, *139*, 6928–6945. [CrossRef]
9. Li, K.; Jiao, M.; Wang, Y.; Wu, Z. CH4 dissociation on NiM(111) (M = Co, Rh, Ir) surface: A first-principles study. *Surf. Sci.* **2013**, *617*, 149–155. [CrossRef]
10. Ray, K.; Sandupatla, A.S.; Deo, G. Activity and stability descriptors of Ni based alloy catalysts for dry reforming of methane: A density functional theory study. *Int. J. Quantum Chem.* **2020**, *121*, e26580. [CrossRef]
11. Sprowl, L.H.; Adam, B.M.; Tucker, J.D.; Árnadóttir, L. First-principles study of the products of CO_2 dissociation on nickel-based alloys: Trends in energetics with alloying element. *Surf. Sci.* **2018**, *677*, 219–231. [CrossRef]
12. Turap, Y.; Wang, I.; Fu, T.; Wu, Y.; Wang, Y. Co–Ni alloy supported on CeO2 as a bimetallic catalyst for dry reforming of methane. *Int. J. Hydrogen Energy* **2020**, *45*, 6538–6548. [CrossRef]
13. Ashok, A.; Kumar, A.; Ponraj, J.; Mansour, S.A.; Tarlochan, F. Effect of Ni incorporation in cobalt oxide lattice on carbon formation during ethanol decomposition reaction. *Appl. Catal. B Environ.* **2019**, *254*, 300–311. [CrossRef]
14. Leba, A.; Yıldırım, R. Determining most effective structural form of nickel-cobalt catalysts for dry reforming of methane. *Int. J. Hydrogen Energy* **2020**, *45*, 4268–4283. [CrossRef]
15. Guo, X.; Liu, H.; Wang, B.; Wang, Q.; Zhang, R. Insight into C + O(OH) reaction for carbon elimination on different types of CoNi(111) surfaces: A DFT study. *RSC Adv.* **2015**, *5*, 19970–19982. [CrossRef]
16. Hao, X.; Wang, Q.; Li, D.; Zhang, R.; Wang, B. The adsorption and dissociation of methane on cobalt surfaces: Thermochemistry and reaction barriers. *RSC Adv.* **2014**, *4*, 43004–43011. [CrossRef]
17. Li, J.; Croiset, E.; Ricardez-Sandoval, L. Methane dissociation on Ni (100), Ni (111), and Ni (553): A comparative density functional theory study. *J. Mol. Catal. A Chem.* **2012**, *365*, 103–114. [CrossRef]
18. Mueller, J.E.; van Duin, A.C.T.; Goddard, W.A. Structures, Energetics, and Reaction Barriers for CHx Bound to the Nickel (111) Surface. *J. Phys. Chem. C* **2009**, *113*, 20290–20306. [CrossRef]
19. Liu, H.; Zhang, R.; Yan, R.; Wang, B.; Xie, K. CH4 dissociation on NiCo (111) surface: A first-principles study. *Appl. Surf. Sci.* **2011**, *257*, 8955–8964. [CrossRef]
20. Chen, S.; Yang, B. Activity and stability of alloyed NiCo catalyst for the dry reforming of methane: A combined DFT and microkinetic modeling study. *Catal. Today* **2021**, *400–401*, 59–65. [CrossRef]
21. Meltzman, H.; Chatain, D.; Avizemer, D.; Besmann, T.M.; Kaplan, W.D. The equilibrium crystal shape of nickel. *Acta Mater.* **2011**, *59*, 3473–3483. [CrossRef]
22. Zhang, W.-B.; Chen, C.; Zhang, S.-Y. Equilibrium Crystal Shape of Ni from First Principles. *J. Phys. Chem. C* **2013**, *117*, 21274–21280. [CrossRef]
23. Liu, J.-X.; Su, H.-Y.; Sun, D.-P.; Zhang, B.-Y.; Li, W.-X. Crystallographic Dependence of CO Activation on Cobalt Catalysts: HCP versus FCC. *J. Am. Chem. Soc.* **2013**, *135*, 16284–16287. [CrossRef] [PubMed]
24. Andersson, M.; Abild-Pedersen, F.; Remediakis, I.; Bligaard, T.; Jones, G.; Engbæk, J.; Lytken, O.; Horch, S.; Nielsen, J.H.; Sehested, J. Structure sensitivity of the methanation reaction: H_2-induced CO dissociation on nickel surfaces. *J. Catal.* **2008**, *255*, 6–19. [CrossRef]

25. Bengaard, H.S.; Nørskov, J.K.; Sehested, J.; Clausen, B.S.; Nielsen, L.P.; Molenbroek, A.M.; Rostrup-Nielsen, J.R. Steam Reforming and Graphite Formation on Ni Catalysts. *J. Catal.* **2002**, *209*, 365–384. [CrossRef]
26. Huang, H.; Yu, Y.; Zhang, M. Structure sensitivity of CH4 formation from successive hydrogenation of C on cobalt: Insights from density functional theory. *Chem. Phys. Lett.* **2019**, *737*, 136824. [CrossRef]
27. Chen, S.; Zaffran, J.; Yang, B. Dry reforming of methane over the cobalt catalyst: Theoretical insights into the reaction kinetics and mechanism for catalyst deactivation. *Appl. Catal. B Environ.* **2020**, *270*, 118859. [CrossRef]
28. Nandula, A.; Trinh, Q.T.; Saeys, M.; Alexandrova, A.N. Origin of Extraordinary Stability of Square-Planar Carbon Atoms in Surface Carbides of Cobalt and Nickel. *Angew. Chem. Int. Ed.* **2015**, *54*, 5312–5316. [CrossRef]
29. Kitakami, O.; Sato, H.; Shimada, Y.; Sato, F.; Tanaka, M. Size effect on the crystal phase of cobalt fine particles. *Phys. Rev. B* **1997**, *56*, 13849–13854. [CrossRef]
30. Pearson, W.B.; Thompson, L.T. The lattice spacings of nickel solid solutions. *Can. J. Phys.* **1957**, *35*, 349–357. [CrossRef]
31. Wang, H.; Miller, J.T.; Shakouri, M.; Xi, C.; Wu, T.; Zhao, H.; Akatay, M.C. XANES and EXAFS studies on metal nanoparticle growth and bimetallic interaction of Ni-based catalysts for CO_2 reforming of CH_4. *Catal. Today* **2012**, *207*, 3–12. [CrossRef]
32. Zhang, H.; Yao, T.; Sun, Z.; Li, Y.; Liu, Q.; Hu, F.; Pan, Z.; He, B.; Xie, Z.; Wei, S. Structural Study on Co−Ni Bimetallic Nanoparticles by X-ray Spectroscopy. *J. Phys. Chem. C* **2010**, *114*, 13596–13600. [CrossRef]
33. Liu, X.; Sun, L.; Deng, W.-Q. Theoretical Investigation of CO_2 Adsorption and Dissociation on Low Index Surfaces of Transition Metals. *J. Phys. Chem. C* **2018**, *122*, 8306–8314. [CrossRef]
34. Peng, G.; Sibener, S.; Schatz, G.C.; Mavrikakis, M. CO_2 Hydrogenation to Formic Acid on Ni(111). *J. Phys. Chem. C* **2012**, *116*, 3001–3006. [CrossRef]
35. Watwe, R.; Bengaard, H.; Rostrup-Nielsen, J.; Dumesic, J.; Nørskov, J. Theoretical Studies of Stability and Reactivity of CHx Species on Ni(111). *J. Catal.* **2000**, *189*, 16–30. [CrossRef]
36. Wang, S.-G.; Cao, D.-B.; Li, Y.-W.; Wang, J.; Jiao, H. CH4 dissociation on Ni surfaces: Density functional theory study. *Surf. Sci.* **2006**, *600*, 3226–3234. [CrossRef]
37. Hammer, B.; Nørskov, J. Electronic factors determining the reactivity of metal surfaces. *Surf. Sci.* **1995**, *343*, 211–220. [CrossRef]
38. Cheng, D.; Barcaro, G.; Charlier, J.-C.; Hou, M.; Fortunelli, A. Homogeneous Nucleation of Graphitic Nanostructures from Carbon Chains on Ni(111). *J. Phys. Chem. C* **2011**, *115*, 10537–10543. [CrossRef]
39. Tan, K.F.; Xu, J.; Chang, J.; Borgna, A.; Saeys, M. Carbon deposition on Co catalysts during Fischer–Tropsch synthesis: A computational and experimental study. *J. Catal.* **2010**, *274*, 121–129. [CrossRef]
40. Robinson, J.; Woodruff, D.P. The local adsorption geometry of CH3 and NH3 on Cu(111): A density functional theory study. *Surf. Sci.* **2002**, *498*, 203–211. [CrossRef]
41. Abild-Pedersen, F.; Greeley, J.; Studt, F.; Rossmeisl, J.; Munter, T.R.; Moses, P.G.; Skúlason, E.; Bligaard, T.; Nørskov, J.K. Scaling Properties of Adsorption Energies for Hydrogen-Containing Molecules on Transition-Metal Surfaces. *Phys. Rev. Lett.* **2007**, *99*, 016105. [CrossRef] [PubMed]
42. Greeley, J.; Mavrikakis, M. Surface and Subsurface Hydrogen: Adsorption Properties on Transition Metals and Near-Surface Alloys. *J. Phys. Chem. B* **2005**, *109*, 3460–3471. [CrossRef] [PubMed]
43. Catapan, R.C.; Oliveira, A.A.M.; Chen, Y.; Vlachos, D.G. DFT Study of the Water–Gas Shift Reaction and Coke Formation on Ni(111) and Ni(211) Surfaces. *J. Phys. Chem. C* **2012**, *116*, 20281–20291. [CrossRef]
44. Zhu, Y.-A.; Chen, D.; Zhou, X.-G.; Yuan, W.-K. DFT studies of dry reforming of methane on Ni catalyst. *Catal. Today* **2009**, *148*, 260–267. [CrossRef]
45. Ren, J.; Guo, H.; Yang, J.; Qin, Z.; Lin, J.; Li, Z. Insights into the mechanisms of CO2 methanation on Ni(111) surfaces by density functional theory. *Appl. Surf. Sci.* **2015**, *351*, 504–516. [CrossRef]
46. Bian, Z.; Kawi, S. Highly carbon-resistant Ni–Co/SiO_2 catalysts derived from phyllosilicates for dry reforming of methane. *J. CO2 Util.* **2017**, *18*, 345–352. [CrossRef]
47. Kresse, G.; Hafner, J. Ab initio molecular dynamics for liquid metals. *Phys. Rev. B* **1993**, *47*, 558–561. [CrossRef]
48. Kresse, G.; Hafner, J. Ab initio molecular-dynamics simulation of the liquid-metal–amorphous-semiconductor transition in germanium. *Phys. Rev. B* **1994**, *49*, 14251–14269. [CrossRef]
49. Kresse, G.; Furthmüller, J. Efficiency of ab-initio total energy calculations for metals and semiconductors using a plane-wave basis set. *Comput. Mater. Sci.* **1996**, *6*, 15–50. [CrossRef]
50. Kresse, G.; Furthmüller, J. Efficient iterative schemes for ab initio total-energy calculations using a plane-wave basis set. *Phys. Rev. B* **1996**, *54*, 11169–11186. [CrossRef]
51. Hammer, B.; Hansen, L.B.; Nørskov, J.K. Improved adsorption energetics within density-functional theory using revised Perdew-Burke-Ernzerhof functionals. *Phys. Rev. B* **1999**, *59*, 7413–7421. [CrossRef]
52. Blöchl, P.E. Projector augmented-wave method. *Phys. Rev. B* **1994**, *50*, 17953–17979. [CrossRef] [PubMed]
53. Kresse, G.; Joubert, D. From ultrasoft pseudopotentials to the projector augmented-wave method. *Phys. Rev. B* **1999**, *59*, 1758–1775. [CrossRef]
54. Methfessel, M.; Paxton, A.T. High-precision sampling for Brillouin-zone integration in metals. *Phys. Rev. B* **1989**, *40*, 3616. [CrossRef] [PubMed]
55. Tyuterev, V.G.; Vast, N. Murnaghan's equation of state for the electronic ground state energy. *Comput. Mater. Sci.* **2006**, *38*, 350–353. [CrossRef]

56. Monkhorst, H.J.; Pack, J.D. Special points for Brillouin-zone integrations. *Phys. Rev. B* **1976**, *13*, 5188. [CrossRef]
57. Tang, W.; Sanville, E.; Henkelman, G. A grid-based Bader analysis algorithm without lattice bias. *J. Phys. Condens. Matter* **2009**, *21*, 084204. [CrossRef]
58. Manz, T.A.; Limas, N.G. Introducing DDEC6 atomic population analysis: Part 1. Charge partitioning theory and methodology. *RSC Adv.* **2016**, *6*, 47771–47801. [CrossRef]
59. Manz, T.A.; Sholl, D.S. Improved Atoms-in-Molecule Charge Partitioning Functional for Simultaneously Reproducing the Electrostatic Potential and Chemical States in Periodic and Nonperiodic Materials. *J. Chem. Theory Comput.* **2012**, *8*, 2844–2867. [CrossRef]
60. Henkelman, G.; Uberuaga, B.P.; Jónsson, H. A climbing image nudged elastic band method for finding saddle points and minimum energy paths. *J. Chem. Phys.* **2000**, *113*, 9901–9904. [CrossRef]
61. Henkelman, G.; Jónsson, H. A dimer method for finding saddle points on high dimensional potential surfaces using only first derivatives. *J. Chem. Phys.* **1999**, *111*, 7010–7022. [CrossRef]
62. Liu, B.; Yuan, F.; Jin, K.; Zhang, Y.; Weber, W.J. Ab initio molecular dynamics investigations of low-energy recoil events in Ni and NiCo. *J. Phys. Condens. Matter* **2015**, *27*, 435006. [CrossRef] [PubMed]

catalysts

MDPI

Article

Promoting Effects of Copper and Iron on Ni/MSN Catalysts for Methane Decomposition

Nur Shamimie Nadzwin Hasnan [1,2], Manoj Pudukudy [3,*], Zahira Yaakob [4], Nur Hidayatul Nazirah Kamarudin [4], Kean Long Lim [2] and Sharifah Najiha Timmiati [2,*]

1 Department of Chemical Science, Faculty of Science and Technology, Universiti Kebangsaan Malaysia, Bangi 43600, Selangor, Malaysia; snadzwin@gmail.com
2 Fuel Cell Institute, Universiti Kebangsaan Malaysia, Bangi 43600, Selangor, Malaysia; kllim@ukm.edu.my
3 Department of Chemistry, Saveetha School of Engineering, Saveetha Institute of Medical and Technical Sciences (SIMATS), Chennai 602105, Tamilnadu, India
4 Research Centre for Sustainable Process Technology (CESPRO), Faculty of Engineering and Built Environment, Universiti Kebangsaan Malaysia, Bangi 43600, Selangor, Malaysia; zahira@ukm.edu.my (Z.Y.); nhnazirah@ukm.edu.my (N.H.N.K.)
* Correspondence: manojpudukudy@gmail.com (M.P.); najiha@ukm.edu.my (S.N.T.)

Abstract: Copper and iron-based bimetallic nickel catalysts supported on Mesostructured Silica Nanoparticles (MSNs) with compositions of 50% Ni–5% Cu/MSN and 50% Ni–5% Fe/MSN were prepared using an impregnation method, and they were compared with a monometallic 50% Ni–MSN catalyst for their activity and stability in methane decomposition reaction. The influence of promoters, such as Cu and Fe, at different reaction temperatures (700 °C, 800 °C and 900 °C) was investigated. The results revealed that the Cu and Fe-promoted catalysts significantly increased the hydrogen yield in methane decomposition compared with the unpromoted catalyst. This could be attributed to the formation of Ni–Cu and Ni–Fe bimetallic alloys in the catalysts, respectively, and this favored the stability of the catalysts. With increasing reaction temperature, the hydrogen yield also increased. However, the hydrogen yield and the lifetime of the nickel catalyst were enhanced upon the addition of iron compared to copper at all the reaction temperatures. The analysis conducted over the spent catalysts validated the formation of multi-walled carbon nanotubes with a bamboo-like internal channel over the catalysts along with a high crystallinity and graphitization degree of the carbon produced.

Keywords: Mesoporous Silica Nanoparticles (MSNs); metal promoters; nickel catalyst; hydrogen; methane decomposition; MWCNTs

1. Introduction

Global warming, a phenomenon caused by increased human activity, has instigated widespread environmental pollution. The emission of greenhouse gases coming from transportation, industrial and other human activities leads to severe environmental pollution. Therefore, an alternative to fossil fuel as a major source of energy for transportation and stationary applications is needed. Many studies have been conducted to develop alternative sources of renewable energy worldwide [1]. Hydrogen energy is a green energy that can be used in stationary and practical applications [2]. Recently, hydrogen has been applied in the field of transportation and has been used in fuel cell technology [3,4].

Hydrogen can be produced from various sources, such as biomass, water, natural gas and hydrocarbon, via different kinds of available methods [5]. Methane decomposition is a practical method to produce hydrogen directly from methane without producing any greenhouse gas [6]. In contrast to conventional steam reforming, methane decomposition requires no further process to purify the produced hydrogen [7]. A variety of carbon structures, such as carbon nanotubes, graphite and graphene, were produced at the end of the reaction [8]. This graphitic carbon can be further utilized for miscellaneous applications, such as a catalyst for several catalytic processes or in hydrogen storage materials [9,10]. Methane decomposition is a highly endothermic process that requires high temperatures of approximately 1200 °C. The reaction can be represented as

$$CH_4 \rightarrow 2H_2 + C \quad \Delta H = 74.87 \text{ kJ/mol}$$

To reduce the reaction temperature, the use of heterogeneous catalysts has been introduced in the reaction. Studies have assessed the use of metal catalysts in methane decomposition [11]. Apart from being low cost, the abundance of transition metals, such as Ni, Cu, Fe and Co, has made them the most used catalyst in methane decomposition [12–15]. Syed-Muhammad et al. summarized studies on metallic catalysts, promoters, metal loading, operational conditions and feedstock mixtures in methane decomposition [16]. They found that these factors play a crucial role in determining the performance in methane decomposition. Ni catalysts supported over several metal oxides, such as silica, magnesia, zirconia and titania, display optimal performance and higher stability in methane decomposition compared with other monometallic catalysts, such as Co, Cu and Fe [17–19]. Hasnan et al. recently found that Ni is the most active metal in methane decomposition [20]. The selection of appropriate catalyst support is crucial in the performance of Ni-based catalysts. Mesoporous silica-based materials are seldom used as catalyst support in methane decomposition, but they are commonly used in other catalytic processes, such as dry reforming [21], CO_2 methanation [22] and drug delivery [23]. The outstanding characteristics of mesoporous silica, such as high surface area and porosity, contribute to adequate active sites available for the reaction, and the presence of a large pore size would help in metal dispersion [24,25].

Apart from that, the influence of the catalyst composition in the bimetallic catalyst also plays an important role in determining its catalytic performance [26]. Pudukudy et al. examined the performance of a Ni catalyst regarding Pd addition in the catalytic decomposition of methane [10]. They found that the incorporation of a co-metal improves the hydrogen yield and stability of the Ni catalyst throughout the 420 min of the reaction. Even though only a small amount of Pd is added, the performance of the Ni-based catalyst is still enhanced compared with that of the unpromoted catalyst. This result agrees with the findings of another study by Pudukudy et al. on a monometallic Ni catalyst, in which Ni/SiO_2 displayed low stability during 300 min of reaction [27]. Saraswat and Pant reported a Ni-based bimetallic catalyst in the catalytic decomposition of methane using Cu and Zn as promoters supported over Ni/MCM-22 [28]. They found that the presence of Cu and Zn prevents nanoparticle agglomeration and enhances the metal dispersion on the catalyst support, thus contributing to the high catalytic performance in methane decomposition. Hence, many studies focused on bimetallic catalysts to develop Ni-based catalysts with improved activity in methane decomposition [29–33].

Cu and Fe are metals commonly studied as promoters in the methane decomposition process. Cunha et al. reported that the formation of a Ni–Cu alloy promotes catalyst stability during the reaction [34]. The addition of Cu inhibited the development of encapsulating carbon, owing to the formation of a Ni–Cu alloy, which segregated the Ni active sites, complicating the reaction of the adsorbed carbon species with each other on the catalyst surface. This result agrees with the findings of Awad et al. [29]. In another study, Bayat et al. examined the activity of a $Ni–Fe/Al_2O_3$ catalyst in methane decomposition and noticed that the introduction of Fe improves the carbon diffusion rate, thus enhancing the catalyst stability, even at high temperatures [32]. Tezel et al. synthesized a Ni–Fe-doped catalyst via a co-impregnation method and obtained 69% methane conversion at 700 °C [35]. They reported that an increase in the reaction temperature improved the initial conversion of methane and enhanced the stability of the catalyst throughout the reaction. This is due to the high amount of Fe present in the catalyst as well as the formation of the cubic structure of the Fe_2O_3 compared to the hexagonal structure, which helps to extend the catalyst lifetime as well as enhance its catalytic stability throughout the reaction [36].

Our previous work examined the effect of Ni loading (5%, 10%, 30% and 50%) over a Mesostructured Silica Nanoparticle (MSN) support and found that a 50% Ni/MSN catalyst exhibited the best results regarding the hydrogen yield and stability throughout 360 min of reaction [37]. This can be ascribed to the enhanced properties of the MSNs that favored the dispersion of the Ni particles on the MSN surface. An increase in the reaction temperature would result in increased activity because methane decomposition is a highly endothermic process [33,36]. However, the stability of the catalyst would be affected as a large amount of carbon is produced [16]. Therefore, this research has been conducted to study the effect of Fe and Cu addition on the activity and stability of a Ni/MSN catalyst in methane decomposition at different reaction temperatures, by considering the high thermal, chemical and mechanical stability of the MSNs. Moreover, the properties of the deposited nanocarbon were also evaluated using various analytical techniques.

2. Results and Discussion

2.1. Catalyst Characterization

The XRD patterns of the fresh 50% Ni/MSN, 50% Ni–5% Cu/MSN and 50% Ni–5% Fe/MSN catalysts are displayed in Figure 1. All the catalysts exhibited sharp diffraction peaks at 2θ values of 37.2°, 43.2°, 62.9°, 75.3° and 79.4°, which were related to the formation of NiO and Ni-based solid solutions for the unpromoted and Cu and Fe-promoted samples, respectively [38,39]. No characteristic peaks for the MSNs were present, and only a small hump was detected in all of the catalysts, which demonstrated the amorphous nature of the MSNs [37]. The absence of any other diffraction peaks further validated the fine dispersion of promoter metallic species in the catalysts. After 5% Cu incorporation in the 50% Ni/MSN sample, a $Ni_xCu_{(1-x)}O$ mixed oxide solid solution could be formed in the catalyst [40]. However, the peaks of $Ni_xCu_{(1-x)}O$ mixed oxide were similar to the peaks of NiO. This indicates that Cu ions are replaced with some Ni ions in its crystal lattice [29,41]. These two species were difficult to differentiate. However, it should be noted that a calcination temperature of 550 °C is sufficient for the formation of $Ni_xCu_{(1-x)}O$ solid solution by their interaction [42]. No other peaks for the formation of phase-segregated CuO and Cu_2O were noticed in the samples. The indistinguishable peaks of the $Ni_xCu_{(1-x)}O$ mixed oxide and NiO, as shown in the XRD patterns, may influence the physicochemical properties of the Ni-based catalysts.

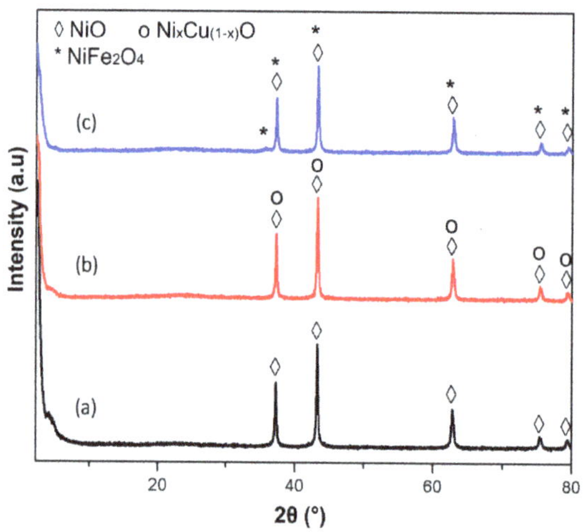

Figure 1. XRD pattern of the fresh catalysts: (**a**) 50% Ni/MSN, (**b**) 50% Ni–5% Cu/MSN and (**c**) 50% Ni–5% Fe/MSN catalysts.

Similar results were observed in the Ni–Fe/MSN catalyst. No peaks associated with the Fe_2O_3 phase were detected in the XRD patterns. The existence of Fe in the catalyst was expected to be in the form of spinel nickel ferrites [38], as both samples exhibited similar diffraction peaks [40], as shown in Figure 1. The crystallite size of the metallic oxides (NiO) in all of the catalysts was calculated using the integral breadth over the high peak intensity diffraction peak at the (200) plane. The crystallite size was calculated to be 25.1 nm, 25.6 nm and 26.5 nm for the 50% Ni/MSN, 50% Ni– 5%Cu/MSN and 50% Ni–5% Fe/MSN catalysts, respectively. The crystallite size of NiO was not altered after the incorporation of metal promoters into the catalysts, which is clear from the intensity of the diffraction peaks.

The surface functional groups present in the catalysts were studied by FTIR spectroscopy. The results presented in Figure 2 show three distinct transmission bands at 457 cm^{-1}, 806 cm^{-1} and 1079 cm^{-1}, which were related to the formation of MSNs [37,43]. These bands can be assigned to the asymmetric and symmetric stretching vibration of the Si–O bonds and the asymmetric stretching vibration of the Si–O–Si bridges, respectively [44]. Another weak band observed at 970 cm^{-1} could be attributed to the stretching vibration of the Si–OH of the silanol groups [10]. All the transmission bands associated with MSNs were maintained in the catalysts after the addition of Ni and promoters. The weak band observed at 566 cm^{-1} was related to the bending vibration of the metal–oxygen bond present in the sample [30]. The presence of a broad transmission band at 3431 cm^{-1} was attributed to the stretching and bending mode vibrations of the hydroxyl group, which were induced by the presence of adsorbed water molecules.

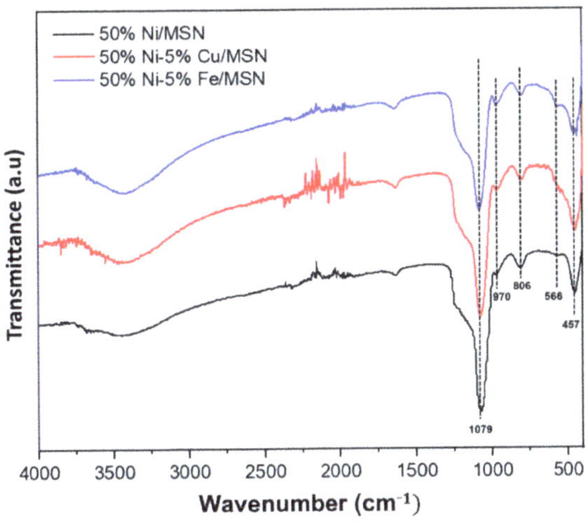

Figure 2. FTIR spectra of 50% Ni/MSN, 50% Ni–5% Cu/MSN and 50% Ni–5% Fe/MSN catalysts.

The morphology of the prepared catalysts was studied using FESEM analysis, and the results are shown in Figure 3. The catalysts had an irregular morphology with polygonal-shaped and non-uniform irregular particles, which were distributed on the MSN support. The 50% Ni/MSN catalyst displayed nanoparticles with sizes ranging from 50 nm to 200 nm. Meanwhile, the Cu- and Fe-promoted Ni/MSN catalysts exhibited the presence of more nano-sized particles ranging from 30 nm to 60 nm and well-developed tetragonal structures with a size ranging from 100 nm to 200 nm. As shown in the figure, more nanoparticles were accumulated in the promoted catalysts, which decreases the surface area of the Ni/MSN catalyst [29]. This could be due to the additional surface coverage of the MSNs after the addition of metal promoters.

BET/BJH measurements were performed to study the textural properties of the catalysts. As shown in Table 1, the MSNs possessed a very high surface area of 1095 m^2/g and a cumulative pore volume of 1.645 cm^3/g, whereas the pore size was found to be 2.9 nm. The surface area and pore volume were drastically decreased upon the loading of metals on the MSN support. This can be associated with the agglomeration of particles on the surface of the MSNs and the blocking of the MSN pores with metal particles, as confirmed by morphological analysis [10,45].

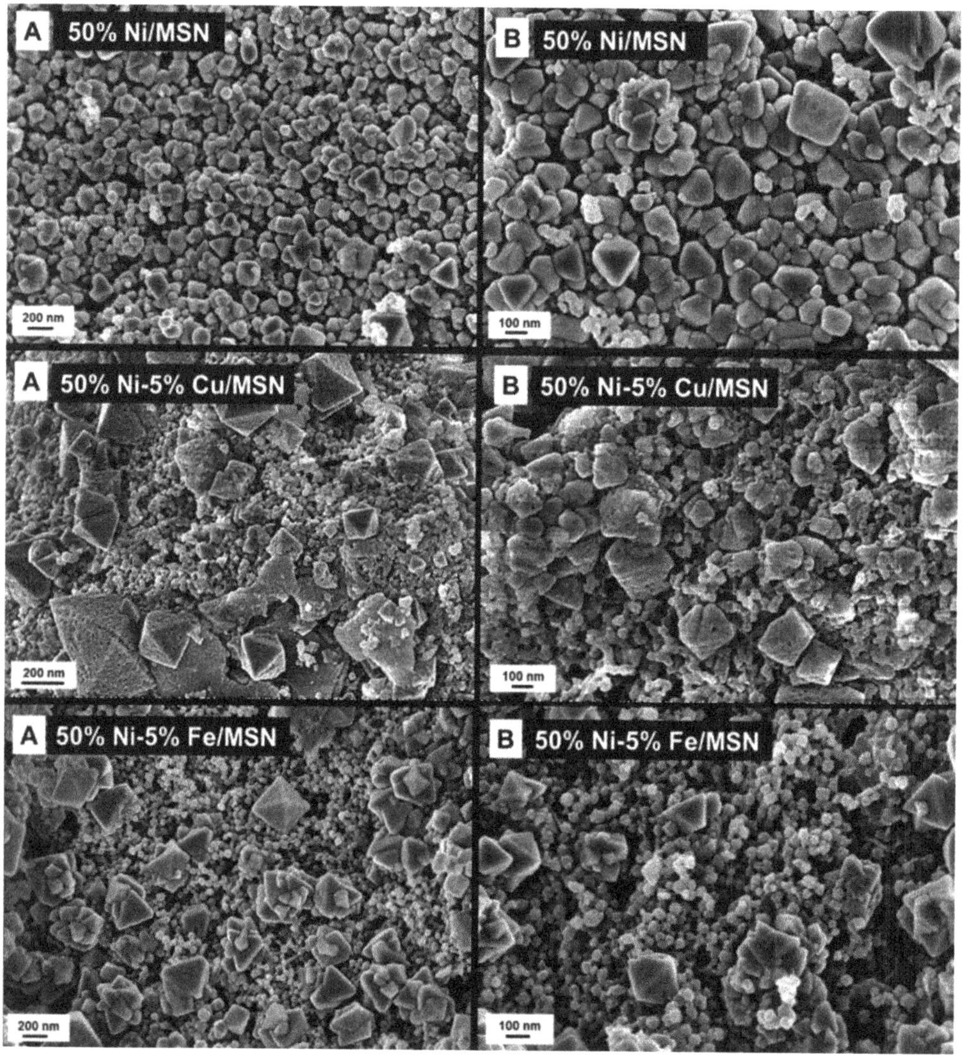

Figure 3. FESEM images of the prepared catalysts at two different magnifications (A and B).

Figure 4 displays the EDX and elemental mapping analyses of the copper and iron-promoted catalysts. The results confirmed the presence of Cu and Fe as promoter metals in the catalyst where the metallic species were dispersed on the surface of the MSN support.

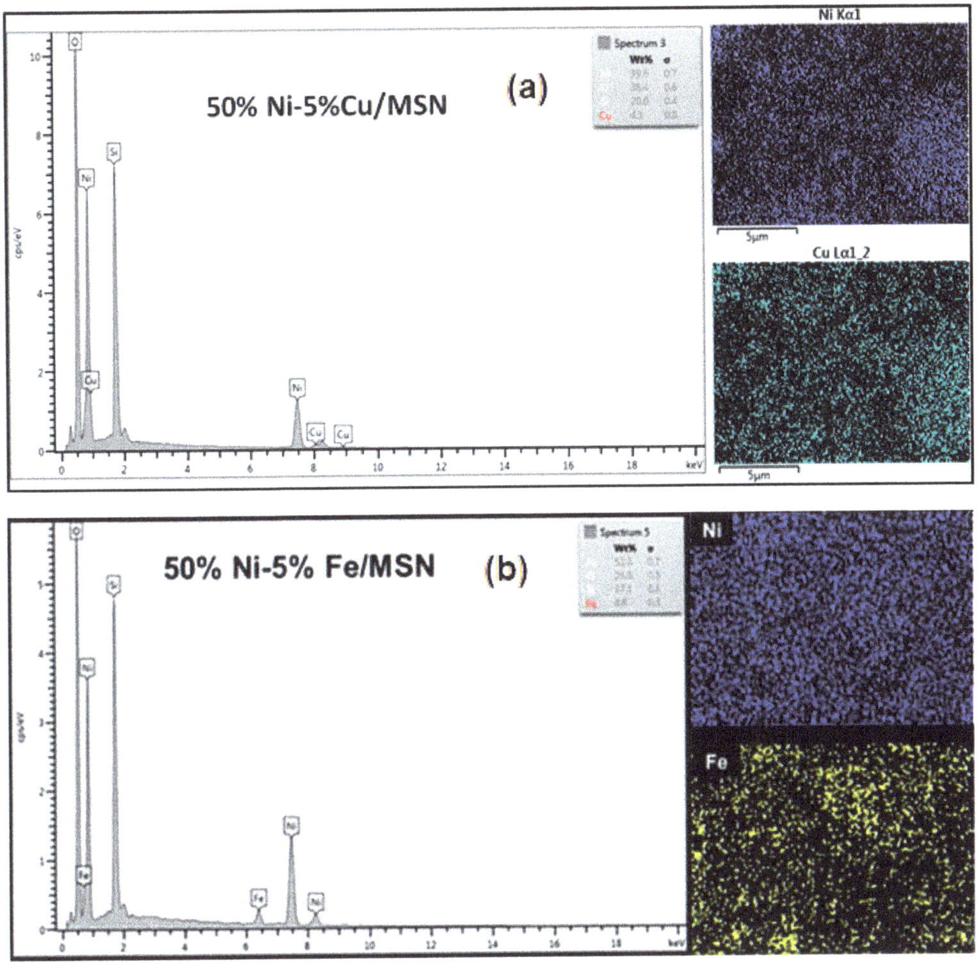

Figure 4. FESEM–EDX spectra and elemental mapping images of the (**a**) Cu and (**b**) Fe-promoted Ni/MSN catalysts.

Table 1. BET/BJH results of the prepared samples.

Sample	BET Surface Area (m^2/g)	Pore Size (nm)	Pore Volume (cm^3/g)
MSN	1095	2.9	1.645
50% Ni/MSN	409	2.3	0.514
50% Ni–5% Cu/MSN	324	2.6	0.601
50% Ni–5% Fe/MSN	319	3.4	0.491

Figure 5 shows the N_2 adsorption/desorption isotherm of the catalysts. The isotherms were type IV isotherms with an H1 hysteresis loop for all of the prepared catalysts, indicating that the particles were uniformly arranged with a cylindrical pore geometry, suggesting uniformity in the pore size distribution of the catalysts. Meanwhile, the 50% Ni/MSN catalyst exhibited type IV isotherms with an H4 hysteresis loop, indicating narrow slit-like pores and hollow spheres with walls composed of ordered mesoporous silica [37]. The inset pore size distribution curves of the catalysts displayed a monomodal pore structure, and the pore size distribution was in the range of 2 to 3.4 nm. This finding was compatible with the mean pore size data provided in Table 1.

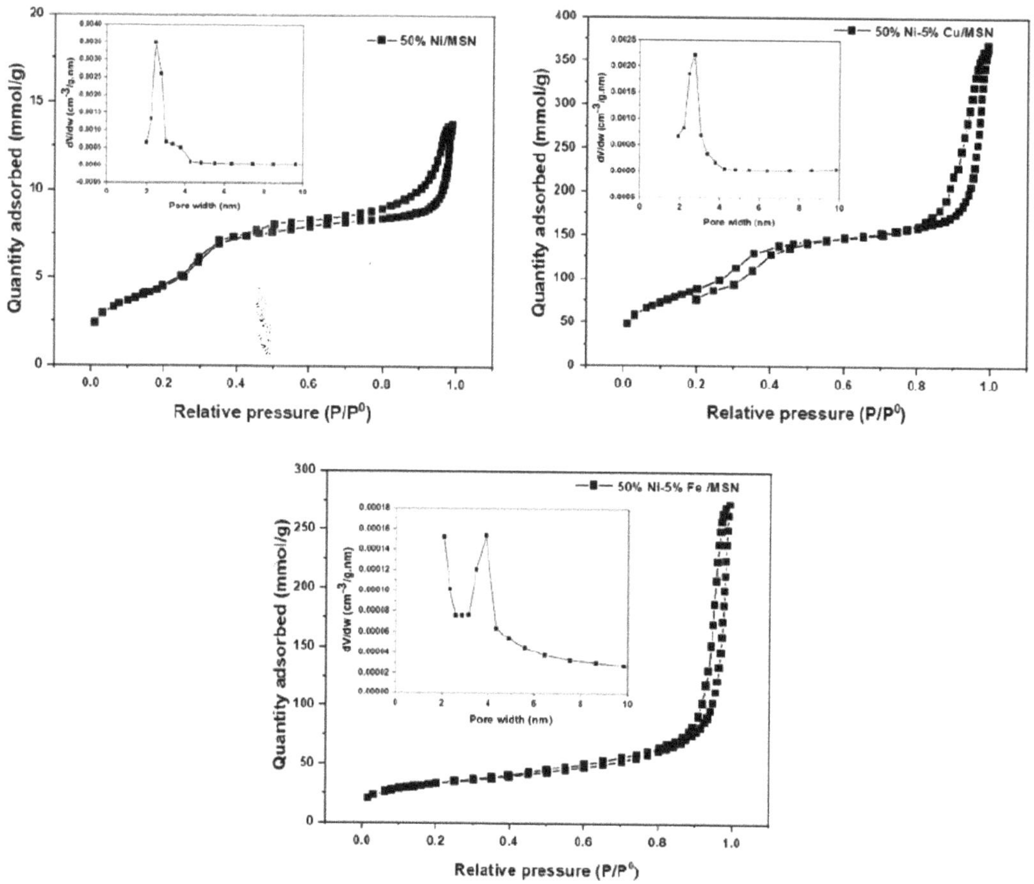

Figure 5. N_2 adsorption–desorption isotherms and pore size distribution curves of 50% Ni/MSN, 50% Ni–5% Cu/MSN and 50% Ni–5% Fe/MSN catalysts.

The reducibility of the unpromoted and Cu and Fe-promoted catalysts was evaluated using TPR analysis. As shown in Figure 6, a clear and well-defined peak centered at 423 °C and one broader reduction hump centered at 577 °C were observed for the 50% Ni/MSN catalyst. The first peak can be associated with the reduction of NiO species with a weak interaction with the MSN support [10]. The latter peak can be ascribed to the reduction of NiO with a strong metal–support interaction [27], as confirmed by Pudukudy et al. [10] who reported that the strong interaction of NiO with the support is developed by the encapsulation of NiO particles inside the SBA-15 mesopore. The 50% Ni–5% Cu/MSN catalyst displayed a small reduction peak at 229 °C, which might be due to the reduction of any copper oxide species in the samples [9,40]. Awad et al. validated the dissociation of H_2 molecules into separate atoms with the help of Cu^0, which leads to the reduction of NiO by the spillover effect [29]. The reduction peaks of the Ni/MSN shifted toward a lower temperature after the addition of Cu, indicating an improvement in the reducibility of the Ni/MSN catalyst.

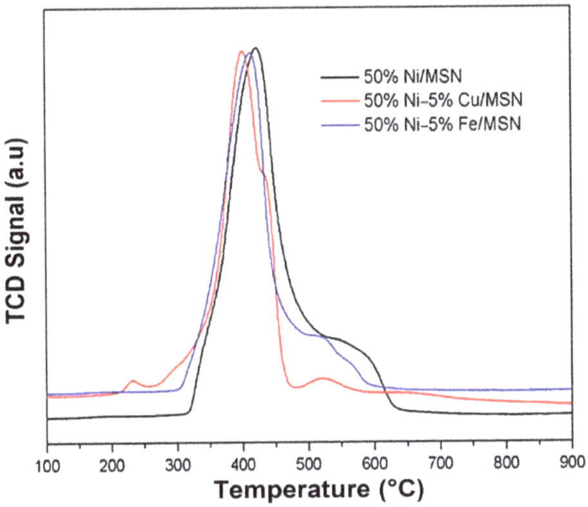

Figure 6. TPR profile of 50% Ni/MSN, 50% Ni–5% Cu/MSN and 50% Ni–5% Fe/MSN catalysts.

The H_2-TPR profile of the 50% Ni–5% Fe/MSN displayed three reduction peaks. The peaks located at 412 °C and 524 °C were attributed to the NiO reduction with weak and strong metal–support interactions, respectively [40]. The last peak centered at 569 °C was related to the reduction of $NiFe_2O_4$ species, which agreed with the XRD results [38,42,46]. It is noteworthy that the reducibility of the Ni/MSN catalyst improved upon the addition of Fe, and the reduction temperature moved towards a lower temperature [36], which is in agreement with Park et al. [47], as they reported the effect of the reduction condition on the reaction over the surface of the catalyst as well as the growth mechanism of the carbon nanotubes that are generated after the reaction.

2.2. *Catalytic Decomposition of Methane*

Methane decomposition is an endothermic process. Thus, the activity would increase as the reaction temperature is increased. Methane decomposition reactions conducted at high temperatures would produce high-quality nanocarbon and enhance the graphitic order of the deposited carbon [48]. The production of multi-walled carbon nanotubes may also increase despite the fact that the aggregation of metal particles may occur at high reaction temperatures [8].

Figure 7 shows the catalytic activity of the prepared catalysts in the methane decomposition. All of the catalysts displayed high initial activity in each of the reaction temperatures. This is attributed to the enhanced textural properties of the MSN support, which provides a high surface area and adequate sites for the active metals initially for the reaction. As can be seen, with an increase in the reaction temperature, the catalytic performance was enhanced for all of the catalysts. However, the Cu and Fe-promoted catalysts exhibited better performance for the reaction compared to the unpromoted catalyst. At 700 °C, the Cu and Fe-promoted Ni/MSN catalysts exhibited higher initial activity compared to the unpromoted catalyst. The highest initial hydrogen yield and final yield measured at 360 min of time on stream were found to be 75% and 73%, 68% and 64% and 64.8% and 61% for the Fe and Cu-promoted and unpromoted catalysts, respectively.

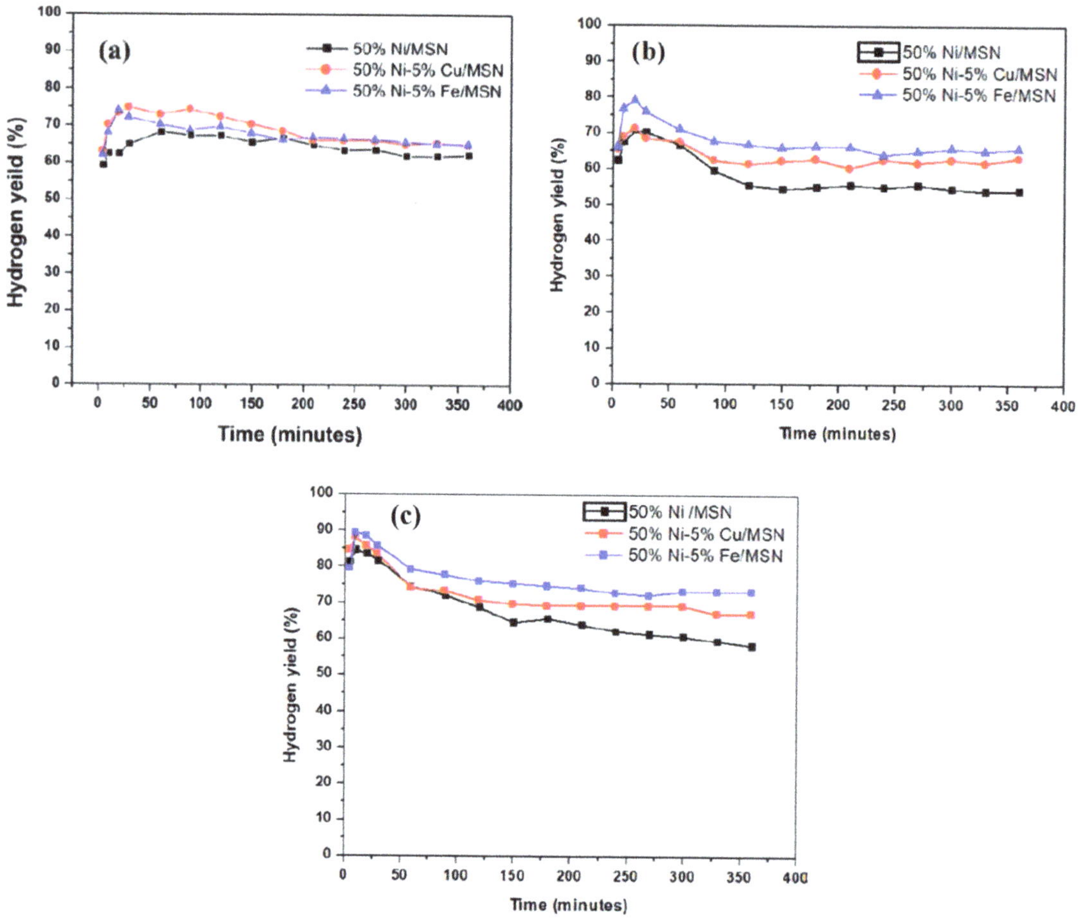

Figure 7. Catalytic performance of the samples for methane decomposition at various reaction temperatures: (**a**) 700 °C, (**b**) 800 °C and (**c**) 900 °C.

However, all the catalysts exhibited good stability throughout the reaction time, whereas the catalytic activity slightly decreased in all of the catalysts. The high stability could be due to the formation of bimetallic alloys in the catalysts. The decreased activity of the catalysts may be due to the deposition of carbon, which lowers the surface area for the menthane molecules to interact with [48].

As the reaction temperature increased to 800 °C, the unpromoted and promoted catalysts showed the same trend with a high hydrogen yield at the initial stage of the reaction. At this temperature, the 50% Ni–5% Fe/MSN catalyst displayed the highest yield (80%) during the first 20 min, followed by the 50% Ni–5% Cu/MSN and 50%-Ni/MSN catalysts with 71% and 69% of the yield, respectively. After 30 min on stream, the activities of the 50% Ni/MSN, 50% Ni–5% Cu/MSN and 50% Ni–5% Fe/MSN catalysts started to decrease slowly until the end of the reaction, and the yield was measured to be 54%, 63% and 66%, respectively. Furthermore, it is worth mentioning that, as seen in Figure 7b, the stability of the Cu and Fe-promoted catalysts was comparatively higher than that of the unpromoted catalyst, as the hydrogen yield difference during the course of the reaction seemed to be minimum for the promoted catalysts. This could be ascribed to the enhanced role of the bimetallic alloys in the samples [30].

A further increase in the reaction temperature to 900 °C increased the hydrogen yield drastically during the first 10 min of the reaction, with a hydrogen yield of 85%, 88% and 91% for the unpromoted and Cu and Fe-promoted catalysts, respectively. As shown in Figure 7c, the activity of the catalysts started to drop after 20 min on stream. The same kinetic curve trend was observed for all of the catalysts. The hydrogen yield of the 50% Ni–5% Cu/MSN catalyst decreased to 74% and remained more or less the same until 360 min on stream, with a final yield of 67%. The 50% Ni–5% Fe/MSN catalyst displayed better activity and stability at 900 °C, with 74% of the final hydrogen yield. This result may be attributed to the Ni–Fe alloy formation, which decreased the rate of decomposition and then enhanced the carbon diffusion rate on the surface of the metal and, therefore, prevented the catalyst deactivation at high reaction temperatures [32,38]. The introduction of Fe into Ni/MSN helped maintain the stability of the catalyst, even at high temperatures [46]. Catalyst deactivation is reported to occur when the carbon diffusion rate is slower than the carbon formation rate, thus leading to the formation of encapsulating carbon [49]. Carbon encapsulation occurs when carbon starts to build on the surface of the catalyst and then encapsulates the metal particles. The high surface area and porosity of the MSNs helped to prevent the agglomeration of metal particles on the MSN support. As a result, all the catalysts achieved adequate stability throughout 360 min of the reaction conducted at high temperatures up to 900 °C. This catalytic performance indicated that the enhanced properties of the promoted catalyst were achieved at higher reaction temperatures owing to the formation of bimetallic alloys [50]. Additionally, at high reaction temperatures, the Ni–Cu or Ni–Fe alloys generated a quasi-liquid condition, then averted the fragmentation process, which led to the generation of small particles as well as the rearrangement of the structural and electronic properties, which prevented the catalysts from agglomerating and then preserved their catalytic stability [51].

2.3. Characterisation of Spent Catalysts

Carbon yield is defined as the ratio of the amount of carbon deposited over the catalyst to the total amount of metallic species present in the catalyst that has been used initially in the decomposition reaction. The yield of carbon obtained after the decomposition reaction at three different reaction temperatures was further calculated using Equation (2), and the results are shown in Figure 8. As can be seen, with an increase in the reaction temperature, the carbon yield increased significantly. The highest carbon yield was obtained for the methane decomposition reaction at 900 °C. The yield was calculated to be 1254%, 1572% and 1896% for the unpromoted Ni/MSN and Cu and Fe-promoted Ni/MSN catalysts at 900 °C, respectively. The increase in the carbon yield could be attributed to the higher conversion of methane at high temperatures over the present catalysts. As expected, the Fe-promoted catalyst exhibited the highest carbon yield as the highest yield of hydrogen was also obtained for the same catalyst. Thus, consistency in the product yield was observed in the methane decomposition reaction.

The morphology and the graphitization degree of the carbon produced by the spent catalysts were further evaluated using XRD, FESEM, Raman and TEM analyses.

Figure 9 displays the XRD patterns of the 50% Ni/MSN, 50% Ni–5% Cu/MSN and 50% Ni–5% Fe/MSN catalysts after the methane decomposition reaction at 900 °C. The well-defined peak centered at 26.3° in all the catalysts was assigned to the presence of graphitic carbon in the spent catalysts. The most intense carbon peak with the highest intensity was achieved by the 50% Ni–5% Fe/MSN catalyst (Figure 8c), which demonstrated that the high crystallinity of nanocarbon was produced with the introduction of Fe at high reaction temperatures. This result agrees with the findings of Torres et al. [48]. Additionally, the d-spacing values of the 26.3° diffraction peaks of the 50% Ni/MSN, 50% Ni–5% Cu/MSN and 50% Ni–5% Fe/MSN spent catalysts were calculated to be 0.3357 nm, 0.3356 nm and 0.3355 nm, respectively. These values were close to the ideal value for a graphite-like structure, which is 0.34 nm, representing a higher quality of carbon produced in this study on the structure and oxidation activity of the solid carbon produced from the catalytic decomposition of methane [52].

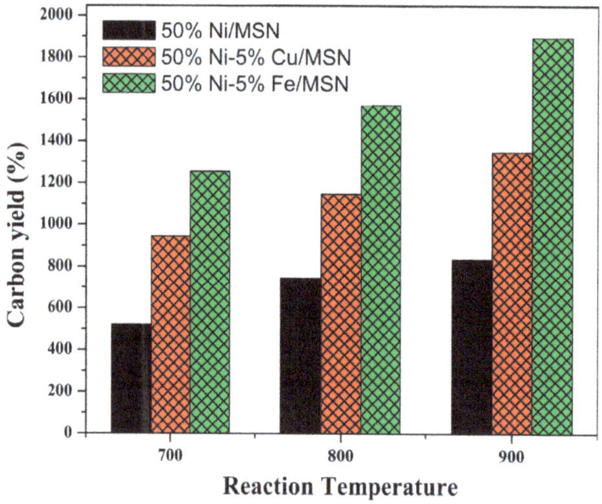

Figure 8. Carbon yield at various reaction temperatures over 50% Ni/MSN and 5% Cu and 5% Fe–promoted 50% Ni/MSN catalysts.

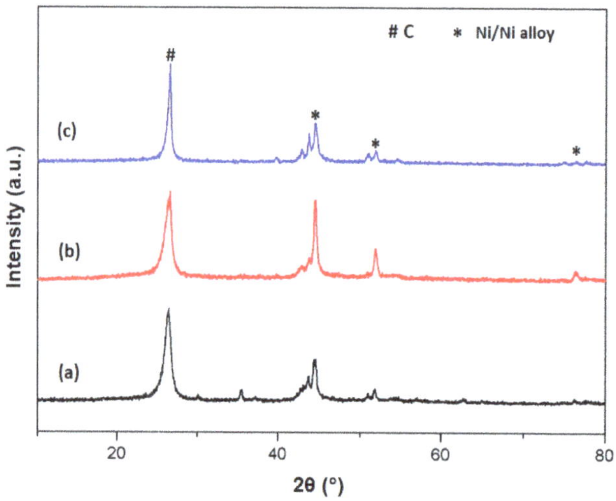

Figure 9. XRD pattern of the spent catalysts: (**a**) 50% Ni/MSN, (**b**) 50% Ni–5% Cu/MSN and (**c**) 50% Ni–5% Fe/MSN at reaction temperature of 900 °C.

The other small peak located at 43.7° also belonged to the graphitic carbon. Furthermore, the diffraction peaks at 44.5° and 51.9° were associated with the presence of metallic Ni and Cu and Fe-based bimetallic nickel alloy species in the catalysts. This result indicates that NiO or metallic oxide species were completely reduced to metallic Ni or bimetallic alloys during the pre-treatment steps of the reaction. The higher intensity of these Ni^0 peaks in the 50% Ni–5% Cu/MSN catalyst (Figure 8b) demonstrated the higher crystallinity of the Ni–Cu alloy particles by the reduction treatment [53]. The other peaks were related to the formation of metallic carbides formed by the reaction of the metals present in the catalyst with the formed carbon at high reaction temperatures, as reported by Pudukudy et al. [54]. The formation of these bimetallic alloys, especially the Ni–Fe alloy, enhanced the hydrogen yield and stability of the Ni catalyst at high temperatures.

Raman spectroscopy was employed to study the crystallinity and graphitization degree of the carbon deposited over the catalysts, and the results are shown in Figure 10. Each catalyst displays two well-defined peaks related to the D-band and G-band. The D-band centered at 1340 cm^{-1} is assigned to the presence of disordered carbon in the graphitic carbon, whereas the G-band centered at 1570 cm^{-1} is attributed to the stretching vibration of in-plane carbon-carbon bonds [53]. The I_D/I_G ratios were found to be 0.82, 0.79 and 0.54 for the 50% Ni/MSN, 50% Ni–5% Cu/MSN and 50% Ni–5% Fe/MSN spent catalysts, respectively. The decreases in the I_D/I_G ratio indicated the improved crystallinity and graphitization degree of the carbon produced at high reaction temperatures.

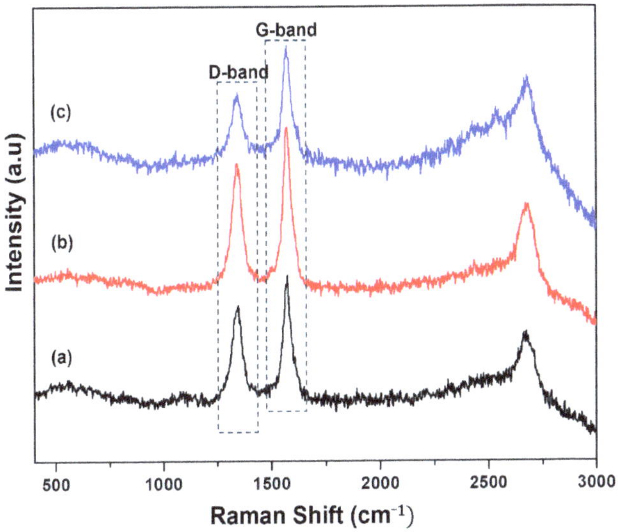

Figure 10. Raman spectra of the spent catalysts: (**a**) 50% Ni/MSN, (**b**) 50% Ni–5% Cu/MSN and (**c**) 50% Ni–5% Fe/MSN at a reaction temperature of 900 °C.

The morphological features of the spent catalysts were investigated by FESEM analysis. As the highest carbon yield was obtained for the Fe-promoted Ni/MSN catalyst, it was further selected for the morphological studies and FESEM micrographs at two different magnifications, as shown in Figure 11. As can be seen, the surfaces of the catalysts were found to be accumulated with the bulk amount of carbon nanotubes (CNTs). The detailed examination of the deposited carbon nanotubes validates the existence of open tips in the CNTs. The lengths of the deposited CNTs were found to vary greatly, whereas the widths were observed to be in the range of 40 nm to 80 nm. The development of the open tips can be attributed to the presence of a strong metal–support interaction in the catalyst. The strong metal–support interaction could be attributed to the encapsulation of metallic nanoparticles on the mesopores of the MSN support. As a result of this, the carbon crystal tends to grow by a base growth mechanism followed by an open tip, while the metal tends to be at the bottom portion of the carbon nanotube [10].

Figure 12 shows the internal morphology of the deposited nanocarbon obtained at 900 °C. The carbon was deposited in the form of a filamentous structure in all of the catalysts. The TEM images showed that the produced filamentous carbon was in the form of multi-walled carbon nanotubes with open tips and diameters ranging from 45 nm to 70 nm. An internal hollow channel diameter of 10 nm to 35 nm was observed for the CNTs, regardless of the role of the promoters in the catalysts. In addition, a bamboo-like internal tube structure was observed for the nanotubes, and some of the metal/alloy nanoparticles were found to be encapsulated inside the carbon nanotubes, which is clearly evident in the TEM images of the Cu-promoted spent catalysts. The metal nanoparticles with a weak metal–support interaction could suffer the encapsulation process.

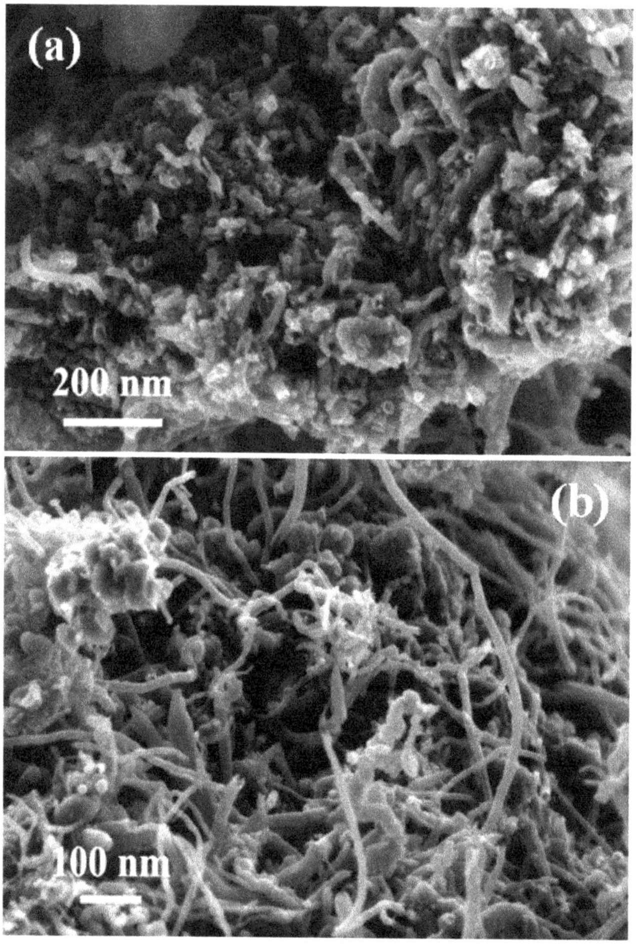

Figure 11. FESEM images of the spent 5%Fe–50%Ni/MSN catalyst obtained after methane decomposition at 900 °C at two different scale bars (**a**) 200 nm and (**b**) 100 nm.

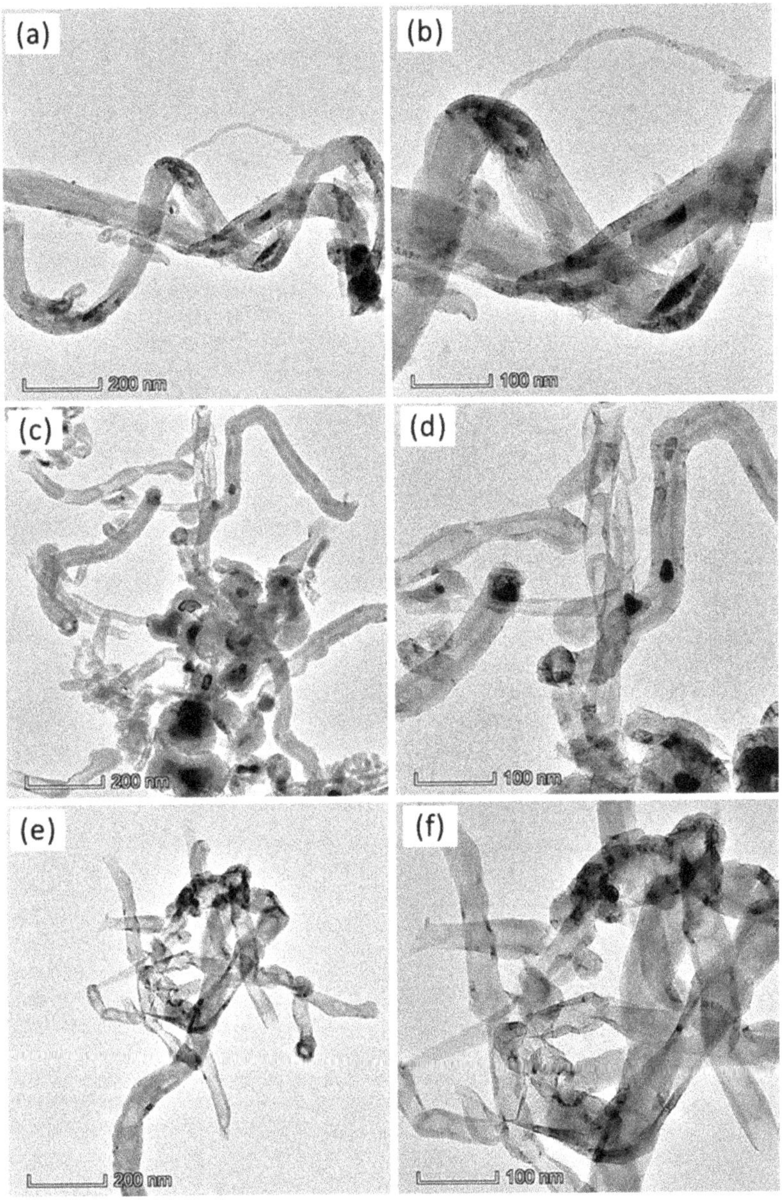

Figure 12. TEM images of nanocarbon obtained over (**a**,**b**) 50% Ni /MSN, (**c**,**d**) 50% Ni–5% Cu/MSN and (**e**,**f**) 50% Ni–5% Fe/MSN at a reaction temperature of 900 °C.

3. Materials and Methods

3.1. Preparation of the Catalysts

The catalyst support, Mesostructured Silica Nanoparticles (MSNs), was synthesized according to our previous work [38]. Firstly, 9.68 g cetyltrimethylammonium bromide (Merck, New York, NY, USA), 58 mL of ammonia solution and 240 mL of ethylene glycol (Merck) were dissolved in 1440 mL of deionized water and then stirred at 50 °C for 30 min. Then, 11.4 mL of tetraethyl orthosilicate (Merck) and 2.1 mL of 3-(triethoxysilyl)-

propylamine (Merck), which act as silica precursors, were added slowly to the mixture, thereby turning the clear mixture into a white suspension. Then, the solution was further stirred at 80 °C for 2 h. Afterwards, the white solution was kept in a refrigerator for separation into two layers. After several days, the white precipitate separated in the solution was collected and washed several times with a distilled water and ethanol mixture at a ratio of 1:1. The obtained as-synthesized MSNs were dried in the oven overnight at 110 °C and then calcined for 3 h at 550 °C.

A co-impregnation method was employed in the synthesis of the metal-loaded MSN catalysts. The desired amount of Ni salt, such as $Ni(NO_3)_2 \cdot 6H_2O$ with 50 wt.% of metallic nickel, in the final sample was weighed and then dissolved in 500 mL of deionized water. The nitrate salts of the co-metal promoters, such as hydrated nitrated salts of Cu and Fe, having 5 wt.% of each metal promoter, were also added. Then, the prepared MSN powder was added slowly to the mixed-metal salt aqueous solution and heated at 90 °C and mixed well with a magnetic stirrer until it became a homogenous paste. After that, the paste was dried at 110 °C overnight and finally calcined at 550 °C for 3 h. The resulting catalysts were in the form of a colored powder and were named 50% Ni/MSN, 50% Ni–5% Cu/MSN and 50% Ni–5% Fe/MSN.

3.2. Characterization of the Catalysts

The physical and chemical properties of the fresh catalyst samples were studied by various methods. X-ray diffraction analysis was used to determine the crystalline structure of the catalysts using a D8 Focus Advance powder diffractometer (Bruker, Billerica, MA, USA) with Cu Kα monochromatic radiation with a wavelength, λ, of 1.5406 Å. The diffraction angle was set from 0 to 80°, and the crystallite sizes were calculated using the Scherrer equation. The functional groups present in the catalysts were determined using Fourier Transform Infrared spectroscopy (FTIR, Thermo Scientific NICOLET 6700 apparatus, Waltham, MA, USA) performed in the range of 400 to 4000 cm^{-1} at room temperature. The nitrogen adsorption–desorption isotherms, together with the surface area, pore size distribution and pore volume, were measured using the BET/BJH method in a Micromeritics ASAP 2020 chemisorption analyzer (Norcross, GA, USA) at −196 °C. The treatment began as the samples were degassed at 300 °C for 3 h before the analysis was carried out. Temperature-programmed reduction (TPR) analysis was carried out to study the reducibility of the catalyst using a Thermo Finnigan TPDRO 1100 model (San Jose, CA, USA). The surface morphology of the catalysts and the spent catalysts was studied using a field emission scanning electron microscope (FESEM) conducted in a Zeiss MERLIN COMPACT machine (Carl Zeiss, Jena, Germany) with an accelerating voltage of 3 kV. The deposited carbon was analyzed by Raman spectroscopy (Thermo Scientific, DXR2xi) and a transmission electron microscope (TEM) (Thermo Fisher Talos 120C model, Waltham, MA, USA).

3.3. Investigation of the Catalytic Performance for Methane Decomposition

The activity of the catalyst was studied for the decomposition of undiluted methane at various reaction temperatures (700–900 °C). The reactions were conducted in a tubular fixed-bed reactor made of stainless steel (height of 60 cm; outer diameter of 3.1 cm; inner diameter of 2.5 cm), which was heated by an electric muffle under atmospheric pressure. The desired weight of the catalyst (1.0 g) was packed in the middle of the reactor. The catalyst was reduced in-situ with hydrogen (150 mL/min) at 600 °C for 90 min. After that, nitrogen gas (150 mL/min) was introduced into the reactor to increase the temperature of the reactor to the desired temperature. The catalytic testing was carried out at atmospheric pressure (0.1 MPa) with a flow of undiluted methane (100 mL/min) for 360 min on stream. The products of the outlet gases were collected in gas bags and were analyzed by a gas chromatograph, GC 2014C (Shimadzu, Osaka, Japan), equipped with a TDX-01 column connected to a thermal conductivity detector (TCD), using helium as the carrier gas. At the end of the experiments, the reactor was flushed with nitrogen (50 mL/min) to cool down the reactor and, thus, the spent catalyst in the form of carbon was taken for further analysis. The hydrogen yield and carbon yield were calculated according to the following Equations (1) and (2)

$$Hydrogen\ yield = \frac{H_2(outlet)}{2 * CH_4(inlet)} * 100 \quad (1)$$

$$Carbon\ yield = \frac{Amount\ of\ carbon\ deposited\ in\ gram}{Total\ amount\ of\ metallic\ species\ presnt\ in\ the\ catalyst\ in\ gram} * 100 \quad (2)$$

4. Conclusions

Ni/MSN catalysts with Cu and Fe-promoters were synthesized by using an impregnation method, and their performance in methane decomposition was investigated at 700 °C, 800 °C and 900 °C. Fine dispersion of the active phase NiO and Cu and Fe-incorporated NiO on the MSN surface with proper metal–support interaction was achieved. Catalytic activity tests revealed that both the hydrogen yield and stability of the Ni/MSN catalyst increased upon the incorporation of Cu and Fe at all of the reaction temperatures. The pronounced performance of the catalyst at high temperatures could be attributed to the enhanced synergistic effects between the metallic nickel species and the MSN support, together with the formation of a bimetallic alloy, which enhanced the stability of the Ni/MSN catalyst, even at high temperatures. The improved performance of the Ni/MSN catalysts in methane decomposition can be ascribed to the high surface area, fine dispersion of the metal species on the surface of the support, an appropriate metal–support interaction and the presence of bimetallic alloy nanoparticles. The characterization studies of the deposited carbon showed that MWCNTs with a high crystallinity and graphitization degree formed over the catalysts after the methane decomposition reaction.

Author Contributions: Conceptualization, N.S.N.H. and M.P.; methodology, N.S.N.H., M.P. and S.N.T.; Validation, N.S.N.H. and M.P.; Formal analysis, N.S.N.H., M.P. and S.N.T.; Investigation, N.S.N.H. and Z.Y.; Data curation, N.S.N.H., M.P., N.H.N.K., K.L.L. and S.N.T.; Writing—original draft preparation, N.S.N.H. and M.P.; Writing—review and editing, N.S.N.H., M.P., S.N.T., Z.Y., N.H.N.K. and K.L.L.; Supervision, M.P. and S.N.T.; Project administration, S.N.T.; Funding acquisition, S.N.T. All authors have read and agreed to the published version of the manuscript.

Funding: The authors gratefully acknowledge the financial supports from University Research Grants [GUP-2022-079] and [PP-SELFUEL-2023].

Data Availability Statement: Data are available upon request.

Acknowledgments: The authors acknowledge the Centre of Research and Instrumentation Management (CRIM) and the Faculty of Science and Technology (FST), Universiti Kebangsaan Malaysia (UKM) for the characterization of materials.

Conflicts of Interest: The authors declare no conflict of interest

References

1. Elavarasan, R.M. The Motivation for Renewable Energy and its Comparison with Other Energy Sources: A Review. *Eur. J. Sustain. Dev. Res.* **2019**, *3*, em0076. [CrossRef] [PubMed]
2. Liu, H.; Xu, L.; Han, Y.; Chen, X.; Sheng, P.; Wang, S.; Huang, X.; Wang, X.; Lu, C.; Luo, H.; et al. Development of a gaseous and solid-state hybrid system for stationary hydrogen energy storage. *Green Energy Environ.* **2021**, *6*, 528–537. [CrossRef]
3. Yang, H.; Han, Y.J.; Yu, J.; Kim, S.; Lee, S.; Kim, G.; Lee, C. Exploring Future Promising Technologies in Hydrogen Fuel Cell Transportation. *Sustainability* **2022**, *14*, 917. [CrossRef]
4. Singla, M.K.; Nijhawan, P.; Oberoi, A.S. Hydrogen fuel and fuel cell technology for cleaner future: A review. *Environ. Sci. Pollut. Res.* **2021**, *28*, 15607–15626. [CrossRef]
5. Balat, H.; Kırtay, E. Hydrogen from biomass—Present scenario and future prospects. *Int. J. Hydrogen Energy* **2010**, *35*, 7416–7426. [CrossRef]
6. Abánades, A.; Rubbia, C.; Salmieri, D. Thermal cracking of methane into Hydrogen for a CO_2-free utilization of natural gas. *Int. J. Hydrogen Energy* **2013**, *38*, 8491–8496. [CrossRef]
7. Fan, Z.; Weng, W.; Zhou, J.; Gu, D.; Xiao, W. Catalytic decomposition of methane to produce hydrogen: A review. *J. Energy Chem.* **2021**, *58*, 415–430. [CrossRef]

8. Pudukudy, M.; Yaakob, Z.; Mazuki, M.Z.; Takriff, M.S.; Jahaya, S.S. One-pot sol-gel synthesis of MgO nanoparticles supported nickel and iron catalysts for undiluted methane decomposition into COx free hydrogen and nanocarbon. *Appl. Catal. B Environ.* **2017**, *218*, 298–316. [CrossRef]
9. Naresh, G.; Vijay Kumar, V.; Anjaneyulu, C.; Tardio, J.; Bhargava, S.K.; Patel, J.; Venugopal, A. Nano size Hβ zeolite as an effective support for Ni and NiCu for COx free hydrogen production by catalytic decomposition of methane. *Int. J. Hydrogen Energy* **2016**, *41*, 19855–19862. [CrossRef]
10. Pudukudy, M.; Yaakob, Z.; Akmal, Z.S. Direct decomposition of methane over Pd promoted Ni/SBA-15 catalysts. *Appl. Surf. Sci.* **2015**, *353*, 127–136. [CrossRef]
11. Kuvshinov, D.G.; Kurmashov, P.B.; Bannov, A.G.; Popov, M.V.; Kuvshinov, G.G. Synthesis of Ni-based catalysts by hexamethylenetetramine-nitrates solution combustion method for co-production of hydrogen and nanofibrous carbon from methane. *Int. J. Hydrogen Energy* **2019**, *44*, 16271–16286. [CrossRef]
12. Al-Fatesh, A.S.; Kasim, S.O.; Ibrahim, A.A.; Al-Awadi, A.S.; Abasaeed, A.E.; Fakeeha, A.H.; Awadallah, A.E. Catalytic methane decomposition over ZrO_2 supported iron catalysts: Effect of WO_3 and La_2O_3 addition on catalytic activity and stability. *Renew. Energy* **2020**, *155*, 969–978. [CrossRef]
13. Somekawa, S.; Yanagi, S.; Yamanaka, T.; Hayashi, H. Preparation of Ni/12CaO·7Al_2O_3 honeycomb-type catalyst for continuous production of hydrogen by methane decomposition. *Inorg. Chem. Commun.* **2022**, *140*, 109401. [CrossRef]
14. Zhang, X.; Zhang, M.; Zhang, J.; Zhang, Q.; Tsubaki, N.; Tan, Y.; Han, Y. Methane decomposition and carbon deposition over Ni/ZrO_2 catalysts: Comparison of amorphous, tetragonal, and monoclinic zirconia phase. *Int. J. Hydrogen Energy* **2019**, *44*, 17887–17899. [CrossRef]
15. Hanifa, N.H.E.; Ismail, M.; Ideris, A. Methane decomposition over Ni supported on palm oil fuel ash (Ni-POFA) catalyst. *Chem. Eng. Res. Des.* **2022**, *178*, 224–231. [CrossRef]
16. Muhammad, A.F.A.S.; Awad, A.; Saidur, R.; Masiran, N.; Salam, A.; Abdullah, B. Recent advances in cleaner hydrogen productions via thermo-catalytic decomposition of methane: Admixture with hydrocarbon. *Int. J. Hydrogen Energy* **2018**, *43*, 18713–18734. [CrossRef]
17. Urdiana, G.; Valdez, R.; Lastra, G.; Valenzuela, M.Á.; Olivas, A. Production of hydrogen and carbon nanomaterials using transition metal catalysts through methane decomposition. *Mater. Lett.* **2018**, *217*, 9–12. [CrossRef]
18. Awadallah, A.; Ahmed, W.; El-Din, M.N.; Aboul-Enein, A. Novel aluminosilicate hollow sphere as a catalyst support for methane decomposition to COx-free hydrogen production. *Appl. Surf. Sci.* **2013**, *287*, 415–422. [CrossRef]
19. Tapia-Parada, K.; Valverde-Aguilar, G.; Mantilla, A.; Valenzuela, M.A.; Hernández, E. Synthesis and characterization of Ni/Ce–SiO_2 and Co/Ce–TiO_2 catalysts for methane decomposition. *Fuel* **2013**, *110*, 70–75. [CrossRef]
20. Hasnan, N.S.N.; Timmiati, S.N.; Lim, K.L.; Yaakob, Z.; Kamaruddin, N.H.N.; Teh, L.P. Recent developments in methane decomposition over heterogeneous catalysts: An overview. *Mater. Renew. Sustain. Energy* **2020**, *9*, 8. [CrossRef]
21. Pizzolitto, C.; Pupulin, E.; Menegazzo, F.; Ghedini, E.; Di Michele, A.; Mattarelli, M.; Cruciani, G.; Signoretto, M. Nickel based catalysts for methane dry reforming: Effect of supports on catalytic activity and stability. *Int. J. Hydrogen Energy* **2019**, *44*, 28065–28076. [CrossRef]
22. Chen, Y.; Qiu, B.; Liu, Y.; Zhang, Y. An active and stable nickel-based catalyst with embedment structure for CO_2 methanation. *Appl. Catal. B Environ.* **2020**, *269*, 118801. [CrossRef]
23. Azali, N.S.; Kamarudin, N.H.N.; Jaafar, J.A.; Timmiati, S.N.; Sajab, M.S. Modification of mesoporous silica nanoparticles with pH responsive polymer poly (2-vinylpyrrolidone) for the release of 5-Florouracil. *Mater. Today Proc.* **2020**, *31*, A12–A17. [CrossRef]
24. Roblero, J.; Pola-Albores, F.; Valenzuela, M.; Rojas-García, E.; Ríos-Valdovinos, E.; Valverde-Aguilar, G. Ni and Ni_3C catalysts supported on mesoporous silica for dry reforming of methane. *Int. J. Hydrogen Energy* **2019**, *44*, 10473–10483. [CrossRef]
25. Wei, Y.; Yang, W.; Yang, Z. An excellent universal catalyst support-mesoporous silica: Preparation, modification and applications in energy-related reactions. *Int. J. Hydrogen Energy* **2022**, *47*, 9537–9565. [CrossRef]
26. Al Mesfer, M.K.; Danish, M.; Shah, M. Synthesis and optimization of hydrotalcite derived Ni-Fe-Cu based catalysts for catalytic methane decomposition process using the design of experiment approach. *J. Taiwan Inst. Chem. Eng.* **2021**, *128*, 370–379. [CrossRef]
27. Pudukudy, M.; Yaakob, Z. Methane decomposition over Ni, Co and Fe based monometallic catalysts supported on sol gel derived SiO_2 microflakes. *Chem. Eng. J.* **2015**, *262*, 1009–1021. [CrossRef]
28. Saraswat, S.K.; Pant, K. Synthesis of carbon nanotubes by thermo catalytic decomposition of methane over Cu and Zn promoted Ni/MCM-22 catalyst. *J. Environ. Chem. Eng.* **2013**, *1*, 746–754. [CrossRef]
29. Awad, A.; Masiran, N.; Salam, M.A.; Vo, D.-V.N.; Abdullah, B. Non-oxidative decomposition of methane/methanol mixture over mesoporous Ni-Cu/Al_2O_3 Co-doped catalysts. *Int. J. Hydrogen Energy* **2019**, *44*, 20889–20899. [CrossRef]
30. Pudukudy, M.; Yaakob, Z.; Akmal, Z.S. Direct decomposition of methane over SBA-15 supported Ni, Co and Fe based bimetallic catalysts. *Appl. Surf. Sci.* **2015**, *330*, 418–430. [CrossRef]
31. Lua, A.C.; Wang, H.Y. Hydrogen production by catalytic decomposition of methane over Ni-Cu-Co alloy particles. *Appl. Catal. B Environ.* **2014**, *156*, 84–93. [CrossRef]
32. Bayat, N.; Rezaei, M.; Meshkani, F. Hydrogen and carbon nanofibers synthesis by methane decomposition over Ni–Pd/Al_2O_3 catalyst. *Int. J. Hydrogen Energy* **2016**, *41*, 5494–5503. [CrossRef]

33. Rategarpanah, A.; Meshkani, F.; Wang, Y.; Arandiyan, H.; Rezaei, M. Thermocatalytic conversion of methane to highly pure hydrogen over Ni–Cu/MgO·Al_2O_3 catalysts: Influence of noble metals (Pt and Pd) on the catalytic activity and stability. *Energy Convers. Manag.* **2018**, *166*, 268–280. [CrossRef]
34. Cunha, A.; Órfão, J.; Figueiredo, J. Methane decomposition on Ni–Cu alloyed Raney-type catalysts. *Int. J. Hydrogen Energy* **2009**, *34*, 4763–4772. [CrossRef]
35. Tezel, E.; Figen, H.E.; Baykara, S.Z. Hydrogen production by methane decomposition using bimetallic Ni–Fe catalysts. *Int. J. Hydrogen Energy* **2019**, *44*, 9930–9940. [CrossRef]
36. Silva, J.A.; Santos, J.B.O.; Torres, D.; Pinilla, J.L.; Suelves, I. Natural Fe-based catalysts for the production of hydrogen and carbon nanomaterials via methane decomposition. *Int. J. Hydrogen Energy* **2021**, *46*, 35137–35148. [CrossRef]
37. Hasnan, N.S.N.; Timmiati, S.N.; Pudukudy, M.; Yaakob, Z.; Lim, K.L.; Taufiq-Yap, Y.H. Catalytic decomposition of methane into hydrogen and carbon nanotubes over mesostructured silica nanoparticle-supported nickel catalysts. *J. Porous Mater.* **2020**, *27*, 369–382. [CrossRef]
38. Bayat, N.; Rezaei, M.; Meshkani, F. Methane decomposition over Ni–Fe/Al_2O_3 catalysts for production of COx-free hydrogen and carbon nanofiber. *Int. J. Hydrogen Energy* **2016**, *41*, 1574–1584. [CrossRef]
39. Pudukudy, M.; Kadier, A.; Yaakob, Z.; Takriff, M.S. Non-oxidative thermocatalytic decomposition of methane into COx free hydrogen and nanocarbon over unsupported porous NiO and Fe_2O_3 catalysts. *Int. J. Hydrogen Energy* **2016**, *41*, 18509–18521. [CrossRef]
40. Li, Y.; Zhang, B.; Xie, X.; Liu, J.; Xu, Y.; Shen, W. Novel Ni catalysts for methane decomposition to hydrogen and carbon nanofibers. *J. Catal.* **2006**, *238*, 412–424. [CrossRef]
41. Lazaro, M.; Echegoyen, Y.; Alegre, C.; Suelves, I.; Moliner, R.; Palacios, J. TiO_2 as textural promoter on high loaded Ni catalysts for methane decomposition. *Int. J. Hydrogen Energy* **2008**, *33*, 3320–3329. [CrossRef]
42. Wang, D.; Zhang, J.; Sun, J.; Gao, W.; Cui, Y. Effect of metal additives on the catalytic performance of Ni/Al_2O_3 catalyst in thermocatalytic decomposition of methane. *Int. J. Hydrogen Energy* **2019**, *44*, 7205–7215. [CrossRef]
43. Sidik, S.; Triwahyono, S.; Jalil, A.; Majid, Z.; Salamun, N.; Talib, N.; Abdullah, T. CO_2 reforming of CH_4 over Ni–Co/MSN for syngas production: Role of Co as a binder and optimization using RSM. *Chem. Eng. J.* **2016**, *295*, 1–10. [CrossRef]
44. Sidik, S.M.; Jalil, A.A.; Triwahyono, S.; Asli, U.A. CO_2 Reforming of Methane Over Ni Supported on Mesostructured Silica Nanoparticles (Ni/Msn): Effect of Ni Loading. *J. Teknol.* **2016**, *78*, 13–18. [CrossRef]
45. Saraswat, S.K.; Pant, K. Synthesis of hydrogen and carbon nanotubes over copper promoted Ni/SiO_2 catalyst by thermocatalytic decomposition of methane. *J. Nat. Gas Sci. Eng.* **2013**, *13*, 52–59. [CrossRef]
46. Wang, W.; Wang, H.; Yang, Y.; Jiang, S. Ni–SiO_2 and Ni–Fe–SiO_2 catalysts for methane decomposition to prepare hydrogen and carbon filaments. *Int. J. Hydrogen Energy* **2012**, *37*, 9058–9066. [CrossRef]
47. Park, S.J.; Kim, K.-D.; Park, Y.S.; Go, K.S.; Kim, W.; Kim, M.; Nho, N.S.; Lee, D.H. Effect of reduction conditions of Mo-Fe/MgO on the formation of carbon nanotube in catalytic methane decomposition. *J. Ind. Eng. Chem.* **2022**, *109*, 384–396. [CrossRef]
48. Torres, D.; Pinilla, J.L.; Suelves, I. Screening of Ni-Cu bimetallic catalysts for hydrogen and carbon nanofilaments production via catalytic decomposition of methane. *Appl. Catal. A Gen.* **2018**, *559*, 10–19. [CrossRef]
49. Nuernberg, G.D.B.; Foletto, E.L.; Campos, C.E.M.; Fajardo, H.V.; Carreño, N.L.V.; Probst, L.F.D. Direct decomposition of methane over Ni catalyst supported in magnesium aluminate. *J. Power Sources* **2012**, *208*, 409–414. [CrossRef]
50. Chai, S.-P.; Zein, S.H.S.; Mohamed, A.R. COx-Free Hydrogen and Carbon Nanofibers Produced from Direct Decomposition of Methane on Nickel-Based Catalysts. *J. Nat. Gas Chem.* **2006**, *15*, 253–258. [CrossRef]
51. Karimi, S.; Bibak, F.; Meshkani, F.; Rastegarpanah, A.; Deng, J.; Liu, Y.; Dai, H. Promotional roles of second metals in catalyzing methane decomposition over the Ni-based catalysts for hydrogen production: A critical review. *Int. J. Hydrogen Energy* **2021**, *46*, 20435–20480. [CrossRef]
52. Chen, J.; Yang, X.; Li, Y. Investigation on the structure and the oxidation activity of the solid carbon produced from catalytic decomposition of methane. *Fuel* **2010**, *89*, 943–948. [CrossRef]
53. Awadallah, A.E.; El-Desouki, D.S.; Aboul-Gheit, N.A.K.; Ibrahim, A.H.; Aboul-Gheit, A.K. Effect of crystalline structure and pore geometry of silica based supported materials on the catalytic behavior of metallic nickel particles during methane decomposition to COx-free hydrogen and carbon nanomaterials. *Int. J. Hydrogen Energy* **2016**, *41*, 16890–16902. [CrossRef]
54. Pudukudy, M.; Yaakob, Z.; Dahani, N.; Takriff, M.S.; Hassan, N.S.M. Production of COx free hydrogen and nanocarbon via methane decomposition over unsupported porous nickel and iron catalysts. *J. Clust. Sci.* **2017**, *28*, 1579–1594. [CrossRef]

catalysts

MDPI

Article

Homogeneous Photo-Fenton Degradation and Mineralization of Model and Simulated Pesticide Wastewaters in Lab- and Pilot-Scale Reactors

Chrysanthi Berberidou [1], Petros Kokkinos [2,3,*], Ioannis Poulios [1,*] and Dionissios Mantzavinos [2]

[1] Department of Chemistry, Aristotle University of Thessaloniki, 54124 Thessaloniki, Greece
[2] Department of Chemical Engineering, University of Patras, Caratheodory 1, University Campus, 26504 Patras, Greece
[3] School of Sciences and Engineering, University of Nicosia, 2417 Nicosia, Cyprus
* Correspondence: pkokkin@upatras.gr (P.K.); poulios@chem.auth.gr (I.P.)

Abstract: The homogeneous photocatalytic degradation of model pesticide clopyralid (CLPR) has been investigated under various experimental setups. Lab-scale experiments under UV-A radiation in an acidic environment showed that the degradation rate generally increased when increasing either Fe^{3+} or H_2O_2 concentration up to a point beyond which (i.e., 100 mg L^{-1} for peroxide or 7 mg L^{-1} for ferric ions) Fenton reagents had little or even detrimental effect on degradation. Thus, there is an optimum concentration of Fenton reagents for maximizing treatment performance, beyond which degradation rates are not enhanced. Excessive concentrations of peroxide and/or catalyst may (i) introduce unnecessary treatment costs, (ii) reduce performance due to scavenging effects, and (iii) raise environmental concerns associated with the disposal of, e.g., high concentrations of iron in the receiving water courses. Switching from UV-A to visible light led to similar rates of degradation, i.e., 86% and 82.2%, respectively, after 90 min of reaction, highlighting the potential of using renewable energy, i.e., natural sunlight, to drive the process. Treatment for 120 min also led to 90% mineralization and quantitative release of nitrogen originally present in the pesticide; this was also accompanied by complete elimination of eco-toxicity to *Vibrio fischeri*. Pilot-scale experiments were performed in a fountain-type reactor using a commercial pesticide formulation containing CLPR. Both the degradation and mineralization rates increased with increasing the intensity of the incident UV-A radiation from 1.88 to 4.03 mW cm^{-2}. Experiments were also conducted with different liquid volumes, i.e., from 3 to 8 L. Illumination of 5 L wastewater resulted in 80% mineralization after 60 min and this only slightly decreased to 73% at 8 L. Overall, the findings underline the promising perspectives of the application of the treatment method in upgrading the quality of water and liquid waste containing pesticides.

Keywords: photocatalysis; photo-Fenton; clopyralid; pesticides; mineralization; toxicity

1. Introduction

The growing global population that is projected to reach 9.7 and 11.0 billion by 2050 and 2100, respectively, generates enormous amount of waste, owing to the production of food, energy, feed, fiber, etc. [1,2].

Pesticide pollution is an issue that negatively impacts on the environment and also concerns human beings and wildlife, being a widespread problem [3–5]. These contaminants are typically chemically stable, resistant to direct decomposition by sunlight, non-biodegradable, and highly toxic [3,6–8]. They are also characterized by relatively high solubility in water, and mobility [9]. Although some common water treatment processes are effective barriers for the abatement of many pesticides, some of them are not readily removed by such technologies and thus advanced treatments are demanded to remove them from the environment [3–5,7,8,10].

Clopyralid (3,6–dichloro-2-pyridine-carboxylic acid) (CLPR) is an auxin-mimic-type herbicide, from the polar and nonvolatile chemical class of pyridine compounds. It is a selective herbicide widely applied to control broadleaf weeds in certain crops and turf and provides control of some brush species on rangeland and pastures [11,12]. CLPR is highly soluble in water, and its solubility increases with pH [7,13,14].

It is known to be particularly stable against hydrolysis and photolysis [13]. It has been frequently detected in the environment and even in drinking water [3,5,11,13]. It is hazardous to a number of endangered plant species and beneficial insects and toxic to certain mammals [15]. Due to its resistance to biodegradation, CLPR is hardly removed by conventional water and wastewater treatments [5,7]. Many studies indicate that a wide range of pesticides are readily degradable by means of heterogenous or homogenous photocatalysis [15]. The pretreatment of wastes containing a hard-to-degrade anthropogenic pollutant is currently one of the most critical challenges of environmental engineering [7]. When a photocatalytic process is applied as a pretreatment, the main objective is the breakdown of organic contents into more biodegradable ones, as well as the removal of toxicity [4,16].

Because of their high reactivity and nonselective oxidation, advanced oxidation processes (AOPs) have been investigated for the treatment of emerging organic pollutants over the past 20 years. They have been proposed as alternative water treatment methods aiming for the mineralization of organic molecules, including pesticides [3,5,8,17–21]. Among different AOPs, heterogenous and homogenous solar photocatalytic detoxification methods have been proposed for the treatment of wastewater containing pesticides [22].

We have previously reported on the heterogenous photocatalytic degradation and mineralization of CLPR in aqueous solutions, employing a wide range of commercial photocatalysts, providing a thorough kinetic analysis in terms of degradation and mineralization, elucidating potential degradation pathways and determining ecotoxicity [15]. We have also previously proposed an integrated system based on solar photocatalysis and constructed wetlands and proven that it has the potential to effectively detoxify wastewater containing CLPR. Three photocatalytic methods under solar light were investigated: the photo-Fenton and the ferrioxalate reagent as well as the combination of photo-Fenton with TiO_2 P25 [22]. We have also shown that reduced graphene oxide (rGO), a low-cost, nontoxic material, may serve as a reliable alternative in the enhancement of TiO_2 photocatalytic efficiency in water processing applications, achieving disinfection and the degradation and mineralization of CLPR [23]. All previous studies of our research group on the abatement of CLPR are summarized in Table 1.

In the present study, we focused on the homogeneous photocatalytic process and studied different factors that affect process efficiency (concentration of H_2O_2 and Fe^{3+} ions, type of irradiation (visible and UV-A)). We also studied the extent of photocatalytic mineralization (DOC and total nitrogen), and the toxicity of CLPR solution. Additionally, we assessed the homogeneous photocatalytic oxidation of simulated wastewater containing CLPR in a pilot scale photoreactor under artificial light and studied the effects of the intensity of the incident UV-A radiation, and of the illuminated volume.

Table 1. Previously published clopyralid studies of our research group, at a glance.

Treatment	Photocatalytic Methods	Level of Implementation	Matrix	Main Outcome	Reference
heterogenous photocatalytic experiments with CLPR solutions containing TiO_2 P25, TiO_2 UV-100, ZnO, TiO_2 Kronos 7000, Kronos 7001 and Kronos 7500 under UV-A irradiation	different operational conditions, e.g., type of photocatalyst, catalyst loading, initial pH and hydrogen peroxide (H_2O_2) concentration	lab	aqueous solutions of clopyralid	TiO_2 mediated photocatalysis has the potential to provide a sustainable solution in the detoxification of wastewater containing recalcitrant pesticides, either alone or in combination with other methods	[16]

Table 1. *Cont.*

Treatment	Photocatalytic Methods	Level of Implementation	Matrix	Main Outcome	Reference
photocatalytic oxidation of CLPR in a bench-scale photocatalytic reactor, equipped with a central UV-A lamp	degradation and mineralization experiments in the presence of three reduced graphene oxide (rGO)/TiO_2 composites containing a 1, 5 or 10 wt% nominal content in rGO and bare P25 for comparison purposes	lab	aqueous solutions of clopyralid	rGO, a low-cost, nontoxic material may serve as a reliable alternative in the enhancement of TiO_2 photocatalytic efficiency in water processing applications	[17]
solar photocatalytic oxidation and constructed wetlands	photo-Fenton and ferrioxalate reagent, and combination of photo-Fenton with TiO_2 P25	pilot	simulated wastewater	the integrated system effectively detoxified wastewater containing clopyralid	[23]

2. Results and Discussion

2.1. Homogeneous Photocatalytic Oxidation of Model CLPR Pesticide in Laboratory Reactor

2.1.1. Effect of H_2O_2 and Fe^{3+} on Degradation Efficiency

The concentration of the reagents in the photo-Fenton method is of vital importance, since it may not only affect the costs associated with the size of the photoreactor, but also the operating costs. The addition of high amounts of hydrogen peroxide can unnecessarily increase operating costs since: (i) reaction rates may be reduced, and (ii) peroxide is usually the single most expensive operating parameter in the photo-Fenton and in similar methods.

Photocatalytic degradation of 40 mg L^{-1} CLPR with photo-Fenton reagent in the presence of 7 mg L^{-1} Fe^{3+} and different H_2O_2 concentrations (200 mg L^{-1}, 150 mg L^{-1}, 100 mg L^{-1}, 75 mg L^{-1}, 50 mg L^{-1}, 25 mg L^{-1}, 0 mg L^{-1}) is shown in Figure 1A. The effect of H_2O_2 concentration on the initial degradation rate is shown in Figure 1B. In the case of UV-A radiation, the rate increases as the H_2O_2 concentration increases to about 100 mg L^{-1}. However, increasing the H_2O_2 concentration from 100 to 200 mg L^{-1} has negative consequences, as at high concentrations H_2O_2 may act as an OH scavenger, as shown below (1):

$$HO\bullet + H_2O_2 \rightarrow H_2O + HO\bullet \quad (1)$$

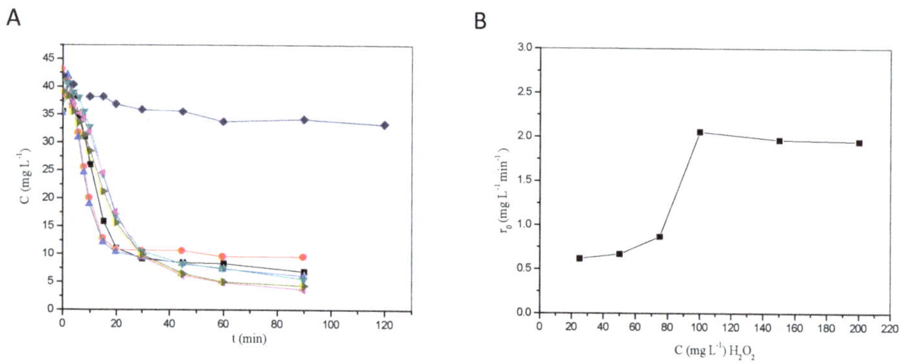

Figure 1. (**A**) Photocatalytic degradation of 40 mg L^{-1} CLPR with photo-Fenton reagent in the presence of 7 mg L^{-1} Fe^{3+} and different H_2O_2 concentrations. [■] 200 mg L^{-1} H_2O_2: (•) 150 mg L^{-1} H_2O_2, (▲) 100 mg L^{-1} H_2O_2, (▼) 75 mg L^{-1} H_2O_2, (◄) 50 mg L^{-1} H_2O_2, (►) 25 mg L^{-1} H_2O_2, (♦) 0 mg L^{-1} H_2O_2. (**B**) Change in the initial degradation rate of CLPR with photo-Fenton reagent in the presence of 7 mg L^{-1} Fe^{3+} and different H_2O_2 concentrations.

In order to study the effect of iron ions on the degradation kinetic, experiments were performed at different initial Fe^{3+} concentrations. As shown in Figure 2A, increasing the Fe^{3+} concentration to 7 mg L^{-1} leads to an increase in the degradation rate. As trivalent iron ions are photo-reduced into divalent ions, which are the major species catalyzing hydrogen peroxide decomposition, the amount of added ferric ions is proportional to the amount of produced hydroxide radicals and hence to the decomposition rate of CLPR. However, increasing Fe^{3+} concentration from 7 to 28 mg L^{-1} did not appear to significantly affect the initial oxidation rate (Figure 2B). This is explained by the fact that for Fe^{3+} concentrations higher than 7 mg L^{-1} the transmission of light is impeded due to the high concentration of iron ions, and consequently it is also impeded in Fe^{3+} photoreduction with oxidation products, thus reducing iron concentration in soluble form, which is capable of causing oxidation.

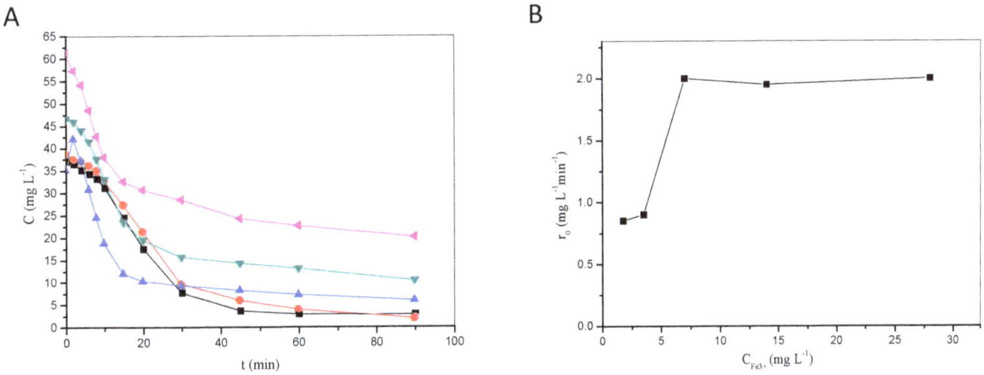

Figure 2. (**A**) Photocatalytic degradation of 40 mg L^{-1} CLPR with photo-Fenton reagent in the presence of 100 mg L^{-1} H_2O_2 and: [■] 2 mg L^{-1} Fe^{3+}, (•) 3.5 mg L^{-1} Fe^{3+}, (▲) 7 mg L^{-1} Fe^{3+}, (▼) 14 mg L^{-1} Fe^{3+}, (◄) 28 mg L^{-1} Fe^{3+}; (**B**) change in the initial degradation rate of CLPR with photo-Fenton reagent in the presence of 100 mg L^{-1} H_2O_2 and different concentrations of Fe^{3+}.

2.1.2. Effect of Irradiation Type

CLPR photocatalytic oxidation under visible radiation and in the presence of the photo-Fenton reagent was promising, since despite the relatively reduced initial degradation rate compared to that under UV-A, it eventually led to similar rates of CLPR degradation after 90 min of irradiation (Figure 3, 82.2% and 86%, respectively). This underscores the significant advantage of photo-Fenton reagent in decomposing organic pollutants into liquid waste under sunlight. Trivalent iron ions also absorb part of the visible spectrum in parallel with ultraviolet radiation.

2.1.3. Mineralization

Under the previously identified optimal conditions, the conversion of organic carbon to inorganic was monitored (Figure 4). Pesticide mineralization in the presence of photo-Fenton reagent was found to be very fast, reaching 90% within 120 min under irradiation (100 mg L^{-1} H_2O_2, 7 mg L^{-1} Fe^{3+}, UV-A). The change in total nitrogen concentration (as a sum of NO_3^-, NH^{4+}, and NO_2^-) over time, as a percentage of the theoretical nitrogen contained in CLPR, under various experimental conditions, is shown in Figure 4. In all cases, ammonia and nitrate anions were the predominantly formed species, while the amounts of NO_2^- were below the detection limit of the method and after prolonged illumination they were oxidized to NO_3^-. The nitrogen conversion of the pyridine ring of the CLPR molecule, under the given experimental conditions, occurs rapidly, as 100% of the total nitrogen of the molecule is released within 90 min of illumination.

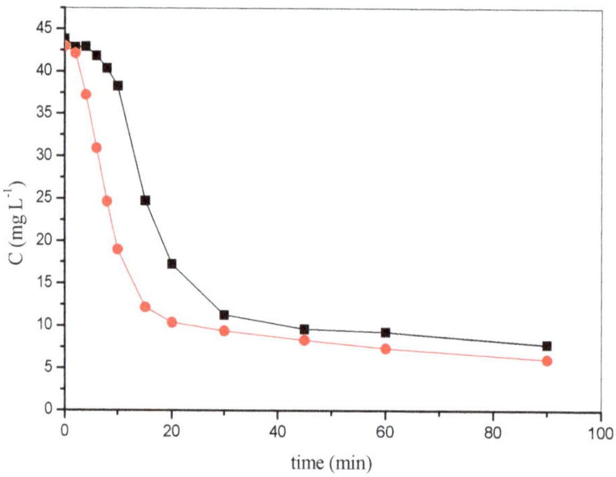

Figure 3. Photocatalytic degradation of 40 mg L^{-1} CLPR with photo-Fenton reagent in the presence of 7 mg L^{-1} Fe^{3+} and 100 mg L^{-1} H_2O_2 at pH: 3.0: (■) visible and (•) UV-A radiation.

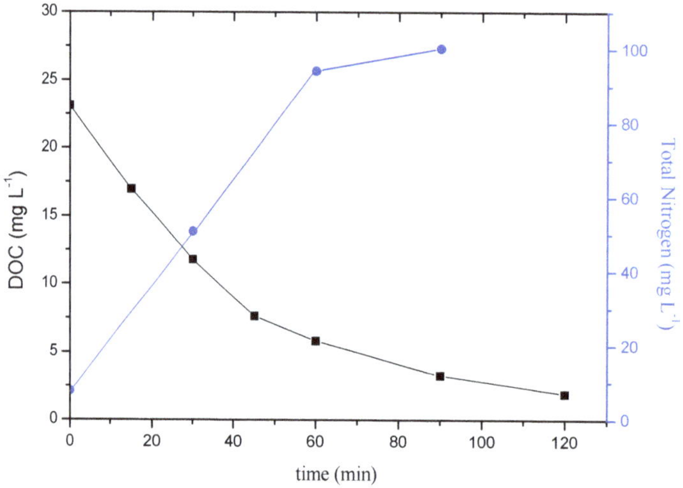

Figure 4. Photocatalytic mineralization of 40 mg L^{-1} CLPR with photo-Fenton reagent in the presence of 100 mg L^{-1} H_2O_2, 7 mg L^{-1} Fe^{3+} and UV-A radiation: (■) DOC and (•) total organic nitrogen.

2.1.4. Determination of Acute Toxicity

To determine the toxicity of the CLPR solution during the homogeneous photocatalytic process, samples were collected at selected time intervals during treatment with photo-Fenton reagent (7 mg L^{-1} Fe^{3+}, 100 mg L^{-1} H_2O_2, pH: 3.0) and UV-A irradiation and their toxicity was assessed on the basis of the inhibition of bioluminescence of *V. fischeri* bacteria (Figure 5). The initial CLPR solution (40 mg L^{-1}) was found to be highly toxic to these microorganisms, causing 89% inhibition of bioluminescence after 15 min of incubation. The profile of the photocatalytically treated samples showed that there is a declining trend in their toxicity, up to 90 min of illumination (7%) while up to 180 min it remains constant. Toxicity is then nullified, which underscores the efficacy of the photo-Fenton reagent in completely eliminating CLPR toxicity under these experimental conditions.

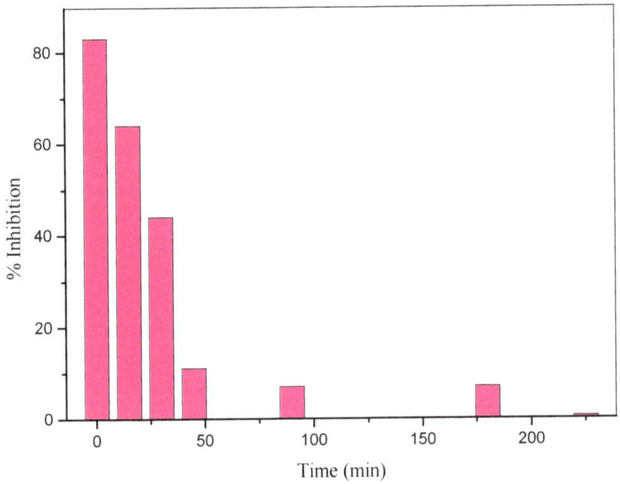

Figure 5. Change of the toxicity of a CLPR solution of initial concentration of 40 mg L^{-1} during photocatalytic oxidation in the presence of photo-Fenton reagent (7 mg L^{-1} Fe^{3+}, 100 mg L^{-1} H_2O_2) and UV-A radiation.

2.2. Homogeneous Photocatalytic Oxidation of Simulated Wastewater Containing CLPR in the Pilot-Scale Photoreactor

2.2.1. Effect of Intensity of Incident UV-A Radiation

The study on the degradation and mineralization of the simulated wastewater was also focused on the homogeneous photocatalytic oxidation with the photo-Fenton reagent (Fe^{3+}/H_2O_2/UV-A, Vis/pH: 3.0) under artificial (UV-A) non-solar illumination. More specifically, the effect of the intensity of the incident UV-A radiation was studied, in the presence of the photo-Fenton reagent (7 mg L^{-1} Fe^{3+}, 100 mg L^{-1} H_2O_2, pH: 3.0). The intensity of the radiation in the pilot photocatalytic system can be changed by varying the distance of the light sources (lamps) from the surface of the photocatalytic treatment tank. Experiments were conducted under five different radiation intensities, and it was found that by progressively increasing the intensity, an increase in both the initial degradation rate (Figure 6A) and the reaction rate constant k (Figure 6B) occurs.

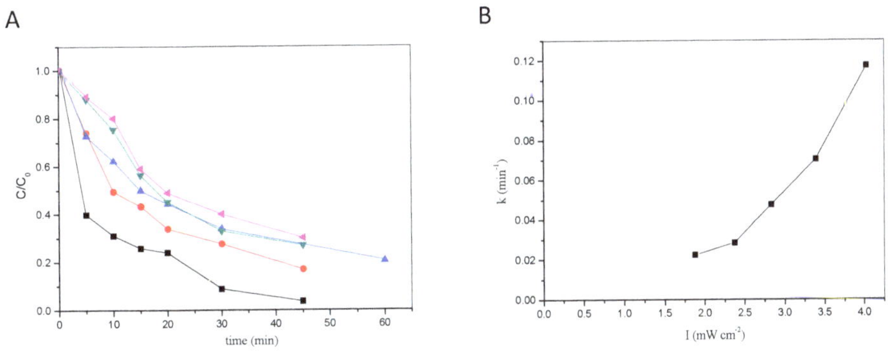

Figure 6. (**A**) Effect of incident UV-A radiation intensity on the degradation of simulated wastewater containing 40 mg L^{-1} CLPR in the presence of the photo-Fenton reagent (7 mg L^{-1} Fe^{3+}, 100 mg L^{-1} H_2O_2, pH: 3.0): [■] I = 4.03 mW cm^{-2}, [•] I = 3.39 mW cm^{-2}, [▲] I = 2.83 mW cm^{-2}, [▼] I = 2.37 mW cm^{-2}, [◄] I = 1.88 mW cm^{-2}. Total volume: 15 L, illuminated volume: 5 L.; (**B**) effect of the intensity of incident UV-A radiation on the reaction rate constant, k.

Moreover, similar conclusions were drawn from the study of the mineralization of the simulated wastewater carried out through the determination of the dissolved organic carbon (DOC) of samples collected at different time intervals (Figure 7A), since the increase in the incident radiation from 1.88 mW cm^{-2} to 4.03 mW cm^{-2} led to a fourfold increase in the rate constant of the mineralization, k_{DOC} (Figure 7B).

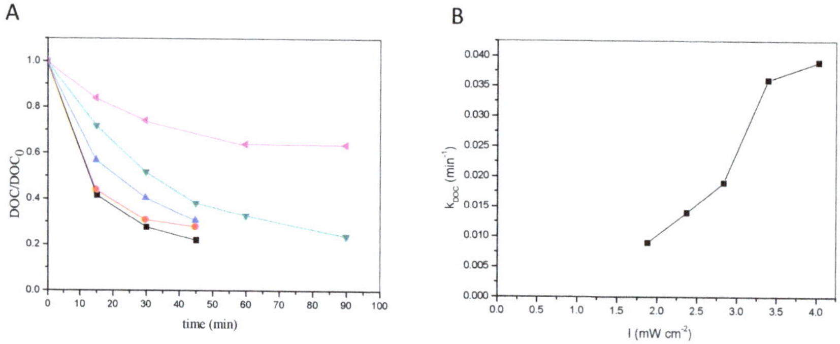

Figure 7. (**A**) Effect of incident UV-A radiation intensity on the mineralization of simulated wastewater containing 40 mg L^{-1} CLPR in the presence of the photo-Fenton reagent (7 mg L^{-1} Fe^{3+}, 100 mg L^{-1} H_2O_2, pH: 3.0): [■] I = 4.03 mW cm^{-2}, [•] I = 3.39 mW cm^{-2}, [▲] I = 2.83 mW cm^{-2}, [▼] I = 2.37 mW cm^{-2}, [◄] I = 1.88 mW cm^{-2}. Total volume: 15 L, illuminated volume: 5 L.; (**B**) effect of intensity of incident UV-A radiation on the reaction constant, k_{DOC}.

2.2.2. Effect of the Illuminated Volume

To assess the effect of the illumination of the simulated wastewater in the homogeneous photocatalytic degradation of CLRP with the photo-Fenton reagent, more experiments were carried out. The volume of illuminated wastewater varies by adding suitable overflow rings that are adapted to the photocatalytic treatment tank and allow it to be filled with 3, 5, or 8 L of wastewater. From Figure 8A, it can be seen that increasing the volume of illuminated wastewater from 3 to 8 L has practically no effect on the degradation of CLPR. Regarding mineralization (Figure 8B), it appears that the illumination of 5 L of the simulated wastewater leads to slightly higher percentages of mineralization after 60 min of illumination (80%) compared to the experiments where the illuminated volume was 3 or 8 L (~73%).

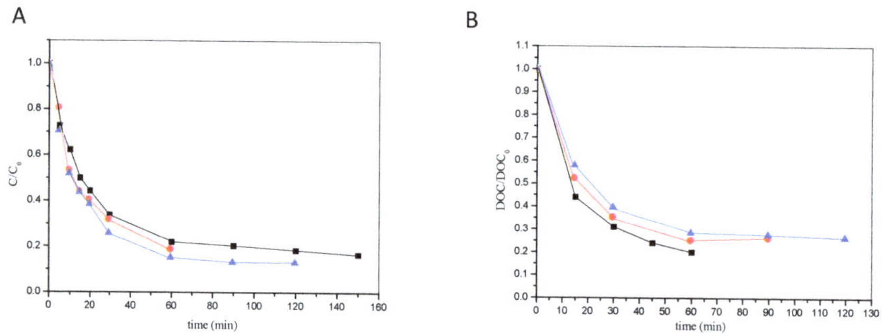

Figure 8. (**A**) Effect of illuminated volume on the degradation efficiency of simulated wastewater containing 40 mg L^{-1} CLPR in the presence of the photo-Fenton reagent (7 mg L^{-1} Fe^{3+}, 100 mg L^{-1} H_2O_2, pH: 3.0): [■] V= 5 L, [•] V = 8 L, [▲] V = 3 L. Intensity of incident UV-A light 3.39 mW cm^{-2}; (**B**) mineralization of simulated wastewater containing 40 mg L^{-1} CLPR in the presence of the photo-Fenton reagent (7 mg L^{-1} Fe^{3+}, 100 mg L^{-1} H_2O_2, pH: 3.0): [■] V = 5 L, [•] V= 8 L, [▲] V = 3 L. Intensity of incident UV-A radiation: 3.39 mW cm^{-2}.

3. Materials and Methods

3.1. Materials

Clopyralid (3,6-dichloro-2-pyridine-carboxylic acid, CAS No. 1702-17-6, Mr: 192 g mol^{-1}, Product No: 36758, PESTANAL®, analytical standard) was a product of Fluka (Sigma-Aldrich Laborchemicalien GmbH, Sellze, Germany) and was used as received. All other reagent-grade chemicals were purchased from Merck and were used without further purification. Doubly deionized water was used throughout the study [14]. For the preparation of simulated wastewater containing clopyralid, used in the pilot scale experiments, appropriate dilutions of a commercial herbicide, Lontrel 100AS® (active ingredient: clopyralid, 10% w/v, Dow Agrosciences, Indianapolis, IN, USA), with tap water were used.

3.2. Photocatalytic Experiments

Photocatalytic experiments were performed in laboratory-scale and pilot-scale reactors. Laboratory-scale experiments were performed in a closed-type glass cylinder reactor of 500 mL capacity under magnetic stirring, as previously described [14]. Internally, the glass cylinder reactor had a smaller borosilicate glass cylinder adapted to house the light source. The borosilicate glass allows radiation of wavelengths higher than 300 nm to pass. Two types of lamps were used: ultraviolet radiation, region A ($320 < hv < 400$), with a maximum at 365 nm (OSRAM DULUX 9W/78, UV-A, Munich, Germany); and visible spectrum radiation ($400 < hv < 550$), with a maximum at 450 nm (OSRAM DULUX S 8/71, Vis, Munich, Germany). After the addition of the pesticide solution to the reactor, its pH was adjusted to 3.0 by adding H_2SO_4 to ensure the presence of Fe^{3+} ions in a soluble form. Subsequently, a sample was collected, Fe^{3+} and H_2O_2 were consecutively added, the solution was illuminated, and samples were collected at certain time intervals and stored in the dark at 25 °C, before any further analysis. Experiments in a pilot-scale reactor were performed as previously described [22], in a unit constituted by three parts: (a) a fountain-type photocatalytic reactor, (b) a reservoir for storing wastewater, which was placed at the bottom part of the reactor, and (c) an Imhoff-type tank. The reactor is illustrated in Figure 9.

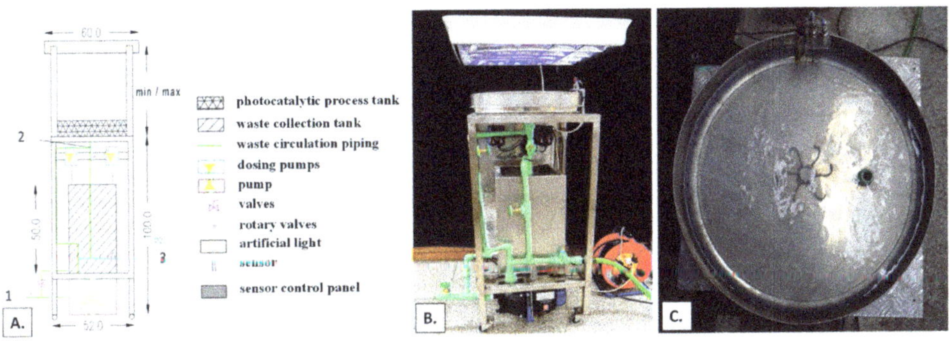

Figure 9. (**A**) Schematic diagram of the fountain-type pilot photocatalytic reactor operating under artificial illumination (UV-A or visible) (front view); (**B**) photo of the fountain-type pilot photocatalytic reactor; (**C**) top view of the photocatalytic treatment tank. The six nozzles through which the waste enters the tank and the hole/pipe through which the waste overflows and returns to the waste collection tank can be distinguished.

3.3. Analytical Procedures

Changes in the concentration of clopyralid were monitored via its characteristic absorption band at 280 nm using a UV-visible spectrophotometer (UV-1700, PharmaSpec, Shimadzu, Kyoto, Japan). Since a linear dependence between the initial concentration of the pesticide and absorption at 280 nm is observed, photodecomposition was monitored spectrophotometrically at this wavelength. Determination of dissolved organic carbon

(DOC) was conducted according to standard methods using a total organic carbon analyzer (Shimadzu VCSH 5000, Kyoto, Japan). Inorganic ions were determined in a Shimadzu system consisting of an LC-10 AD pump, a CDD-6A conductometric detector (0.25 μL flow-cell) and a CTO-10A column oven, as previously described [15]. Chromatographic conditions have also previously been described in detail [15,22].

3.4. Toxicity Analysis

Toxicity analysis was performed using the Microtox Model 500 Analyzer (Azur Environmental, Carlsbad, CA, USA), by assessing the bioluminescence of the marine bacteria *V. fischeri*. Freeze-dried bacteria, reconstitution solution, diluent (2% *w/v* NaCl), and adjustment solution (non-toxic 22% *w/v* NaCl) were obtained from Azur. Change in the toxicity of a CLPR solution of initial concentration of 40 mg L^{-1} during photocatalytic oxidation in the presence of photo-Fenton reagent (7 mg L^{-1} Fe^{3+}, 100 mg L^{-1} H_2O_2) and UV-A radiation was studied. The inhibition of bioluminescence, compared with a toxic-free control to provide the percentage of inhibition, was calculated using Microtox calculation software.

4. Conclusions

Concerning the homogeneous photocatalytic oxidation of CLPR pesticide in a laboratory photocatalytic reactor, the following conclusions were drawn:

- There usually is an optimum concentration of Fenton reagents to maximize treatment performance, beyond which degradation rates are not enhanced. This has been demonstrated in this work with lab-scale photocatalytic experiments. Excessive concentrations of peroxide and/or catalyst may (i) introduce unnecessary treatment costs, (ii) reduce performance due to scavenging effects, and (iii) raise environmental concerns associated with the disposal of, e.g., high concentrations of iron in the receiving water courses.
- Interestingly, photocatalytic oxidation under visible light resulted in comparable degradation rates to UV-A radiation; this pinpoints the potential of using renewable energy, i.e., natural sunlight, to drive the process.
- Moreover, the proposed process was capable of converting the organic carbon of CLPR to carbon dioxide, as well as the nitrogen content of the pyridine ring to NO_3^- and NH_4^+. Mineralization was accompanied by detoxification as has been demonstrated in experiments on marine bacteria.
- The process was tested in a pilot-scale, fountain-type reactor treating a commercial formulation of the pesticide under consideration. The results were promising, resulting in considerable degradation and mineralization, whose rates increased with increasing light intensity; an increase of the incident radiation from 1.88 mW cm^{-2} to 4.03 mW cm^{-2} led to a quadrupling of the reaction rate constant k_{DOC}.
- The increase in the volume of the illuminated wastewater does not practically affect the initial rate of its degradation during the photocatalytic oxidation with the photo-Fenton reagent. Regarding mineralization, illumination of 5 L of simulated wastewater leads to slightly higher mineralization rates after 60 min of illumination (80%), compared to experiments where the illuminated volume was 3 or 8 L (~73%). Thus, wastewater volumes up to 8 L could be treated without efficiency loss, which shows the potential of the process for large-scale applications.

Author Contributions: Conceptualization, I.P.; methodology, C.B., I.P. and D.M.; validation, C.B.; formal analysis, C.B.; data curation, C.B. and P.K.; writing—original draft preparation, P.K.; writing—review and editing, P.K., C.B., I.P. and D.M.; supervision, I.P. and D.M.; project administration, I.P.; funding acquisition, I.P. All authors have read and agreed to the published version of the manuscript.

Funding: The study was implemented within the framework of the research project entitled 'A novel method for detoxification and reuse of wastewater containing pesticides by solar photocatalysis and constructed wetlands' (project No: 957) of the Action ARISTEIA of the Operational Program 'Education and Lifelong Learning' (Action's Beneficiary: General Secretariat for Research and Technology) and is co-financed by the European Social Fund (ESF) and the Greek State.

Data Availability Statement: Not applicable.

Conflicts of Interest: The authors declare no conflict of interest.

References

1. Appa, R.M.; Naidu, B.R.; Venkateswarlu, D.; Hanafiah, M.M.; Lakkaboyana, S.K.; Lakshmidevi, J.; Venkateswarlu, K. Water extract of pomegranate ash–I2 as sustainable system for external oxidant/metal/catalyst-free oxidative iodination of (hetero)arenes. *Green Chem. Lett. Rev.* **2021**, *14*, 700–712. [CrossRef]
2. Venkateswarlu, K. Ashes from organic waste as reagents in synthetic chemistry: A review. *Environ. Chem. Lett.* **2021**, *19*, 3887–3950. [CrossRef]
3. Semitsoglou-Tsiapou, S.; Templeton, M.R.; Graham, N.J.D.; Hernández Leal, L.; Martijn, B.J.; Royce, A.; Kruithof, J.C. Low pressure UV/H_2O_2 treatment for the degradation of the pesticides metaldehyde, clopyralid and mecoprop—Kinetics and reaction product formation. *Water Res.* **2016**, *91*, 285–294. [CrossRef]
4. Carboneras Contreras, M.B.; Fourcade, F.; Assadi, A.; Amrane, A.; Jesus Fernandez-Morales, F. Electro Fenton removal of clopyralid in soil washing effluents. *Chemosphere* **2019**, *237*, 124447. [CrossRef] [PubMed]
5. Rajah, Z.; Guiza, M.; Solís, R.R.; Rivas, F.J.; Ouederni, A. Catalytic and photocatalytic ozonation with activated carbon as technologies in the removal of aqueous micropollutants. *J. Photochem. Photobiol. A* **2019**, *382*, 111961. [CrossRef]
6. Huang, X.; Fong, S.; Deanovic, L.; Young, T.M. Toxicity of herbicides in highway runoff. *Environ. Toxicol. Chem.* **2005**, *24*, 2336–2340. [CrossRef] [PubMed]
7. Santos, G.O.S.; Eguiluz, K.I.B.; Salazar-Banda, G.R.; Saez, C.; Rodrigo, M.A. Biodegradability improvement of clopyralid wastes through electrolysis using different diamond anodes. *Environ. Res.* **2020**, *188*, 109747. [CrossRef]
8. Millán, M.; García-Orozco, V.M.; Lobato, J.; Fernández-Marchante, C.M.; Roa-Morales, G.; Linares-Hernández, I.; Natividad, R.; Rodrigo, M.A. Toward more sustainable photovoltaic solar electrochemical oxidation treatments: Influence of hydraulic and electrical distribution. *J. Environ. Manag.* **2021**, *285*, 112064. [CrossRef] [PubMed]
9. Solís, R.R.; Rivas, F.J.; Gimeno, O.; Pérez-Bote, J.L. Photocatalytic ozonation of pyridine-based herbicides by N-doped titania. *J. Chem. Technol. Biotechnol.* **2016**, *91*, 1998–2008. [CrossRef]
10. Rajah, Z.; Guiza, M.; Solís, R.R.; Becheikh, N.; Rivas, F.J.; Ouederni, A. Clopyralid degradation using solar-photocatalytic/ozone process with olive stone activated carbon. *J. Environ. Chem. Eng.* **2019**, *7*, 102900. [CrossRef]
11. Sojić, D.V.; Anderluh, V.B.; Orcić, D.Z.; Abramović, B.F. Photodegradation of clopyralid in TiO_2 suspensions: Identification of intermediates and reaction pathways. *J. Hazard. Mater.* **2009**, *168*, 94–101. [CrossRef]
12. Muñoz-Morales, M.; Sáez, C.; Cañizares, P.; Rodrigo, M.A. Improvement of electrochemical oxidation efficiency through combination with adsorption processes. *J. Environ. Manag.* **2020**, *262*, 110364. [CrossRef]
13. Tizaoui, C.; Mezughi, K.; Bickley, R. Heterogeneous photocatalytic removal of the herbicide clopyralid and its comparison with UV/H_2O_2 and ozone oxidation techniques. *Desalination* **2011**, *273*, 197–204. [CrossRef]
14. Gu, B.; Meldrum, B.; McCabe, T.; Phillips, S. Enhancing concentration and mass sensitivities for liquid chromatography trace analysis of clopyralid in drinking water. *J. Sep. Sci.* **2012**, *35*, 185–192. [CrossRef]
15. Berberidou, C.; Kitsiou, V.; Karahanidou, S.; Lambropoulou, D.A.; Kouras, A.; Kosma, C.I.; Albanis, T.A.; Poulios, I. Photocatalytic degradation of the herbicide clopyralid: Kinetics, degradation pathways and ecotoxicity evaluation. *J. Chem. Technol. Biotechnol.* **2016**, *91*, 2510–2518. [CrossRef]
16. Pérez, J.F.; Llanos, J.; Sáez, C.; López, C.; Cañizares, P.; Rodrigo, M.A. On the design of a jet-aerated microfluidic flow-through reactor for wastewater treatment by electro-Fenton. *Sep. Purif. Technol.* **2019**, *208*, 123–129. [CrossRef]
17. Sedlazeck, K.P.; Vollprecht, D.; Müller, P.; Mischitz, R.; Gill, J.; Trois, W.; Maunz, I.; Frate, R.; Mann, O.; Wruss, K. Decomposition of dissolved organic contaminants by combining a boron-doped diamond electrode, zero-valent iron and ultraviolet radiation. *Chemosphere* **2019**, *217*, 897–904. [CrossRef] [PubMed]
18. Carboneras, M.B.; Cañizares, P.; Rodrigo, M.A.; Villaseñor, J.; Fernandez-Morales, F.J. Improving biodegradability of soil washing effluents using anodic oxidation. *Bioresour. Technol.* **2018**, *252*, 1–6. [CrossRef]
19. Lara-Ramos, J.A.; Saez, C.; Machuca-Martínez, F.; Rodrigo, M.A. Electro-ozonizers: A new approach for an old problem. *Sep. Purif. Technol.* **2020**, *241*, 116701. [CrossRef]
20. Kokkinos, P.; Venieri, D.; Mantzavinos, D. Advanced oxidation processes for water and wastewater viral disinfection. A systematic review. *Food Environ. Virol.* **2021**, *13*, 283–302. [CrossRef]
21. Pandit, V.U.; Arbuj, S.S.; Pandit, Y.B.; Naik, S.D.; Rane, S.B.; Mulik, U.P.; Gosavi, S.W.; Kale, B.B. Solar light driven dye degradation using novel organo-inorganic (6, 13-pentacenequinone-TiO_2) nanocomposite. *RSC Adv.* **2015**, *5*, 10326–10331. [CrossRef]
22. Berberidou, C.; Kitsiou, V.; Lambropoulou, D.A.; Antoniadis, A.; Ntonou, E.; Zalidis, G.C.; Poulios, I. Evaluation of an alternative method for wastewater treatment containing pesticides using solar photocatalytic oxidation and constructed wetlands. *J. Environ. Manag.* **2017**, *195 Pt 2*, 133–139. [CrossRef] [PubMed]
23. Berberidou, C.; Kyzas, G.Z.; Paspaltsis, I.; Sklaviadis, T.; Poulios, I. Photocatalytic disinfection and purification of water employing reduced graphene oxide/TiO_2 composites. *J. Chem. Technol. Biotechnol.* **2019**, *94*, 3905–3914. [CrossRef]

catalysts MDPI

Article

Catalytic Conversion of Sugars into Lactic Acid via a RuO_x/MoS_2 Catalyst

Zongling Li [1,2], Pengfei Wu [2], Jifeng Pang [2], Xianquan Li [2,3], Shangru Zhai [1,*] and Mingyuan Zheng [2,4,*]

1 Faculty of Light Industry and Chemical Engineering, Dalian Polytechnic University, Dalian 116034, China
2 CAS Key Laboratory of Science and Technology on Applied Catalysis, Dalian Institute of Chemical Physics (DICP), Chinese Academy of Sciences (CAS), 457 Zhongshan Road, Dalian 116023, China
3 University of Chinese Academy of Sciences, Beijing 100049, China
4 Dalian National Laboratory for Clean Energy, Chinese Academy of Sciences, Dalian 116023, China
* Correspondence: zhaisr@dlpu.edu.cn (S.Z.); myzheng@dicp.ac.cn (M.Z.)

Abstract: The catalytic transformation of sugars into lactic acid has shown great potential for the scalable utilization of renewable biomass. Herein, RuO_x/MoS_2 catalysts were synthesized with the assistance of CaO for the one-pot conversion of glucose to lactic acid. Under the reaction conditions of 120 °C and 1 MPa O_2, a 96.6% glucose conversion and a 54.3% lactic acid yield were realized in the one-pot catalytic reaction, with relatively high stability after four successive cycles. This catalytic system was also effective for the conversion of many other carbohydrate substrates, such as fructose, xylose and cellulose (selectivity 68.9%, 78.2% and 50.6%, respectively). According to catalyst characterizations and conditional experiments, the highly dispersed RuO_x species on the surface of MoS_2, together with OH^-, promoted isomerization, retro-aldol condensation, dehydration and hydration reactions, resulting in a relatively high lactic acid yield for sugar conversions.

Keywords: glucose; catalyst; lactic acid; ruthenium; calcium

1. Introduction

As the only abundant renewable carbon resource, the efficient conversion of biomass into fuel and chemicals has attracted significant attention [1,2]. Due to their abundance and wide availability, cellulose and sugars have been used in central research on biomass conversion and converted into a range of oxygen-containing compounds [3–6]. Among these attractive chemicals, lactic acid (LA) has emerged as one of the most promising due to its fast-growing market [7–9]. Currently, LA is widely used in the food and beverage sector as well as the cosmetics and pharmaceutics industries. Additionally, LA is also an important raw material for the production of bioplastics, polylactide in particular, which is regarded as one of the top ten industrially important chemicals by the United States Department of Energy [10,11].

At present, the fermentation of starch is the most common method for LA production, providing more than 90% of LA in market [12,13]. However, the bacterial fermentation method has the disadvantages of multiple treatment steps and low productivity. Alternatively, the catalytic conversion method of LA production from biomass has attracted increasing research interest during recent years thanks to its versatile feedstocks and high reaction efficiency [14]. To alleviate the condensation or decomposition reactions, the catalytic reaction is usually conducted under alcohol medium to produce esters or in water solution with metal salts to stabilize the final LA. Typically, Sn(IV)-beta or Sn(IV) grafted onto mesoporous silica support catalyzed the conversion of carbohydrates in an alcohol medium into a ca. 60–75% yield of alkyl lactate [15–19]. Alternatively, homogeneous metal salts were applied to crack sugars and cellulosic biomass into LA under hydrothermal conditions and yielded > 60% LA from the lignocellulose-based biomass [20–24]. The carbohydrates, including glucose, cellulose and starch, are feasibly transformed into LA in the presence of NaOH and/or $Ca(OH)_2$ but suffer from low LA selectivity and fierce reaction conditions [25]. Therefore, a number of works have investigated heterogeneous catalysts to promote the conversion of sugars into LA. For instance, Murzin et al. developed a Sn(salen)/octylmethyl imidazolium bromide catalyst and achieved a 58% yield of LA with 68% glucose conversion at 473 K [26]. Zhang et al. reported that the introduction of indium and tin into Beta zeolites is able to promote the transformation of glucose to LA, affording a LA yield of 53% at 443 K [27]. Nakajima et al. synthesized a new type of solid acid-base bifunctional catalyst $YNbO_4$ through a simple co-precipitation method, which obtained a 36% LA yield at 413 K for 5 h [28]. For instance, Choudhary et al. prepared a CTAB-capped Cu/MgO catalyst for the conversion of biomass-derived sugars in the presence of NaOH and obtained a 70% yield of LA at a mild temperature of 393 K [6]. Wang et al. employed the polymerizates of imidazole and epichlorohydrin ([IMEP]Cl) as catalysts to convert glucose to LA in water in the presence of 50 mM NaOH and yielded 63% LA with 99% glucose conversion at 373 K [29]. In spite of these impressive results, developing novel catalysts for the conversion of sugars to LA is still a hotly researched topic.

Ru-based catalysts have been widely used in the conversion of biomass-derived chemicals to organic acids due to the unique chemical properties of Ru/RuO_x [30,31]. Additionally, MoS_2 is a promising catalyst or stable support used in hydrothermal conditions [32,33]. Herein, a series of RuO_x/MoS_2 catalysts were synthesized for the conversion of sugars into LA with the assistance of CaO in aqueous solutions. Different catalysts and supports were first screened to yield the best RuO_x/MoS_2 catalyst. The influence of pivotal reaction parameters on the LA yield was investigated. Then, the optimized catalysts were systematically characterized to unveil the role of Ru species in the reaction and the reaction networks was proposed according to conditional experiments.

2. Results and Discussion

2.1. Catalytic Conversion of Glucose via Different Catalysts

In view of the versatile acidic and metallic properties of Ru-based catalysts in biomass conversions [30,34], a variety of 1 wt% Ru catalysts supported on different carriers were screened in glucose conversion for LA production. As shown in Table 1, under a mild condition of 2 h reaction at 120 °C under the pressure of 1.0 MPa O_2, the glucose conversion ranged from 30% to 98%, and the LA selectivity varied greatly with the carbon balance changed within a large range of 40–85%, according to the HPLC, GC-MS and HPLC-MS analysis of the products. Via the activated carbon (AC)- and TS-1-supported RuO_x catalysts (Table 1, Entries 1–2), fructose was the major product, with 30–50% selectivity and high conversions (>72%) of glucose, but LA selectivity was merely ca. 10%. This indicates that these two catalysts have notable activity in glucose isomerization but are much less active for retro-aldol reaction and C-C bond breakage [35]. On the other hand, besides the low yields of LA, the $1RuO_x$/MWW and $1RuO_x$/HPA catalysts (Table 1, Entries 3–4) exhibited distinguished selectivity (ca. 25%) for gluconic acid formation, reflecting the typical activity for catalytic oxidation usually observed via a noble metal catalyst [36,37]. With respect to the LA formation, the most remarkable performance was observed in the catalysts of $1RuO_x/TiO_2$, $1RuO_x$/BN and $1RuO_x/MoS_2$ (Table 1, Entries 5–7), which provided 30–46% LA selectivity and 65–95% glucose conversion. The carbon balance in the reaction v these RuO_x catalysts changed within the large range of 40–85%, suggesting that certain side reactions often took place. The reactant solution was clear and colorless after the reaction. The HPLC analysis showed that no 5-hydroxylmethylfurfural (a product of fructose dehydration, often serving as a precursor for polymerization to form humins) was detected in the products, indicating the formation of heavier products was not the main cause of the carbon loss. The analysis of the gas phase product showed that CO_2 was generated in the reaction. We speculated that the CO_2 production largely accounted for the loss of carbon balance in the reaction. According to the highest LA selectivity and the high thermal stability of the MoS_2-supported metal catalysts previously observed [38], the RuO_x/MoS_2 system was chosen for further investigation.

Table 1. Product distribution for glucose conversion via different Ru-based catalysts.

Entry [a]	Catalysts	Con. (%)	Sel. (%)				
			L-Lactic Acid	Glycolic Acid	Glyceric Acid	Gluconic Acid	Fructose
1	$1RuO_x/AC$	98.3	9.3	5.1	6.3	0	32.8
2	$1RuO_x$/TS-1	72.1	17.4	0	0	16.2	50.0
3	$1RuO_x$/MWW	94.5	17.0	5.6	0	25.8	6.7
4	$1RuO_x$/HPA	30.1	12.6	0	0	29.3	0
5	$1RuO_x/TiO_2$	94.6	35.7	4.5	0	2.4	30.2
6	$1RuO_x$/BN	73.6	30.4	2.1	0	0	15.5
7	$1RuO_x/MoS_2$	64.9	46.3	5.5	14.0	2.3	0

[a] Reaction conditions: 20 mL 1% glucose, 0.10 g catalyst, 120 °C, 1.0 MPa O_2, 2 h; con., sel., MWW and BN are abbreviations for conversion, selectivity, zeolite with MWW structure and boron nitride, respectively.

Alkaline environments greatly contribute to the catalytic conversion of bio-based compounds to LA [39]. Therefore, two types of frequently used base, i.e., NaOH and CaO, with different alkalinities, were added into the reaction system for glucose conversion. As shown in Table 2, Entries 1 and 2, the addition of the base obviously improved the conversion of glucose, while the selectivity to LA was only increased under the assistance of CaO. According to the reaction results and previous studies [40–42], we speculated that CaO provided alkalinity and Ca^{2+} cations for the reaction. The glucose isomerization and the followed retro-aldol reaction were catalyzed by the suitable base, forming C3 intermediates as the precursor of LA. At the same time, the soluble Ca^{2+} combined with part of the generated LA to form more stable calcium lactate to a certain extent, which promoted the forward reaction. Via the $1RuO_x/MoS_2$+CaO catalyst, the LA selectivity was improved to 50.5% with a 98.9% glucose conversion. To investigate the role of different sectors in the catalytic conversion, CaO, MoS_2 and MoS_2+CaO were evaluated in the catalytic system, respectively. As shown in Table 2, Entries 3–5, the presence of MoS_2 merely showed a remarkable selectivity to gluconic acid but not to the LA product (38.8% vs. 9.2%). On the other hand, CaO afforded 29.5% yield of LA along with 15.2% gluconic acid formation. Combining the MoS_2+CaO in the reaction simply obtained a compromised performance but without the synergistic effect on LA or gluconic acid production. In another case, Ru was loaded on CaO to detect the collaboration between them. Over the $1RuO_x$/CaO catalyst (Table 2, Entry 6), LA selectivity was very similar to that of CaO, but the fructose selectivity was increased by more than 25%, similar to that via $1RuO_x/TiO_2$ (Table 1, entry 5). Thus, from these conditional experiments, it can be learned that merely through the RuO_x/MoS_2+CaO catalyst, synergistic effects were observed, which realized a higher yield of LA and glucose conversion simultaneously. Note that the carrier CaO was partially hydrolyzed during the reaction process, which may induce the supported RuO_x species to being dissolved in aqueous solution. In contrast, the MoS_2 is very stable under hydrothermal conditions, providing a potential stable support to load Ru species [43]. Thus, the RuO_x/MoS_2 catalyst was optimized for conversions under different conditions.

The effect of Ru loading on catalytic performance was finally investigated. As shown in Table 2, Entries 7–9, the LA yield was maximized to 53.7% via the $2RuO_x/MoS_2$ catalyst. At lower Ru loadings (0.5% and 1%), the main by-product was C_6 acids, indicating the weak activity of catalysts in retro-aldol condensation reactions. In contrast, at a high Ru loading of 5%, glyceric acid and fructose was formed largely due to the hydration of dihydroxyacetone and/or glyceraldehyde and depressed retro-aldol condensation. Thus, a reasonable Ru content, i.e., 2%, should be loaded on MoS_2 to balance the retro-aldol condensation and hydration reactions for high LA yields.

Table 2. Effects of different components via catalysts on the glucose conversions.

Entry [a.]	Catalyst	Con. (%)	Sel. (%)				
			L-Lactic Acid	Glycolic Acid	Glyceric Acid	Gluconic Acid	Fructose
1 [b]	$1RuO_x/MoS_2$+ NaOH	75.5	37.9	1.9	2.1	27.7	3.1
2 [c]	$1RuO_x/MoS_2$+ CaO	98.9	50.5	7.5	6.7	6.3	0
3	CaO	100	29.5	20.0	7.1	15.2	2.7
4	MoS_2	70.1	9.2	0.6	0	38.3	0
5	MoS_2+CaO	100	25.0	16.1	10.7	31.3	4.2
6	$1RuO_x$/CaO	97.4	30.5	7.2	0	0	29.3
7	$0.5RuO_x$/CaO	80.7	50.5	6.7	9.7	15.2	0
8	$2RuO_x$/CaO	96.6	54.3	5.4	4.6	6.3	0
9	$5RuO_x$/CaO	99.2	49.3	10.6	1.1	1.7	30.2

[a] Reaction conditions: 20 mL 1% glucose, 0.10 g catalyst, 120 °C, 1.0 MPa O_2, 2 h; con. and sel. are abbreviations for conversion and selectivity, respectively. [b] 0.10 mol/L NaOH. [c] 0.10 g CaO.

For the optimized $2RuO_x$/CaO catalyst, the reaction parameters, including reaction temperature, time, CaO dosage and catalyst stability, were investigated for the conversion of glucose to LA. As shown in Figure 1a, the selectivity to LA displayed a volcano curve as a function of temperature and was maximized to 54.3% at 120 °C with the conversion of glucose maintained at 95% at temperatures between 100 °C and 160 °C. When the reaction temperature was higher than the optimal point of 120 °C, the yield of LA decreased dramatically with a concomitantly increased yield of acetic acid. The reason for this should be attributed to the random decomposition of glucose and the oxidative decarbonylation of LA due to the metastable of glucose and LA at high temperatures. The influence of reaction time is shown in Figure 1b. At the 0.5 h reaction, the glucose conversion reached 93.1% with ca. 30% LA selectivity. Prolonging the reaction time to 2 h led to the increased LA yield, mainly due to the cascade conversion of the key intermediate of fructose to the downstream product. It is worth noting that even further extending the reaction time, the product compositions remained almost unchanged, confirming the good stability of LA in the present reaction system. The reaction atmosphere was also tested, as seen in Figure S1. Although the conversion of glucose to LA did not need the participation of O_2, the presence of O_2 herein encouraged the synthesis of LA in comparison to that in the N_2 atmosphere. The enhanced performance should be related to the catalyst state under the oxidation conditions, e.g., RuO_x species.

As the key catalytic component in this catalytic system, the effect of CaO dosage was investigated and is shown in Figure 1c. As expected, LA formation is very dependent on the amount of CaO, reaching a maximum with a 0.10 g addition of CaO. Concurrently, the undesired glucose oxidation route was greatly depressed by the competition of base-catalyzed glucose isomerization and retro-aldol condensation reaction. Excessive CaO dosage played negative roles in LA production, which should be attributed to the low saturated solubility of $Ca(OH)_2$ (~0.025 M at room temperature) and uncontrollable basicity. The reusability of RuO_x/MoS_2 catalyst was evaluated, and results are shown in Figure 1d. After four cycles, the glucose conversion slightly decreased from 96% to 76% with a relatively stable LA selectivity of ca. 45%, indicating the relatively high stability of the catalyst. After the second cycle, the key intermediate of fructose emerged, which can be attributed to the coverage of the Ru species with a strongly interacting Ca species.

2.2. Active Sites for Glucose Conversions

To verify the stability of support and the optimized catalyst under oxygen-containing hydrothermal reaction conditions, TGA was carried out to evaluate the decomposition temperature of the MoS_2 and the supported $2RuO_x/MoS_2$ in air. As shown in Figure 2a, the first weight loss stage from room temperature to 54 °C corresponds to the desorption of physically adsorbed water. The second and third weight losses distributed at around 390 °C and 720 °C could be attributed to the decomposition of the surficial and complete oxidation of MoS_2, respectively [44]. Figure 2b illustrates that the introduction of the Ru species has no discernible impact on the thermal stability of the sample. Thus, at low reaction temperatures between 100 °C and 160 °C, the catalyst structure will not suffer from oxidative deterioration, which provides solid guarantee for the high stability of catalysts in the reaction.

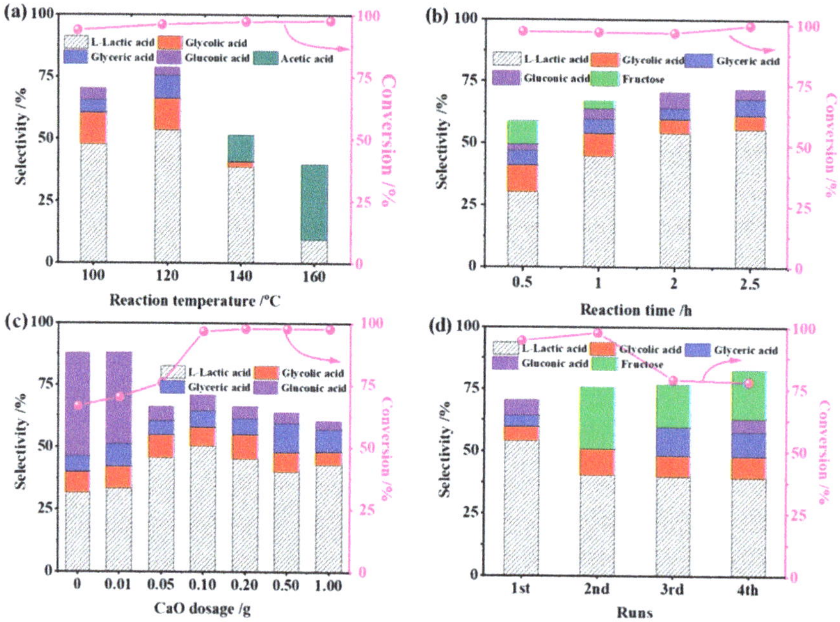

Figure 1. Catalytic conversion of glucose at different (**a**) reaction temperatures, (**b**) reaction times, (**c**) amounts of CaO and (**d**) stabilities of catalysts (typical reaction conditions: 20 mL 1% glucose, 0.10 g catalyst, 120 °C, 1.0 MPa O_2, 2 h. One condition was altered while other factors were kept the same in each case).

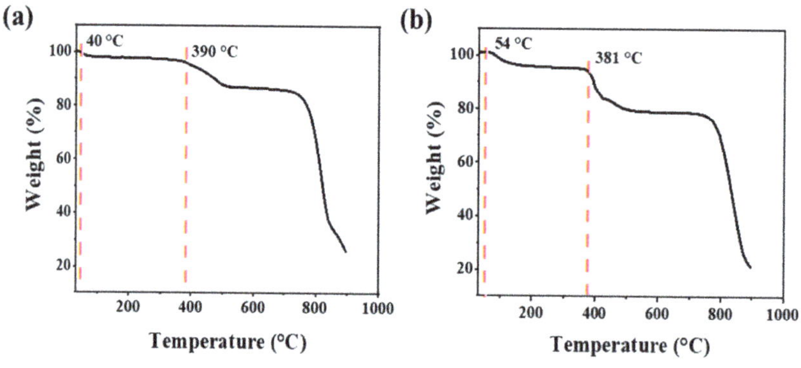

Figure 2. TGA curves of (**a**) MoS_2 and (**b**) $2RuO_x/MoS_2$ catalyst.

The XRD patterns of the fresh and used catalysts are displayed in Figure 3a. For the fresh $2RuO_x/MoS_2$ catalyst, the diffraction peaks at 14.4°, 32.2°, 39.6°, 49.8°, 58.8° (labeled with clubs), and 37.4° and 53.8° (labeled with spades) were attributed to MoS_2 and CaO, respectively. No characteristic peaks corresponding to RuO_x were detected, indicating the high dispersion of the Ru species. For the used catalyst, new peaks attributed to the $CaCO_3$

phase appeared, which may be caused by the reaction of CaO with CO_2 generated in the reaction. Raman spectra in Figure 3b further reveal the structure of catalysts. Both MoS_2 and $2RuO_x/MoS_2$ showed two characteristic peaks at 409 cm^{-1} and 383 cm^{-1}, corresponding to the E^1_{2g} and A_{1g} characteristic vibration modes of MoS_2-2H phase, respectively [45]. Similar to the XRD patterns, the RuO_x species were not detected, confirming the high dispersion of the Ru species.

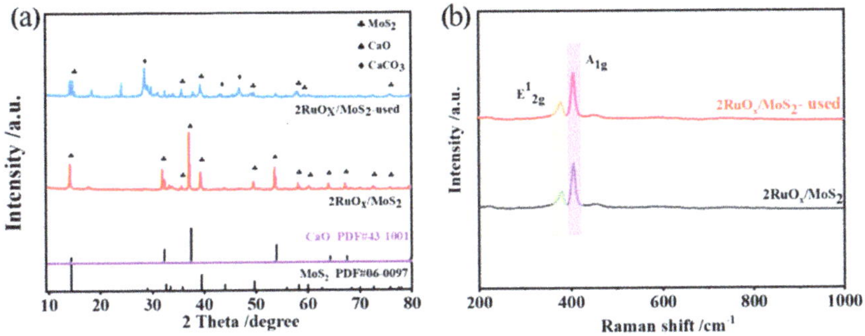

Figure 3. (**a**) XRD patterns and (**b**) Raman spectra of the fresh and used $2RuO_x/MoS_2$+CaO catalyst.

To disclose the metal dispersion and morphology of catalysts, the SEM and TEM images of the optimized catalyst are compiled in Figure 4. The carrier MoS_2 showed a multilayer smooth sheet morphology with a thickness of about 100 nm. The highly dispersed of RuO_x species are clearly shown in the high-resolution TEM images of Figure 4c,d. Even up to 2 wt% of the metal loading onto the MoS_2 with a low BET surface area of < 10 m^2/g (Figure S2), the average size of RuO_x particles was centered at 0.53 nm. Meanwhile, the elemental mapping in Figure 4e–j reveals that the Ru particles were uniformly distributed on the catalyst surface. This is highly consistent with the results obtained via the Pt/MoS_2 catalysts reported previously, owing to the uniquely strong interaction between MoS_2 and noble metals [38]. The used catalyst was also characterized and is shown in Figure S3. According to the SEM image, a large number of amorphous particles attributed to Ca species were randomly distributed on the layered MoS_2. In addition, the TEM and EDS analyses verified the remarkable structural stability of $2RuO_x/MoS_2$. The intrinsic structures of MoS_2 and RuO_x were completely maintained through the catalytic reaction, similar to that of the fresh catalyst. The ICP-OES measurements in Table 3 further highlighted the fact that Ru leaching rarely occurred under hydrothermal reaction conditions. Above all, despite Ru sites being partially covered by Ca species, the $2RuO_x/MoS_2$ catalyst was still able to produce a reasonable yield of LA in the subsequent cycling tests thanks to its excellent stability.

Table 3. Elemental analysis of catalysts and product solution by ICP-OES.

Catalysts	Ru Loading [a] /%	Mo Amount /%	Leached Amount/% [b]	
			Ru	Mo
$2RuO_x/MoS_2$	2.3	60.0	2.4×10^{-5}	9.5×10^{-4}
$2RuO_x/MoS_2$ -used	2.0	56.2	2.9×10^{-6}	5.5×10^{-6}

[a] The Ru loading of the catalysts. [b] The mass ratio of elements leached into product solution.

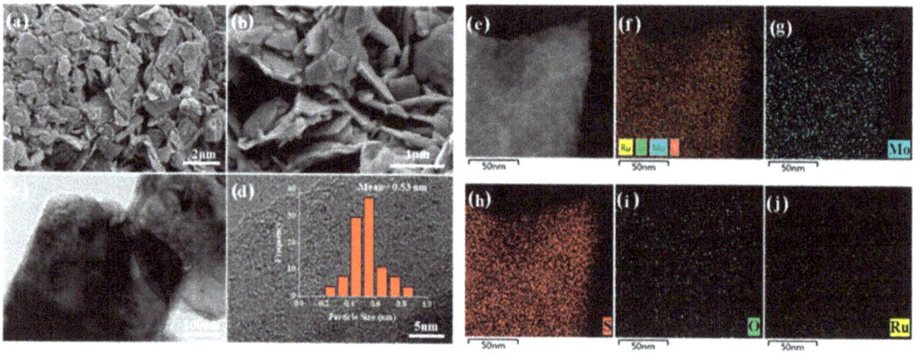

Figure 4. (**a**,**b**) SEM images, (**c**,**d**) TEM images and (**e–j**) element mapping of the $2RuO_x/MoS_2$ catalyst.

The chemical states of Mo and Ru species on the catalysts were investigated by XPS (the whole spectrum is shown in Figure S4). As shown in Figure 5a, for the catalyst of $2RuO_x/MoS_2$, three valence states of Mo species were identified, including the dominant Mo^{4+} (229.09 eV and 232.28 eV) and a small amount of Mo^{5+} (233.45 eV) and Mo^{6+} (235.56 eV) [46,47]. Since no MoO_3 related signals were observed by Raman characterizations, it was speculated that the high-valence Mo species was distributed in the bulk phase of MoS_2. Through the catalytic reaction, the proportion of Mo^{6+} increased by 8.95% after reaction, which may be caused by the partial oxidation of the carrier under the oxidizing atmosphere. For the Ru spectra in Figure 5b, the signal of Ru 3d overlaps with that of C ls. By means of deconvolution analysis, the Ru^{4+} species in the form of RuO_2 (280.93 eV and 286.08 eV) were observed [48]. After the reaction, characteristic peaks assigned to RuO_2 vanished, which should be attributed to the tendentiously deposition of $CaCO_3$ onto the metal sites of the $2RuO_x/MoS_2$ catalyst, agreeing with our hypothesis that the Ru species were covered to decrease the activity, rather than the deactivation of $2RuO_x/MoS_2$ catalysts.

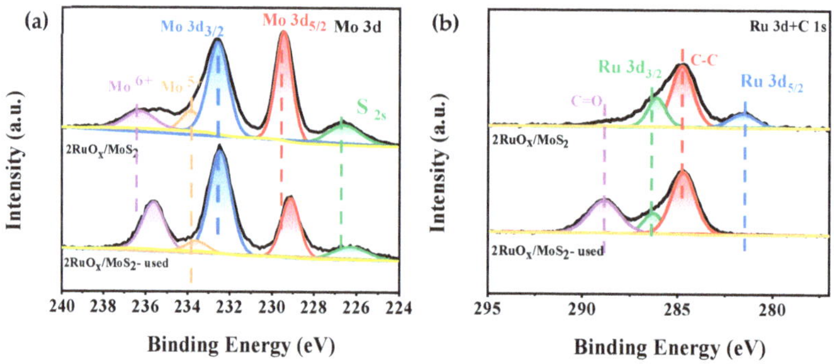

Figure 5. XPS spectra of the fresh and used $2RuO_x/MoS_2$ catalyst (**a**) Mo 3d, (**b**) Ru 3d and C1s.

2.3. Reaction Networks

Various C5 and C6 sugars, polyols and polyhydroxy acids were employed as feedstocks to evaluate the possible reaction networks via this catalytic system. As shown in Table 4, Entries 1–3, the RuO_x/MoS_2 catalyst with the assistance of CaO showed a good catalytic performance in the conversion of sugars, including glucose, fructose and xylose. As expected, because the obstacle of glucose isomerization was eliminated, the feedstock of fructose provided a higher selectivity to LA than glucose (68.9% vs. 54.3%). Interestingly, xylose exhibited the best performance for LA production, possibly due to the fewer side reactions occurring after the C5 reactant being degraded into C3 and C2 moieties. Moreover, it was difficult to effectively convert both polyols and gluconic acid into LA. It is noteworthy that the selectivity of LA reached 50.6% in the 8 h reaction at 160 °C when cellulose was used as the reactant (Table 4, Entry 7). This suggests that the RuO_x/MoS_2+ CaO catalyst has a potential to treat recalcitrant feedstock.

Table 4. Catalytic performance of RuO_x/MoS_2 catalyst in the presence of CaO with different feedstocks [a].

Entry	Reagent	Con. /%	Sel./%				
			L-Lactic Acid	Glycolic Acid	Glyceric Acid	Gluconic Acid	Fructose
1	Glucose	96.6	54.3	5.4	4.6	6.3	0
2	Fructose	97.3	68.9	6.2	5.0	0	0
3	Xylose	100	78.2	17.6	0	0	0
4	Sorbitol	51.8	35.5	27.5	0	0	0
5	Mannitol	100	9.3	8.4	0	0	0
6	Gluconic acid	58.1	12.9	3.2	4.8	0	0
7 [b]	Cellulose	20.5	50.6	4.8	2.1	0	0
8 [c]	Glucose	76.2	43.5	2.6	1.9	0	0

[a] Reaction conditions: 20 mL 1% reagent, 0.10 g catalyst, 0.10 g CaO, 120 °C, 1.0 MPa O_2, 2 h; [b] 20 mL H_2O, 0.16g reagent, 160 °C, 8 h; [c] 20 mL 5 wt% reagent, 0.50 g catalyst, 0.50 g CaO; con. and sel. are abbreviations for conversion and selectivity, respectively.

In addition, the activity of the catalyst was tested in glucose conversion at a high concentration (5 wt%). As shown in Table 4, Entry 8, the selectivity of LA was slightly decreased to 43.5%, providing LA productivity of 8.3 g/L/h, which is notably higher than that of the standard of LA production via industrial fermentation (5 g/L/h) [49].

The above results demonstrate again that the base-catalyzed and RuO_x/MoS_2-catalyzed retro-aldol condensation route together contributed to the LA formation. CaO dissolved in water to release OH^- and Ca^{2+}. OH^- catalyzed the isomerization of glucose to fructose, followed by degradation into dihydroxyacetone and glyceraldehyde through retro-aldol condensation. Afterward, glyceraldehyde was further catalytically dehydrated by the elimination reaction under the actions of OH^- to generate the pyruvaldehyde intermediate [40,42]. Finally, obtained pyruvaldehyde can form metal chelates with Ca^{2+} and transform into LA and/or lactate via the 1,2-hydride shift [41]. On the other hand, the Ru species on the RuO_x/MoS_2 in an oxidation state played the crucial role in promoting retro-aldol condensation reactions.

Finally, the reaction pathway is briefly proposed in Scheme 1. Under the mild conditions of 120 °C and 1MPa O_2, the CaO and Ru species catalyzed the isomerization of glucose into fructose. Afterward, the retro-aldol condensation occurred via the RuO_x species with dihydroxyacetone and glyceraldehyde production. Finally, the obtained glyceraldehyde was transformed into LA via dehydration and hydration reactions via the Ru species. Simultaneously, side reactions, such as glucose direct oxidation and the direct conversion of dihydroxyacetone, occurred, resulting in the by-products of gluconic acid, glyceric acid, glycolic acid and acetic acid.

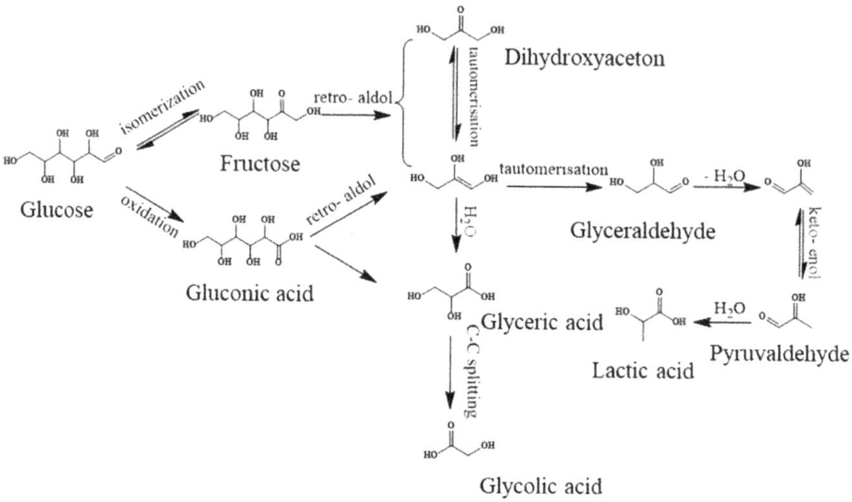

Scheme 1. Proposed reaction pathways for the catalytic conversion of glucose to lactic acid via the RuO_x/MoS_2 catalyst in the presence of CaO [42,49].

3. Materials and Methods

3.1. Materials

All reagents were purchased commercially without further purification. Glucose (99%) and calcium oxide were purchased from Tianjin Damao Chemical Reagent Co., Ltd. (Tianjin, China). Sodium hydroxide was purchased from Tianjin Yongda Chemical Reagent Co. Ltd. Activated carbon (AC) was purchased from Norit Corporation. Calcium carbonate was purchased from Xilong Science Co., Ltd. (Shantou, China). Standard sample of lactic acid (LA) with 95% purity was purchased from Sigma-Aldrich Co., Ltd. (St. Louis, MO, USA). Ruthenium chloride hydrate, molybdenum disulfide, hydroxyapatite (HPA), titanium dioxide, titanium silicalite-1 (TS-1), Boron nitride (BN) and all other chemicals were analytical reagents and purchased from Shanghai Aladdin Biochemical Technology Co., Ltd. (Shanghai, China).

3.2. Catalysts Preparation

RuO_x/support catalysts were prepared using the incipient wetness impregnation method with MoS_2, AC, TiO_2, TS-1, HPA and BN as supports. Typically, 2 g of support was mixed with 0.26–2.6mL aqueous solutions of $RuCl_3 \cdot 3H_2O$ (3.792% g/mL) and maintained at 25 °C for 12 h. Then, the sample was dried at 60 °C overnight and calcined at 350 °C for 4 h in air. The theoretical contents of Ru were controlled to remain at 0.5–5 wt%. The cost of preparing 1 g of $1RuO_x/MoS_2$ catalyst was approximately CNY 1.1 (USD 0.16).

3.3. Catalysts Characterizations

X-ray powder diffraction (XRD, PANalytical B. V., Almelo, The Netherlands) patterns of catalysts were collected on a PANalytical X'pert Pro-1 equipped with Cu-Kα monochromatic radiation (0.1541 nm), operating at 40 kV and 40 mA. A continuous mode was used in the 2θ range from 10° to 80° at a scanning speed of 5°/min.

Thermogravimetric (TG, TA Instruments, Newcastle, DE, USA) analysis was conducted with SDT Q600 equipment under an air flow of 100 mL/min from 30 to 900 °C at a heating rate of 10 °C/min.

Scanning electron microscopy (SEM, HITACHI, Tokyo, Japan) images were obtained by using a Carl Zeiss Merlin Compact scanning electron microscope.

High-angle annular dark-field scanning-transmission electron microscopy (HAADF-STEM, JEOL, Tokyo, Japan) images and energy-dispersive X-ray spectroscopy (EDS, JEOL, Tokyo, Japan) elemental mapping were obtained by using a JEM-2100F microscope, operated at a voltage of 200 kV.

The contents of Ru and Mo in the synthesized catalysts and the leaching amount in the reaction solution were determined by an inductively coupled plasma optical emission spectrometer (ICP-OES, PerkinElmer, Waltham, MA, USA) on a Perkin Elmer ICP-OES 7300DV.

X-ray photoelectron spectra (XPS Thermo Fisher Scientific, Waltham, MA, USA) were recorded on a Thermo Scientific K-Alpha+ spectrometer with a monochromatic Al-Kα X-ray source as the excitation source. The binding energies were corrected by using the C1s peak at 284.4 eV as a reference.

3.4. Catalytic Conversion of Glucose

The catalytic conversion of glucose was carried out in a high-pressure batch reactor. Typically, 20 mL of 1% glucose aqueous solutions, 0.10 g of catalyst and 0.10 g of CaO were placed into the reactor. Then, the reactor was charged in O_2 five times to remove the air. Finally, the oxygen pressure in the reactor was increased to 1.0 MPa and sealed for the reaction at 120 °C for 2 h.

After reaction, the product was cooled to room temperature and analyzed with a high-performance liquid chromatograph HPLC, Agilent 1200, (Agilent Technologies, Palo Alto, CA, USA) equipped with a Rezex ROA-Orgenic Acid H^+ column and a differential refractive index detector with 5 mM sulfuric acid aqueous as the mobile phase. In detail, the GC-MS (Agilent Technologies, Palo Alto, CA, USA) and HPLC-MS methods (Agilent Technologies, Palo Alto, CA, USA)were used to identify the products in reaction solutions or identify the products after methyl esterification treatment [31]. Then, the standard samples were used to further confirm and quantitatively analyze the products on the HPLC instrument.

Glucose conversion and product selectivity were calculated using following equations:

$$\text{Conversion } (\%) = (1 - \frac{\text{moles of unconverted glucose}}{\text{moles of initial glucose}}) \times 100\% \tag{1}$$

$$\text{Selectivity } (\%) = \frac{\text{moles of product}}{\text{theoretical moles of product}} \times 100\% \tag{2}$$

4. Conclusions

A series of RuO_x/MoS_2 catalysts were synthesized for the conversion of sugars into LA with the assistance of CaO. Unlike the conventional base-catalyst system, the $2RuO_x/MoS_2$ catalyst exhibited enhanced catalytic activity and stability. Under the low reaction temperature of 120 °C with low alkalinity, the LA yield reached 54.3% at a 96.6% glucose conversion, even in the recycle runs. The RuO_x species promoted the isomerization of glucose to fructose, and then the fructose was cracked into the C3 compound of glyceraldehyde via the RuO_x species via the retro-aldol condensation reaction. Finally, the C3 intermediate was further converted into LA via dehydration and hydration reactions. Although the finally yield of LA still needs to be improved, this work still provides valuable information for the development of the novel heterogeneous catalyst for carbohydrates conversions. The future study may focus on how to further enhance the LA selectivity and the reaction efficiency, particularly in the cellulosic feedstock conversion, which involves the catalyst development as well as a kinetic reaction investigation.

Supplementary Materials: The following are available online at https://www.mdpi.com/article/10.3390/catal13030545/s1: Figure S1. Results of catalytic conversion of glucose at different reaction atmospheres (reaction conditions: 20 mL 1% glucose, 0.10 g catalyst, 0.10 g CaO. 120 °C, 1.0 MPa O_2, 2 h); Figure S2. Nitrogen adsorption–desorption isotherms BET surface area of $2RuO_x/MoS_2$ catalyst;

Figure S3. TEM and EDS images of the used $2RuO_x/MoS_2$; Figure S4. The survey spectra of fresh and used $2RuO_x/MoS_2$ catalysts.

Author Contributions: Data curation, Z.L.; formal analysis Z.L., X.L., P.W. and J.P.; supervision, S.Z. and M.Z.; writing—original draft, Z.L.; writing—review and editing, P.W., J.P. and M.Z. All authors have read and agreed to the published version of the manuscript.

Funding: This work was supported by the National Natural Science Foundation of China (21721004, 21676039, 22108274), "Transformational Technologies for Clean Energy and Demonstration", Strategic Priority Research Program of the Chinese Academy of Sciences (XDA 21060200).

Data Availability Statement: The data is included in the article or Supplementary Materials.

Conflicts of Interest: The authors declare no conflict of interest.

References

1. Nwosu, U.; Wang, A.; Palma, B.; Zhao, H.; Khan, M.A.; Kibria, M.; Hu, J. Selective biomass photoreforming for valuable chemicals and fuels: A critical review. *Renew. Sustain. Energy Rev.* **2021**, *148*, 111266. [CrossRef]
2. Li, C.; Zhao, X.; Wang, A.; Huber, G.W.; Zhang, T. Catalytic Transformation of Lignin for the Production of Chemicals and Fuels. *Chem. Rev.* **2015**, *115*, 11559–11624. [CrossRef] [PubMed]
3. Artz, J.; Palkovits, R. Cellulose-based platform chemical: The path to application. *Curr. Opin. Green Sustain. Chem.* **2018**, *14*, 14–18. [CrossRef]
4. Gao, X.; Zhong, H.; Yao, G.; Guo, W.; Jin, F. Hydrothermal conversion of glucose into organic acids with bentonite as a solid-base catalyst. *Catal. Today* **2016**, *274*, 49–54. [CrossRef]
5. Wang, F.-F.; Liu, J.; Li, H.; Liu, C.-L.; Yang, R.-Z.; Dong, W.-S. Conversion of cellulose to lactic acid catalyzed by erbium-exchanged montmorillonite K10. *Green. Chem.* **2015**, *17*, 2455–2463. [CrossRef]
6. Choudhary, H.; Nishimura, S.; Ebitani, K. Synthesis of high-value organic acids from sugars promoted by hydrothermally loaded Cu oxide species on magnesia. *Appl. Catal. B* **2015**, *162*, 1–10. [CrossRef]
7. Mäki-Arvela, P.; Simakova, I.L.; Salmi, T.; Murzin, D.Y. Production of Lactic Acid/Lactates from Biomass and Their Catalytic Transformations to Commodities. *Chem. Rev.* **2014**, *114*, 1909–1971. [CrossRef] [PubMed]
8. Chenebault, C.; Moscoviz, R.; Trably, E.; Escudié, R.; Percheron, B. Lactic acid production from food waste using a microbial consortium: Focus on key parameters for process upscaling and fermentation residues valorization. *Bioresour. Technol.* **2022**, *354*, 127230. [CrossRef]
9. Nagarajan, D.; Chen, C.-Y.; Ariyadasa, T.U.; Lee, D.-J.; Chang, J.-S. Macroalgal biomass as a potential resource for lactic acid fermentation. *Chemosphere* **2022**, *309*, 136694. [CrossRef]
10. Becker, J.; Lange, A.; Fabarius, J.; Wittmann, C. Top value platform chemicals:Bio-based production of organic acids. *Curr. Opin. Biotechnol.* **2015**, *36*, 168–175. [CrossRef] [PubMed]
11. Bozell, J.J.; Petersen, G.R. Technology development for the production of biobased products from biorefinery carbohydrates—The US Department of Energy's "Top 10" revisited. *Green Chem.* **2010**, *12*, 539–554. [CrossRef]
12. Wang, Y.; Cao, W.; Luo, J.; Qi, B.; Wan, Y. One step open fermentation for lactic acid production from inedible starchy biomass by thermophilic Bacillus coagulans IPE22. *Bioresour. Technol.* **2019**, *272*, 398–406. [CrossRef]
13. Trakarnpaiboon, S.; Srisuk, N.; Piyachomkwan, K.; Yang, S.-T.; Kitpreechavanich, V. l-Lactic acid production from liquefied cassava starch by thermotolerant Rhizopus microsporus: Characterization and optimization. *Process Biochem.* **2017**, *63*, 26–34. [CrossRef]
14. Li, S.; Deng, W.; Li, Y.; Zhang, Q.; Wang, Y. Catalytic conversion of cellulose-based biomass and glycerol to lactic acid. *J. Energy Chem.* **2019**, *32*, 138–151. [CrossRef]
15. Holm, M.S.; Saravanamurugan, S.; Taarning, E. Conversion of Sugars to Lactic Acid Derivatives Using Heterogeneous Zeotype Catalysts. *Science* **2010**, *328*, 602–605. [CrossRef] [PubMed]
16. de Clippel, F.; Dusselier, M.; Van Rompaey, R.; Vanelderen, P.; Dijkmans, J.; Makshina, E.; Giebeler, L.; Oswald, S.; Baron, G.V.; Denayer, J.F.M.; et al. Fast and Selective Sugar Conversion to Alkyl Lactate and Lactic Acid with Bifunctional Carbon–Silica Catalysts. *J. Am. Chem. Soc.* **2012**, *134*, 10089–10101. [CrossRef] [PubMed]
17. Guo, Q.; Fan, F.; Pidko, E.A.; van der Graaff, W.N.P.; Feng, Z.; Li, C.; Hensen, E.J.M. Highly Active and Recyclable Sn-MWW Zeolite Catalyst for Sugar Conversion to Methyl Lactate and Lactic Acid. *ChemSusChem* **2013**, *6*, 1352–1356. [CrossRef]
18. Pang, J.; Zheng, M.; Li, X.; Song, L.; Sun, R.; Sebastian, J.; Wang, A.; Wang, J.; Wang, X.; Zhang, T. Catalytic conversion of carbohydrates to methyl lactate using isolated tin sites in SBA-15. *ChemistrySelect* **2017**, *2*, 309–314. [CrossRef]
19. Sun, P.; Liu, C.; Wang, H.; Liao, Y.; Li, X.; Liu, Q.; Sels, B.F.; Wang, C. Rational Positioning of Metal Ions to Stabilize Open Tin Sites in Beta Zeolite for Catalytic Conversion of Sugars. *Angew. Chem. Int. Ed.* **2023**, *62*, e202215737. [CrossRef]
20. Kong, L.; Li, G.; Wang, H.; He, W.; Ling, F. Hydrothermal catalytic conversion of biomass for lactic acid production. *J. Chem. Technol. Biot.* **2008**, *83*, 383–388. [CrossRef]
21. Wang, Y.; Deng, W.; Wang, B.; Zhang, Q.; Wan, X.; Tang, Z.; Wang, Y.; Zhu, C.; Cao, Z.; Wang, G.; et al. Chemical synthesis of lactic acid from cellulose catalysed by lead(II) ions in water. *Nat. Commun.* **2013**, *4*, 2141. [CrossRef] [PubMed]

22. Bayu, A.; Yoshida, A.; Karnjanakom, S.; Kusakabe, K.; Hao, X.; Prakoso, T.; Abudula, A.; Guan, G. Catalytic conversion of biomass derivatives to lactic acid with increased selectivity in an aqueous tin(ii) chloride/choline chloride system. *Green Chem.* **2018**, *20*, 4112–4119. [CrossRef]
23. Xu, S.; Wu, Y.; Li, J.; He, T.; Xiao, Y.; Zhou, C.; Hu, C. Directing the Simultaneous Conversion of Hemicellulose and Cellulose in Raw Biomass to Lactic Acid. *ACS Sustain. Chem. Eng.* **2020**, *8*, 4244–4255. [CrossRef]
24. Kutrakul, N.; Liu, A.; Ratchahat, S.; Posoknistakul, P.; Laosiripojana, N.; Wu, K.C.W.; Sakdaronnarong, C. Highly selective catalytic conversion of raw sugar and sugarcane bagasse to lactic acid over $YbCl_3$, $ErCl_3$, and $CeCl_3$ Lewis acid catalysts without alkaline in a hot-compressed water reaction system. *Chem. Eng. Res. Des.* **2022**, *187*, 549–569. [CrossRef]
25. Yan, X.; Jin, F.; Tohji, K.; Kishita, A.; Enomoto, H. Hydrothermal conversion of carbohydrate biomass to lactic acid. *AIChE J.* **2010**, *56*, 2727–2733. [CrossRef]
26. Mäki-Arvela, P.; Aho, A.; Murzin, D.Y. Heterogeneous Catalytic Synthesis of Methyl Lactate and Lactic Acid from Sugars and Their Derivatives. *ChemSusChem* **2020**, *13*, 4833–4855. [CrossRef]
27. Xia, M.; Dong, W.; Shen, Z.; Xiao, S.; Chen, W.; Gu, M.; Zhang, Y. Efficient production of lactic acid from biomass-derived carbohydrates under synergistic effects of indium and tin in In–Sn-Beta zeolites. *Sustain. Energy Fuels* **2020**, *4*, 5327–5338. [CrossRef]
28. Kim, M.; Ronchetti, S.; Onida, B.; Ichikuni, N.; Fukuoka, A.; Kato, H.; Nakajima, K. Lewis Acid and Base Catalysis of $YNbO_4$ toward Aqueous-Phase Conversion of Hexose and Triose Sugars to Lactic Acid in Water. *ChemCatChem* **2020**, *12*, 350–359. [CrossRef]
29. Wang, X.; Song, Y.; Huang, C.; Liang, F.; Chen, B. Lactic acid production from glucose over polymer catalysts in aqueous alkaline solution under mild conditions. *Green Chem.* **2014**, *16*, 4234–4240. [CrossRef]
30. Liang, G.; Wang, A.; Li, L.; Xu, G.; Yan, N.; Zhang, T. Production of Primary Amines by Reductive Amination of Biomass-Derived Aldehydes/Ketones. *Angew. Chem. Int. Ed.* **2017**, *56*, 3050–3054. [CrossRef]
31. Song, L.; Wang, R.; Che, L.; Jiang, Y.; Zhou, M.; Zhao, Y.; Pang, J.; Jiang, M.; Zhou, G.; Zheng, M.; et al. Catalytic Aerobic Oxidation of Lignocellulose-Derived Levulinic Acid in Aqueous Solution: A Novel Route to Synthesize Dicarboxylic Acids for Bio-Based Polymers. *ACS Catal.* **2021**, *11*, 11588–11596. [CrossRef]
32. Diao, X.; Ji, N.; Li, X.; Rong, Y.; Zhao, Y.; Lu, X.; Song, C.; Liu, C.; Chen, G.; Ma, L.; et al. Fabricating high temperature stable Mo-Co_9S_8/Al2O3 catalyst for selective hydrodeoxygenation of lignin to arenes. *Appl. Catal. B* **2022**, *305*, 121067. [CrossRef]
33. Anjum, M.A.R.; Jeong, H.Y.; Lee, M.H.; Shin, H.S.; Lee, J.S. Efficient Hydrogen Evolution Reaction Catalysis in Alkaline Media by All-in-One MoS_2 with Multifunctional Active Sites. *Adv. Mater.* **2018**, *30*, 1707105. [CrossRef]
34. Kobayashi, H.; Matsuhashi, H.; Komanoya, T.; Hara, K.; Fukuoka, A. Transfer hydrogenation of cellulose to sugar alcohols over supported ruthenium catalysts. *Chem. Commun.* **2011**, *47*, 2366–2368. [CrossRef]
35. Zheng, M.; Pang, J.; Sun, R.; Wang, A.; Zhang, T. Selectivity control for cellulose to diols:Dancing on eggs. *ACS Catal.* **2017**, *7*, 1939–1954. [CrossRef]
36. Wan, Y.; Zhang, L.; Chen, Y.; Lin, J.; Hu, W.; Wang, S.; Lin, J.; Wan, S.; Wang, Y. One-pot synthesis of gluconic acid from biomass-derived levoglucosan using a Au/Cs2.5H0.5PW12O40 catalyst. *Green Chem.* **2019**, *21*, 6318–6325. [CrossRef]
37. Zhang, Q.; Wan, Z.; Yu, I.K.M.; Tsang, D.C.W. Sustainable production of high-value gluconic acid and glucaric acid through oxidation of biomass-derived glucose: A critical review. *J. Clean. Prod.* **2021**, *312*, 127745. [CrossRef]
38. Wang, J.; Fang, W.; Hu, Y.; Zhang, Y.; Dang, J.; Wu, Y.; Chen, B.; Zhao, H.; Li, Z. Single atom Ru doping 2H-MoS_2 as highly efficient hydrogen evolution reaction electrocatalyst in a wide pH range. *Appl. Catal. B Environ.* **2021**, *298*, 120490. [CrossRef]
39. Yan, X.; Jin, F.; Tohji, K.; Moriya, T.; Enomoto, H. Production of lactic acid from glucose by alkaline hydrothermal reaction. *J. Mater. Sci.* **2007**, *42*, 9995–9999. [CrossRef]
40. Wang, Z.; Mo, C.; Xu, S.; Chen, S.; Deng, T.; Zhu, W.; Wang, H. $Ca(OH)_2$ induced a controlled-release catalytic system for the efficient conversion of high-concentration glucose to lactic acid. *Mol. Catal.* **2021**, *502*, 111406. [CrossRef]
41. Lux, S.; Siebenhofer, M. Synthesis of lactic acid from dihydroxyacetone:Use of alkaline-earth metal hydroxides. *Catal. Sci. Technol.* **2013**, *3*, 1380–1385. [CrossRef]
42. Esposito, D.; Antonietti, M. Chemical Conversion of Sugars to Lactic Acid by Alkaline Hydrothermal Processes. *ChemSusChem* **2013**, *6*, 989–992. [CrossRef]
43. Li, X.; Han, S.; Qiao, Z.; Zeng, X.; Cao, D.; Chen, J. Ru monolayer island doped MoS2 catalysts for efficient hydrogen evolution reaction. *Chem. Eng. J.* **2023**, *453*, 139803. [CrossRef]
44. Li, X.; Pang, J.; Luo, W.; Zhao, Y.; Pan, X.; Zheng, M. Catalytic Conversion of Tetrahydrofurfuryl Alcohol over Stable Pt/MoS_2 Catalysts. *Catal. Lett.* **2021**, *151*, 2734–2747. [CrossRef]
45. Li, H.; Zhang, Q.; Yap, C.C.R.; Tay, B.K.; Edwin, T.H.T.; Olivier, A.; Baillargeat, D. From Bulk to Monolayer MoS_2: Evolution of Raman Scattering. *Adv. Funct. Mater.* **2012**, *22*, 1385–1390. [CrossRef]
46. Song, W.; Lai, W.; Lian, Y.; Jiang, X.; Yang, W. Sulfated ZrO_2 supported CoMo sulfide catalyst by surface exsolution for enhanced hydrodeoxygenation of lignin-derived ethers to aromatics. *Fuel* **2020**, *263*, 116705. [CrossRef]
47. Wang, G.; Zhang, G.; Ke, X.; Chen, X.; Chen, X.; Wang, Y.; Huang, G.; Dong, J.; Chu, S.; Sui, M. Direct Synthesis of Stable 1T-MoS2 Doped with Ni Single Atoms for Water Splitting in Alkaline Media. *Small* **2022**, *18*, e2107238. [CrossRef] [PubMed]

48. Zhu, F.; Xue, J.; Zeng, L.; Shang, J.; Lu, S.; Cao, X.; Abrahams, B.F.; Gu, H.; Lang, J. One-pot pyrolysis synthesis of highly active Ru/RuOX nanoclusters for water splitting. *Nano Res.* **2022**, *15*, 1020–1026. [CrossRef]
49. Jin, F.; Zhou, Z.; Enomoto, H.; Moriya, T.; Higashijima, H. Conversion Mechanism of Cellulosic Biomass to Lactic Acid in Subcritical Water and Acid–base Catalytic Effect of Subcritical Water. *Chem. Lett.* **2004**, *33*, 126–127. [CrossRef]

catalysts

MDPI

Communication

Synthesis of a Series of Methyl Benzoates through Esterification with a Zr/Ti Solid Acid Catalyst

Xiaofeng Yu, Chunjie Shi, Yueling Cheng, Yejing Zhu, Renyuan Song * and Shengfei Hu

School of Materials and Chemical Engineering, Bengbu University, Bengbu 233000, China
* Correspondence: sry@bbc.edu.cn

Abstract: Methyl benzoate (MB) compounds are prepared by reacting various benzoic acids with methanol using an acidic catalyst. In this study, the solid acids of zirconium metal solids fixed with various substances were studied. We determined that zirconium metal catalysts with fixed Ti had the best activity. The catalytic synthesis of a series of MB compounds using titanium zirconium solid acids was studied. The direct condensation of benzoic acid and methanol using a metallic Lewis acid without other auxiliary Bronsted acids is reported for the first time.

Keywords: methyl benzoate; solid acid; catalysis

1. Introduction

Methyl benzoate (MB) compounds are an important class of chemical products. Most of these compounds have low toxicity and are mainly used in scents and as solvents [1–14]. For example, MB has a fruity flavor that can be used in pineapple, strawberry, and cherry syrups, and it as acts as a solvent in resin rubber. Additionally, *p*-methoxybenzoic acid is also used as a flavoring, and methyl p-bromobenzoate is the main raw material of pemetrexed disodium, an antitumor drug, and diglitin A [15], an antifungal compound [16–18].

The synthesis of MB is catalyzed by various inorganic and organic acids, such as sulfuric, phosphoric, and *p*-toluenesulfonic acid [19–25]. However, these acids are not recoverable, and a large amount of water is required to clean them; therefore, a large amount of wastewater is produced during this process, polluting the environment. Sulfuric acid is often used in industry because of its low price; however, the yield of sulfuric acid is not high because of its oxidation. Solid acids are now being developed for use in ester synthesis reactions as an alternative to common acids [6,26].

Solid acids are both strongly acidic and insoluble in organic solvents [26,27]. Therefore, solid acids are recoverable and can be used multiple times. Herein, we present our synthesis of a series of reusable supported solid acids for use in the synthesis of MB compounds. Beena Tyagi's research group reported the use of zirconate sulfate as a catalyst for methyl palmitate [28,29]. A zirconia-based phyto-nanocatalyst was used to catalyze novel non-edible Bischofia javanica seed oil into methyl ester [30]. In this study, zirconium with various carriers was used as a catalyst in the synthesis of methyl benzoates. Pure zirconium compounds are solids, such as zirconia, which can be used in the form of solid spheres [30]. The addition of a carrier leads to zirconium forming a cage structure more conducive to catalysis.

Zirconium solid acid was synthesized using catalysts with different supports, and the catalyst with the highest activity was selected. This catalyst catalyzed the esterification of methanol and benzoic acid with different functional groups. The catalytic formation of methyl benzoate using Lewis Al_2O_3 has been reported, requiring the auxiliary catalyst $MeSO_3H$ [31]. The catalytic generation of methyl benzoate with Lewis $Fe_2(SO_4)_3 \cdot xH_2O$ has been reported. The auxiliary catalyst concentrated H_2SO_4 is needed [32]. The use of metal Lewis as a catalyst for the synthesis of MB without adding other acids has not been reported.

2. Results

2.1. Characteristics of the Catalyst

Zirconium solid acid and its carriers were synthesized in a one-to-one atomic molar ratio. These zirconium solid acids were synthesized from various raw materials (Table 1). In order for these atoms to mix well, the raw materials used must be completely liquid. The ingredients chosen are either inherently liquid or very soluble in water. The selection of raw materials is shown in Table 1. The solid form of the raw material was mixed into a certain concentration of water solution before use.

Table 1. Various raw materials for zirconium solid acids.

No.	Raw Material/g		Raw Material/g	
1	$ZrOCl_2·8H_2O$	10 g	$(C_4H_9O)_4Ti$	10.56 g
2	$ZrOCl_2·8H_2O$	10 g	$CdCl_2$	5.69 g
3	$ZrOCl_2·8H_2O$	10 g	$FeCl_3$	5.03 g
4	$ZrOCl_2·8H_2O$	10 g	-	-
5	$ZrOCl_2·8H_2O$	10 g	$AlCl_3$	4.14 g
6	$ZrOCl_2·8H_2O$	10 g	$(C_2H_5O)_4Si$	6.46 g
7	$ZrOCl_2·8H_2O$	10 g	$ZnCl_2$	4.23 g

2.1.1. XRD

Figure 1 shows the X-ray diffraction (XRD) patterns of the solid acids (Table 1). The XRD curves indicate that the catalysts had a large number of amorphous phases and show obvious amorphous diffraction peaks (steamed bun peaks) in the range of 2θ = 10°–40°. Zinc oxide [33] and titanium oxide [34] have been reported to have good crystal forms. The prepared catalysts had poor crystal forms.

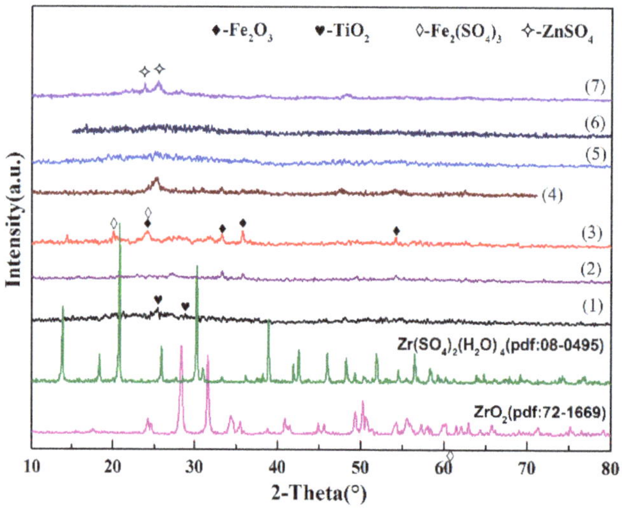

Figure 1. XRD curves of the catalysts in Table 1.

2.1.2. FT-IR

Above is the infrared spectrum of Zr/Ti (entry 10, Table 2, abbreviated as ZT10). The infrared multiplet absorption peak at 1040–1180 cm^{-1} (Figure 2) was attributed to the S=O double bond. It was observed that the $[SO_4]^{2-}$ was bonded to the catalyst. The peak at 1400 cm^{-1} was attributed to the vibrational absorption of the covalent S=O double bond, which is the characteristic absorption of a solid superacid. This finding indicates the successful synthesis of the solid superacid. The other absorption peaks belonging to the solid catalysts are not detailed.

Table 2. Synthesis of methyl *p*-toluenate and methyl *p*-chlorobenzoate using different catalysts [a,b].

Entry	Zr:C [c]	Molar Ratio	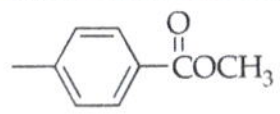	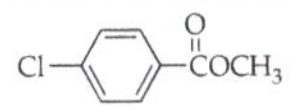
1	Zr:Ti	1:1	55.9%	52.7%
2	Zr:Cd	1:1	49.9%	35.8%
3	Zr:Fe	1:1	50.4%	50.4%
4	Zr	-	10.0%	7.2%
5	Zr:Al	1:1	9.0%	14.6%
6	Zr:Si	1:1	19.9%	21.5%
7	Zr:Zn	1:1	51.2%	50.1%
8	Zr:Ti	1:0.5	50.4%	50.3%
9	Zr:Ti	0.5:1	43.1%	45.5%
10	Zr:Ti	1.5:1	66.1%	68.1%
11	Zr:Ti	1.2:1	60.3%	62.1%
12	Zr:Ti	2:1	40.5%	45.2%

[a] Reaction condition: acid (2.0 mmol), CH_3OH (15 mL), Cat. (10 wt%), Oil bath (120 °C), 6 h. [b] Isolated yields.
[c] C = Carrier atom.

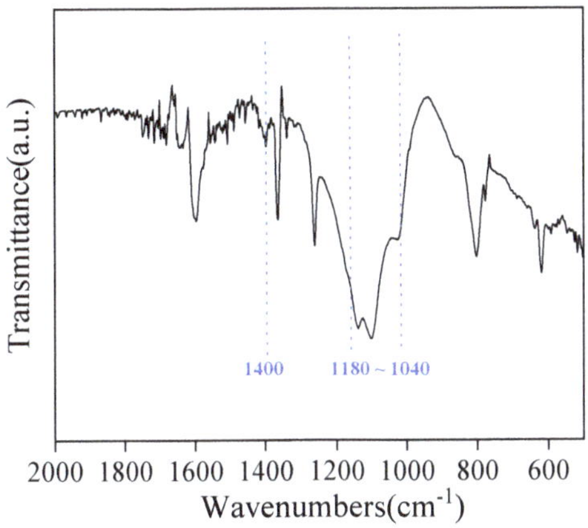

Figure 2. FT−IR spectra of ZT10.

2.1.3. SEM

The following are the SEM pictures. Diagram A shows a 1 μm electron microscopy of ZT10. B shows a 5 μm electron microscopic image of ZT10. C presents an electron microscopy of Zr (entry 4, Table 2, abbreviated as Zr0) at 1 μm (Figure 3). D shows a 5 μm electron microscope image of Zr0. Titanium metal was used as a carrier to form a honeycomb solid of zirconium metal. In the catalytic process, methyl alcohol and benzoic acid can enter the honeycomb to react, effectively improving the catalytic activity.

2.1.4. EDS

It can be seen from Figure 4 that the surface of ZT10 contained C, Ti, O, Zr, S, etc. The content of Zr was particularly high, which indicates that zirconium plays a major role in the synthesized catalyst. Sulfur, oxygen, and metals were partly in the form of sulfate.

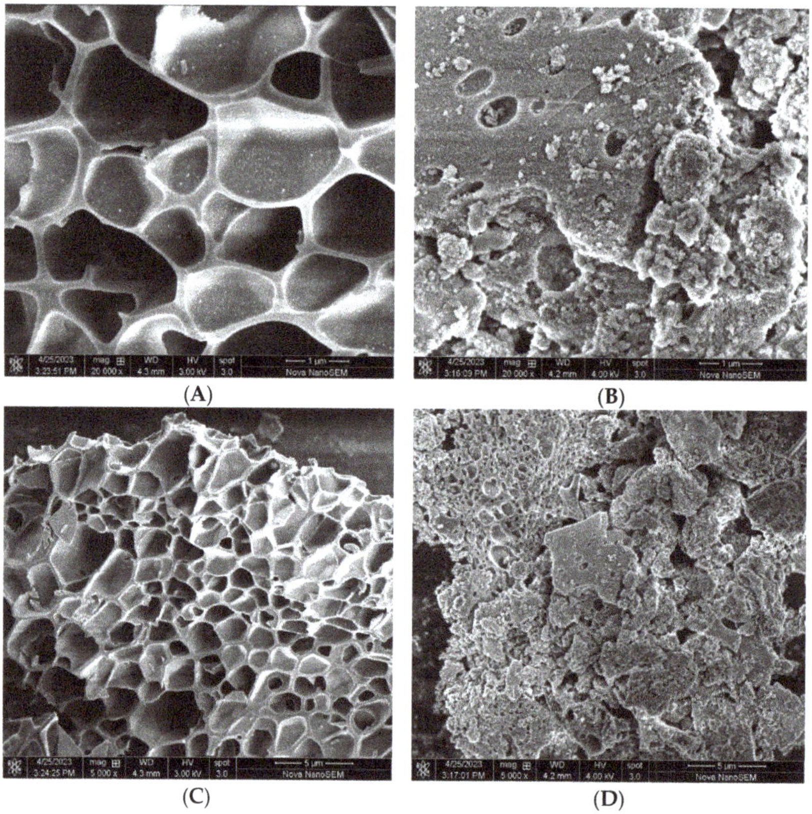

Figure 3. SEM images: (**A**) (ZT10, 1 μm), (**B**) (Zr0, 1 μm), (**C**) (ZT10, 5 μm), (**D**) (Zr0, 5 μm).

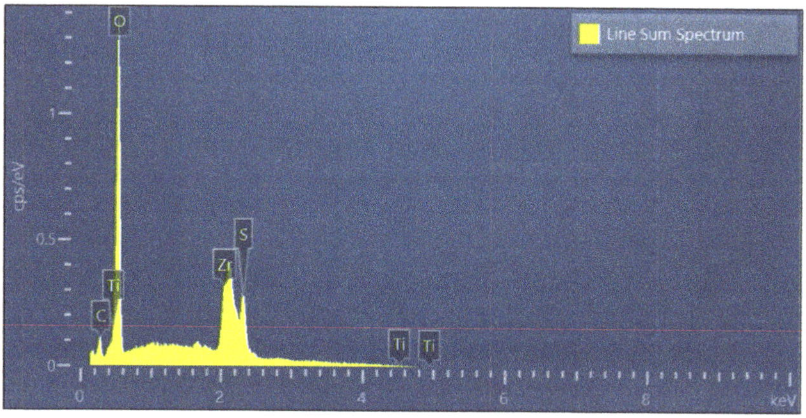

Figure 4. EDS image of ZT10.

Figure 5 below shows the TGA of catalyst ZT10. The results show that there was a significant weightlessness at 84.32 °C. This was due to the evaporation of free water on the surface of the sample. The main weight loss temperature was 735.28 °C. At this temperature C, S, and O atoms were recombined to form small molecules such as CO_2, CO, SO_2, etc. The evaporation of these small molecules was the reason for the mass loss.

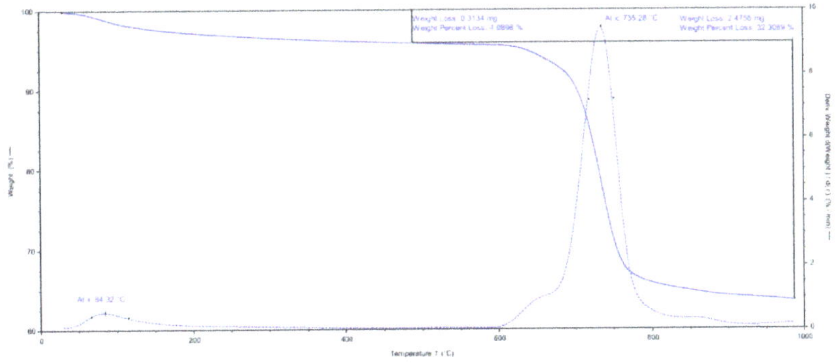

Figure 5. TGA of ZT10.

2.2. *Comparison of Catalyst Activity*

These solid acids catalyzed reactions of methanol with *p*-chlorobenzoic acid and *p*-methylbenzoic acid. Titanium, as a supporting compound, exhibited the highest catalytic activity among the solid acids. The catalytic activities of zirconium and titanium mixtures with different atomic molar ratios were observed. The zirconium-titanium atomic molar ratio of 1.2 exhibited the highest catalytic activity. Zirconium alone was found to have a poor catalytic effect. The honeycomb structure of ZT10 results in higher catalytic activity. Zinc-zif-7 (CAS:909531-29-9) with a three-dimensional network structure was used to catalyze the reaction between benzoic acid and methyl ester (under the same conditions as in Table 2) with a yield of only 34.1%. This may be due to the fact that the central metal zinc atom is less acidic than zirconium.

The mechanism of esterification may be ZT10 binding with carboxylate. Fast dissociation of hydrogen ions was observed. The hydrogen ion activated the carbonyl group. Methanol electrophiles formed the carbonyl group. The water molecules were removed to form methyl benzoate.

2.3. *Properties and Activity of Preferred Catalyst*

The strength of ZT10 was determined using the Hammett method. The solid acid could significantly change the color of 2,4-dinitrotoluene ($H_0 = -13.75$) and 2,4-dinitrofluorobenzene ($H_0 = -14.52$) but not that of 1,3,5-trinitrobenzene ($H0 = -16.02$). The catalyst was an acid with an acid strength range of $-16.02 < H_0 < -14.52$. This indicates that the solid acid was a solid superacid.

A series of catalytic studies were carried out using the preferred solid acid with a Zr–Ti atomic ratio of 1.2:1. The catalytic results are summarized in Table 3. The yields of the compounds (**1**, **4**, **7**, **10**, **13**) indicate that the presence of a large group in the ortho position decreased the yield. Orthosteric effects play a dominant role in the esterification process. However, the steric hindrance effect at the para-position was small (**16**, **17**, **18**). Therefore, MB with strong electron-absorbing groups on the benzene ring was obtained in a low yield. For example, the yield of MB with a nitro group in the ortho position was only 20.1%, and the compound with a trifluoromethyl group in the ortho position was obtained with only 29.1% yield. This is because of high steric hindrance and strong electron absorption effects, which make esterification difficult. For the same functional group, the para-substituted product yields were higher than those of the relative and intermediate sites, regardless of whether it was an electron-withdrawing or electron-donating group (e.g., **1**, **2**, **3**; **4**, **5**, **6**; **19**, **20**, **21**). In the case of the benzene ring with two carboxyl groups, the reaction activity was low because of the large steric hindrance and strong electric absorption of the carboxyl group. The two carboxyl groups each underwent esterification, and moderate yields were also obtained for **31**, which had a large steric hindrance from three methoxy groups, and **32**, which had a perfluorinated substitution.

Table 3. Synthesis of methyl benzoate with substituents [a,b].

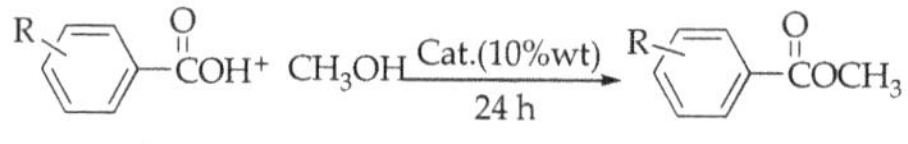

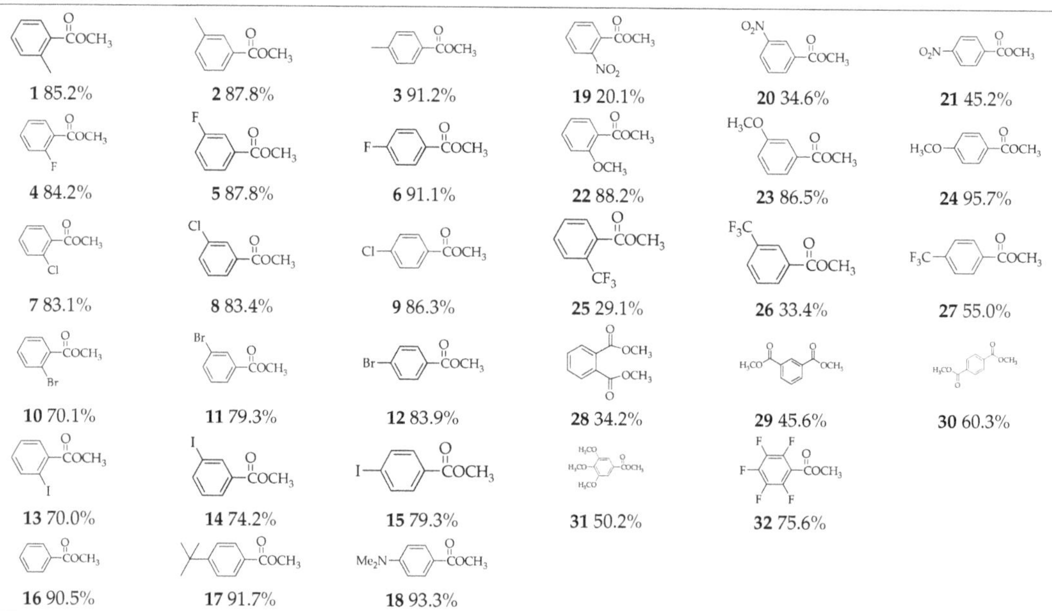

1 85.2%	**2** 87.8%	**3** 91.2%	**19** 20.1%	**20** 34.6%	**21** 45.2%
4 84.2%	**5** 87.8%	**6** 91.1%	**22** 88.2%	**23** 86.5%	**24** 95.7%
7 83.1%	**8** 83.4%	**9** 86.3%	**25** 29.1%	**26** 33.4%	**27** 55.0%
10 70.1%	**11** 79.3%	**12** 83.9%	**28** 34.2%	**29** 45.6%	**30** 60.3%
13 70.0%	**14** 74.2%	**15** 79.3%	**31** 50.2%	**32** 75.6%	
16 90.5%	**17** 91.7%	**18** 93.3%			

[a] Reaction condition: acid (2.0 mmol), CH_3OH (15 mL), Cat. (10 wt%) (Zr:Ti = 1.2:1), oil bath (120 °C), 24 h.
[b] Isolated yields.

As previously reported, direct metal–Lewis acid interaction has not been found to catalyze the reaction of direct benzoic acid and methanol to form methyl benzoate. Dehydration esterification is a reversible reaction. It requires high energy and the removal of the resulting water molecules to move the reaction forward. In the catalytic process, methanol is strongly reflux. Methanol can be supplemented when the amount of methanol is too low. The methanol that does not condense will remove the water and move the reaction forward.

2.4. Properties and Activity of Recovered Catalyst

Because the esterification reaction is heterogeneous, the catalyst and reactants can be directly separated via solid–liquid separation after the initial reaction. After the reaction liquid was filtered and discarded, new reactants were added. The activity of the second reaction was similar to that of the first. Taking compound **7** as an example, the catalyst afforded an 84.1% yield in the second reaction.

Nitrogen adsorption of the recovered and original catalysts was also studied. The specific surface area of the catalyst was 242 m^2/g, and the pore size was approximately 2 nm. The recovered and primary catalysts exhibited comparable pore structures and specific surface areas (Table 4).

Table 4. Surface properties of the catalyst and the regenerated one.

Samples	Surface Area (m^2/g)	Pore Volume ($cm^{-3}g^{-1}$)	Pore Diameter (nm)
Catalyst (zirconium: titanium = 1.2:1)	242	0.210	2.32
Regenerated Catalyst (zirconium: titanium = 1.2:1)	248	0.205	2.34

3. Materials and Methods

3.1. Synthesis of the Catalyst

Dichlorooxyzirconium, cadmium chloride, aluminum chloride, tetraethoxysilane iron chloride, zinc chloride, tetrabutyl titanate, and methanol were purchased from Sinopharm Chemical Reagent Co., Ltd. (Shanghai, China) Benzoic acid and other benzene compounds were purchased from Aladdin Reagent Co., Ltd. (Shanghai, China) Deionized water (18.2 MΩ·cm) was used in the catalyst synthesis. All materials were used as received. ^{1}H, ^{19}F, and ^{13}C NMR spectra (Figures S1–S32) were measured on a Bruker AVANCE (400 MHz, Bruker, Billerica, MA, USA) spectrometer using TMS as the internal standard and $CDCl_3$ as a solvent.

The zirconium catalyst was synthesized with cadmium as a carrier. Zirconium oxychloride was prepared as a 50% aqueous solution. Cadmium chloride was prepared as 30% aqueous solutions. The ratio of the solutions was calculated according to the molar ratio, and quantitative zirconium oxychloride and aqueous chloride solutions were mixed by weight. The mixture was placed in an oven at 100 °C until it reached a constant weight. The dried mixture was added to a 2 mol/L sulfuric acid solution and soaked for 24 h. After soaking, the products were dried in the oven at 120 °C until the weight was constant. The mixture was bagged and heated in a Muffle furnace at 550 °C for 12 h.

The zirconium catalyst was synthesized with cadmium as a carrier. Zirconium oxychloride was prepared as a 50% aqueous solution. Zinc chloride was prepared as 50% aqueous solutions. The ratio of the solutions was calculated according to the molar ratio, and quantitative zirconium oxychloride and aqueous chloride solutions were mixed by weight. The mixture was placed in an oven at 100 °C until it reached a constant weight. The dried mixture was added to a 2 mol/L sulfuric acid solution and soaked for 24 h. After soaking, the products were dried in the oven at 120 °C until the weight was constant. The mixture was bagged and heated in a Muffle furnace at 550 °C for 12 h.

The zirconium catalyst was synthesized with aluminum as a carrier. Zirconium oxychloride was prepared as a 50% aqueous solution. Aluminum chloride was prepared as 60% aqueous solutions. The ratio of the solutions was calculated according to the molar ratio, and quantitative zirconium oxychloride and aqueous chloride solutions were mixed by weight. The mixture was placed in an oven at 100 °C until it reached a constant weight. The dried mixture was added to a 2 mol/L sulfuric acid solution and soaked for 24 h. After soaking, the products were dried in the oven at 120 °C until the weight was constant. The mixture was bagged and heated in a Muffle furnace at 550 °C for 12 h.

A zirconium oxychloride solution was prepared using silicon as the support catalyst at 50 wt% based on the molar ratio. After rapid stirring, proportional ethyl orthosilicate was added. The mixture was stirred and placed in the oven at 100 °C until the weight was constant. The dried mixture was added to a 2 mol/L sulfuric acid solution and soaked for 24 h. The mixed solution was placed in the oven at 120 °C until the weight was constant. The mixture was bagged and heated in a Muffle furnace at 550 °C for 12 h.

A zirconium oxychloride solution was prepared using titanium as the support catalyst at 50 wt% based on the molar ratio. A proportional amount of tetrabutyl titanate was added after rapid stirring. After thorough mixing, it was placed in the oven to bake at 100 °C until the weight was constant. The dried mixture was added to a 2 mol/L sulfuric acid solution and soaked for 24 h. The mixed solution was placed in the oven at 120 °C until the weight was constant. The mixture was bagged and heated in a Muffle furnace at 550 °C for 12 h.

3.2. General Catalytic Process

Taking *p*-methylbenzoic acid as an example, 0.27 g of (2 mmol) *p*-methylbenzoic acid, 15.00 mL of methanol, and 0.027 g of catalyst were added into a 25 mL single-mouth flask successively and vigorously stirred. In an oil bath at 120 °C, the mixture was strongly refluxed. If there is too little methanol during reflux, a small amount can be added. The reaction was stopped after 24 h, and the catalyst was removed. The product was isolated using column chromatography with petroleum ether/ethyl acetate (5:1 *v*/*v*) as the mobile phase.

3.3. Recovery Catalyst Catalytic Process

The reaction was completed and the reaction liquid was poured out. A small amount of residual solid from methanol washing was used as a catalyst. Additionally, 0.27 g of (2 mmol) *p*-methylbenzoic acid and 15.00 mL of methanol were then added into a 25 mL single-mouth flask with the used catalyst and vigorously stirred. In an oil bath at 120 °C, the mixture was strongly refluxed. If there was too little methanol during reflux, a small amount was added. The reaction was stopped after 24 h, and the catalyst was removed. The product was isolated using column chromatography with petroleum ether/ethyl acetate (5:1 *v*/*v*) as the mobile phase.

After the catalyst was recovered, it was baked at 100 °C until a constant weight was reached for nitrogen adsorption.

4. Conclusions

Zirconium solid acid catalysts, supported by various compounds, have been synthesized, and their catalytic activities for synthesizing MB compounds were studied. The experimental results show that titanium was the best carrier for zirconium. The catalyst (ZT10) prepared with a Zr: Ti molar ratio of 1.2:1 exhibited the best catalytic effect.

ZT10 catalyzed the synthesis of MB from benzoic acid with a substituted group and methanol. The experimental results demonstrate that this catalyst could catalyze the reactions of benzoic acid and methanol, regardless of whether the benzene ring of benzoic acid contains electron-donating groups, strong electron-absorbing groups, small steric hindrance groups, or large steric hindrance groups. We also demonstrated that the catalyst could be easily separated from a reaction mixture and reused with no impact on its performance.

Supplementary Materials: The following supporting information can be downloaded at: https://www.mdpi.com/article/10.3390/catal13050915/s1; Figure S1: ^{1}H and ^{13}C NMR of methyl 2-methylbenzoate; Figure S2: ^{1}H and ^{13}C NMR of methyl 3-methylbenzoate; Figure S3: ^{1}H and ^{13}C NMR of methyl 4-methylbenzoate; Figure S4: ^{1}H, ^{13}C and ^{19}F NMR of methyl methyl 2-fluorobenzoate; Figure S5: ^{1}H, ^{13}C and ^{19}F NMR of methyl methyl 3-fluorobenzoate; Figure S6: ^{1}H, ^{13}C and ^{19}F NMR of methyl methyl 4-fluorobenzoate; Figure S7: ^{1}H and ^{13}C NMR of methyl 2-chlorobenzoate; Figure S8: ^{1}H and ^{13}C NMR of methyl 3-chlorobenzoate; Figure S9: ^{1}H and ^{13}C NMR of methyl 4-chlorobenzoate; Figure S10: ^{1}H and ^{13}C NMR of methyl 2-bromobenzoate; Figure S11: ^{1}H and ^{13}C NMR of methyl 3-bromobenzoate; Figure S12: ^{1}H and ^{13}C NMR of methyl 4-bromobenzoate; Figure S13: ^{1}H and ^{13}C NMR of methyl 2-iodobenzoate; Figure S14: ^{1}H and ^{13}C NMR of methyl 3-iodobenzoate; Figure S15: ^{1}H and ^{13}C NMR of methyl 4-iodobenzoate; Figure S16: ^{1}H and ^{13}C NMR of methyl benzoate; Figure S17: ^{1}H and ^{13}C NMR of methyl 4-*Tert*-Butylbenzoate; Figure S18: ^{1}H and ^{13}C NMR of methyl 4-cyanobenzoate; Figure S19: ^{1}H and ^{13}C NMR of Methyl 2-nitrobenzoate; Figure S20: ^{1}H and ^{13}C NMR of Methyl 3-nitrobenzoate; Figure S21: ^{1}H and ^{13}C NMR of Methyl 4-nitrobenzoate; Figure S22: ^{1}H and ^{13}C NMR of Methyl 2-methoxybenzoate; Figure S23: ^{1}H and ^{13}C NMR of Methyl 3-methoxybenzoate; Figure S24: ^{1}H and ^{13}C NMR of Methyl 4-methoxybenzoate; Figure S25: ^{1}H, ^{13}C and ^{19}F NMR of Methyl 2-trifluoromethylbenzoate; Figure S26: ^{1}H, ^{13}C and ^{19}F NMR of Methyl 3-trifluoromethylbenzoate; Figure S27: ^{1}H, ^{13}C and

^{19}F NMR of Methyl 4-trifluoromethylbenzoate; Figure S28: ^{1}H and ^{13}C NMR of o-Dimethyl phthalate; Figure S29: ^{1}H and ^{13}C NMR of m-Dimethyl terephthalate; Figure S30: ^{1}H and ^{13}C NMR of p-Dimethyl terephthalate; Figure S31: ^{1}H and ^{13}C NMR of methyl 3,4,5-trimethoxybenzoate; Figure S32: ^{1}H, ^{13}C and ^{19}F NMR of Methyl 2,3,4,5,6-pentafluorobenzoate. References [35–41] are cited in the Supplementary Materials.

Author Contributions: Conceptualization, writing—original draft preparation, X.Y. and C.S.; methodology, S.H.; formal analysis, Y.C. and Y.Z.; writing—review and editing, R.S. All authors have read and agreed to the published version of the manuscript.

Funding: This research was funded by the Natural Science Foundation of Anhui Provincial Natural Science Foundation of China (2008085ME174), Higher Education Institutions of Anhui Province (KJ2019A0856), and the Bengbu University-funded project of scientific research startups for high-level talents (BBXY2018KYQD20).

Conflicts of Interest: The authors declare no conflict of interest.

References

1. Kapoor, D.; Bhardwaj, S.; Sharma, N.R. Fragrance Stimulation Mechanisms of Flowers and their Regulation under Environmental Constraints. *J. Plant Growth Regul.* **2023**, *42*, 60–82. [CrossRef]
2. Tian, Y.-F.; Chen, Y.-X.; Tong, X.; Hou, S.; Zhao, M.-M.; Feng, Y.-Z. Flavor differences of soybean and defatted soybean fermented soy sauce and its correlation with the enzyme profiles of the kojis. *J. Sci. Food Agric.* **2023**, *103*, 606–615. [CrossRef]
3. El Mansouri, A.-E.; Lachhab, S.; Oubella, A.; Mehdi, A.; Neyts, J.; Jochmans, D.; Chiu, W.; Vangeel, L.; De Jonghe, S.; Morjani, H.; et al. Synthesis, characterization, molecular docking, and anticancer activities of new 1,3,4-oxadiazole-5-fluorocytosine hybrid derivatives. *J. Mol. Struct.* **2023**, *1272*, 134135. [CrossRef] [PubMed]
4. Ning, S.; Zheng, L.; Gao, Q.; Shi, L.; Liu, Y.; Sun, C.; Zhang, Z.; Xiang, J. New electrotriggers: P-methoxycarbonylbenzyl (pMCB) as an electroremovable protecting group for carboxylic acids, phosphoric acids and alcohols. *Green Chem.* **2022**, *24*, 5632–5636. [CrossRef]
5. Li, J.-G.; Peng, Y.-Q. Efficient Methyl Esterification Using Methoxyl Silica Gel as a Novel Dehydrating Reagent. *J. Chin. Chem. Soc.* **2010**, *57*, 305–308. [CrossRef]
6. Travis, B.R.; Sivakumar, M.; Hollist, G.O.; Borhan, B. Facile Oxidation of Aldehydes to Acids and Esters with Oxone. *Org. Lett.* **2003**, *5*, 1031–1034. [CrossRef]
7. Talukdar, D.; Sharma, K.; Bharadwaj, S.K.; Thakur, A.J. VO (acac)$_2$: An Efficient Catalyst for the Oxidation of Aldehydes to the Corresponding Acids in the Presence of Aqueous H_2O_2. *Synlett* **2013**, *24*, 963–966. [CrossRef]
8. Han, J.; Gu, F.; Li, Y. N-Doped Sub-3 nm Co Nanoparticles as Highly Efficient and Durable Aerobic Oxidative Coupling Catalysts. *Chem.-Asian J.* **2016**, *11*, 2594–2601. [CrossRef]
9. Lerebours, R.; Wolf, C. Chemoselective Nucleophilic Arylation and Single-Step Oxidative Esterification of Aldehydes Using Siloxanes and a Palladium−Phosphinous Acid as a Reaction Switch. *J. Am. Chem. Soc.* **2006**, *128*, 13052–13053. [CrossRef]
10. Subramanian, K.; Yedage, S.L.; Bhanage, B.M. An Electrochemical Method for Carboxylic Ester Synthesis from N-Alkoxyamides. *J. Org. Chem.* **2017**, *82*, 10025–10032. [CrossRef] [PubMed]
11. Liu, M.; Zhang, Z.; Yan, J.; Liu, S.; Liu, H.; Liu, Z.; Wang, W.; He, Z.; Han, B. Aerobic Oxidative Cleavage and Esterification of C(OH)–C Bonds. *Chem* **2020**, *6*, 3288–3296. [CrossRef]
12. Mao, F.; Qi, Z.; Fan, H.; Sui, D.; Chen, R.; Huang, J. Heterogeneous cobalt catalysts for selective oxygenation of alcohols to aldehydes, esters and nitriles. *RSC Adv.* **2017**, *7*, 1498–1503. [CrossRef]
13. Chng, L.L.; Yang, J.; Ying, J.Y. Efficient Synthesis of Amides and Esters from Alcohols under Aerobic Ambient Conditions Catalyzed by a Au/Mesoporous Al_2O_3 Nanocatalyst. *ChemSusChem* **2015**, *8*, 1916–1925. [CrossRef] [PubMed]
14. Liu, C.; Wang, J.; Meng, L.; Deng, Y.; Li, Y.; Lei, A. Palladium-Catalyzed Aerobic Oxidative Direct Esterification of Alcohols. *Angew. Chem. Int. Ed.* **2011**, *50*, 5144–5148. [CrossRef]
15. Wang, K.; Xu, Y.; Wang, J. Palladium-Catalyzed Cyclizative Borylation of Allenyl Ketones through Carbene Boryl Migratory Insertion: Access to Densely Substituted Furyl Boronates. *Chem.-A Eur. J.* **2023**, *29*, e2022036972023. [CrossRef]
16. Xiao, Z.; Zhang, S.; Zhu, J.; Niu, Y.; Xiong, W.; Chen, F. Identification of key aromas of grapefruit juice and study of their contributions to the enhancement of sweetness perception. *Eur. Food Res. Technol.* **2023**, *249*, 537–551. [CrossRef]
17. Chen, X.; He, Z.; He, L.; Li, C.; Tao, H.; Wang, X.; Liu, L.; Zeng, X.; Ran, G. Effects of perilla seed oil addition on the physicochemical properties, sensory, and volatile compounds of potato blueberry flavored yogurt and its shelf-life prediction. *LWT* **2023**, *173*, 114383. [CrossRef]
18. Ji, G.; Liu, G.; Li, B.; Tan, H.; Zheng, R.; Sun, X.; He, F. Influence on the aroma substances and functional ingredients of apple juice by lactic acid bacteria fermentation. *Food Biosci.* **2023**, *51*, 102337. [CrossRef]
19. Keana, J.F.W.; Cai, S.X. New reagents for photoaffinity labeling: Synthesis and photolysis of functionalized perfluorophenyl azides. *J. Org. Chem.* **1990**, *55*, 3640–3647. [CrossRef]

20. De Oliveira, C.S.; Lira, B.F.; Dos Santos Falcão-Silva, V.; Siqueira-Junior, J.P.; Barbosa-Filho, J.M.; De Athayde-Filho, P.F. Synthesis, Molecular Properties Prediction, and Anti-staphylococcal Activity of N-Acylhydrazones and New 1,3,4-Oxadiazole Derivatives. *Molecules* **2012**, *17*, 5095–5107. [CrossRef] [PubMed]
21. Štefane, B.; Kočevar, M.; Polanc, S. Ceric (IV) Ammonium nitrate mediated transesterification and esterification. *Synth. Commun.* **2002**, *32*, 1703–1707. [CrossRef]
22. Martínez, R.; Zamudio, G.J.N.; Pretelin-Castillo, G.; Torres-Ochoa, R.O.; Medina-Franco, J.L.; Pinzón, C.I.E.; Miranda, M.S.; Hernández, E.; Alanís-Garza, B. Synthesis and antitubercular activity of new N-[5-(4-chlorophenyl)-1,3,4-oxadiazol-2-yl]-(nitroheteroaryl)carboxamides. *Heterocycl. Commun.* **2019**, *25*, 52–59. [CrossRef]
23. Roy, S.; Samanta, D.; Kumar, P.; Maji, T.K. Pure white light emission and charge transfer in organogels of symmetrical and unsymmetrical π-chromophoric oligo-p-(phenyleneethynylene) bola-amphiphiles. *Chem. Commun.* **2018**, *54*, 275–278. [CrossRef] [PubMed]
24. Stephens, M.D.; Yodsanit, N.; Melander, C. Potentiation of the fosmidomycin analogue FR 900098 with substituted 2-oxazolines against Francisella novicida. *MedChemComm* **2016**, *7*, 1952–1956. [CrossRef] [PubMed]
25. Gianni, J.; Pirovano, V.; Abbiati, G. Silver triflate/p-TSA co-catalysed synthesis of 3-substituted isocoumarins from 2-alkynylbenzoates. *Org. Biomol. Chem.* **2018**, *16*, 3213–3219. [CrossRef]
26. Sekewael, S.J.; Pratika, R.A.; Hauli, L.; Amin, A.K.; Utami, M.; Wijaya, K. Recent Progress on Sulfated Nanozirconia as a Solid Acid Catalyst in the Hydrocracking Reaction. *Catalysts* **2022**, *12*, 191. [CrossRef]
27. He, X.; Guo, X.; Xia, G.; Xu, R.; Wu, Y.; Luan, X. A Green Route to Methyl Formate from CO_2-Derived Formamides over Solid Base Catalysts. *Catalysts* **2023**, *13*, 487. [CrossRef]
28. Saravanan, K.; Tyagi, B.; Shukla, R.S.; Bajaj, H.C. Esterification of palmitic acid with methanol over template-assisted mesoporous sulfated zirconia solid acid catalyst. *Appl. Catal. B Environ.* **2015**, *172*, 108–115. [CrossRef]
29. Saravanan, K.; Tyagi, B.; Shukla, R.S.; Bajaj, H.C. Solvent free synthesis of methyl palmitate over sulfated zirconia solid acid catalyst. *Fuel* **2016**, *165*, 298–305. [CrossRef]
30. Ameen, M.; Zafar, M.; Ramadan, M.F.; Ahmad, M.; Makhkamov, T.; Bokhari, A.; Mubashir, M.; Chuah, L.F.; Show, P.L. Conversion of novel non-edible Bischofia javanica seed oil into methyl ester via recyclable zirconia-based phyto-nanocatalyst: A circular bioeconomy approach for eco-sustenance. *Environ. Technol. Innov.* **2023**, *30*, 103101. [CrossRef]
31. Sharghi, H.; Sarvari, M.H. $Al_2O_3/MeSO_3H$ (AMA) as a new reagent with high selective ability for monoesterification of diols. *Tetrahedron* **2003**, *59*, 3627–3633. [CrossRef]
32. Xu, Q.-H.; Liu, W.-Y.; Chen, B.-H.; Ma, Y.-X. An improved method for the esterification of aromatic acids with ethanol and methanol. *Synth. Commun.* **2001**, *31*, 2113–2117. [CrossRef]
33. Guo, H.; Yang, H.; Huang, J.; Tong, J.; Liu, X.; Wang, Y.; Qiao, W.; Han, J. Theoretical and experimental insight into plasma-catalytic degradation of aqueous p-nitrophenol with graphene-ZnO nanoparticles. *Sep. Purif. Technol.* **2022**, *295*, 121362. [CrossRef]
34. Guo, H.; Wang, Y.; Yao, X.; Zhang, Y.; Li, Z.; Pan, S.; Han, J.; Xu, L.; Qiao, W.; Li, J.; et al. A comprehensive insight into plasma-catalytic removal of antibiotic oxytetracycline based on graphene-TiO_2-Fe_3O_4 nanocomposites. *Chem. Eng. J.* **2021**, *425*, 130614. [CrossRef]
35. Chavan, S.P.; Dantale, S.W.; Govande, C.A.; Venkatraman, M.S.; Praveen, C. Titanosilicate (TS-1) Catalyzed Oxidation of Aromatic Aldehydes to Esters. *Synlett* **2002**, *2*, 0267–0268. [CrossRef]
36. Jaoui, M.; Kleindienst, T.E.; Lewandowski, M.; Edney, E.O. Identification and Quantification of Aerosol Polar Oxygenated Compounds Bearing Carboxylic or Hydroxyl Groups. 1. Method Development. *Anal. Chem.* **2004**, *76*, 4765–4778. [CrossRef]
37. Gopinath, R.; Barkakaty, B.; Talukdar, B.; Patel, B.K. Peroxovanadium-Catalyzed Oxidative Esterification of Aldehydes. *J. Org. Chem.* **2003**, *68*, 2944–2947. [CrossRef]
38. El-Damasy, A.K.; Jin, H.; Seo, S.H.; Bang, E.-K.; Keum, G. Design, synthesis, and biological evaluations of novel 3-amino-4-ethynyl indazole derivatives as Bcr-Abl kinase inhibitors with potent cellular antileukemic activity. *Eur. J. Med. Chem.* **2020**, *207*, 112710. [CrossRef]
39. Jiang, Y.; Pan, S.; Zhang, Y.; Yu, J.; Liu, H. Copper-Catalyzed Decarboxylative Methylation of AromaticCarboxylic Acids with $PhI(OAc)_2$. *Eur. J. Org. Chem.* **2014**, *10*, 2027–2031. [CrossRef]
40. Senaweera, S.M.; Singh, A.; Weaver, J.D. Photocatalytic Hydrodefluorination: Facile Access to Partially Fluorinated Aromatics. *J. Am. Chem. Soc.* **2014**, *136*, 3002–3005. [CrossRef]
41. Ross, J.P.; Couture, P.; Warkentin, J. Nucleophilic aromatic substitution with dialkoxycarbenes. *Can. J. Chem.* **1997**, *75*, 1331–1335. [CrossRef]

catalysts MDPI

Article

Efficient Synthesis of Furfural from Corncob by a Novel Biochar-Based Heterogeneous Chemocatalyst in Choline Chloride: Maleic Acid–Water

Linsong Yang [1,†], Yucheng Li [1,†], Yuqi Wu [1,†], Yucai He [1,2,*] and Cuiluan Ma [3]

1 School of Pharmacy, Changzhou University, Changzhou 213164, China; linsongyang@cczu.edu.cn (L.Y.); s21091055082@smail.cczu.edu.cn (Y.L.); wyqup12@163.com (Y.W.)
2 State Key Laboratory of Bioreactor Engineering, School of Biochemical Engineering, East China University of Science and Technology, Shanghai 200237, China
3 State Key Laboratory of Biocatalysis and Enzyme Engineering, School of Life Sciences, Hubei University, Wuhan 430062, China; macuiluan@163.com
* Correspondence: yucaihe@cczu.edu.cn or heyucai2001@163.com
† These authors contributed equally to this work.

Abstract: The use of plentiful and renewable feedstock for producing chemicals is fundamental for the development of sustainable chemical processes. Using fish scale as a biobased carrier, a novel biochar SO_4^{2-}/SnO_2-FFS heterogeneous chemocatalyst was prepared to catalyze furfural production from xylose-rich corncob-hydrolysates obtained from acid hydrolysis of corncob in a deep eutectic solvent (DES)–water system. By characterizing the physical as well as chemical properties of SO_4^{2-}/SnO_2-FFS by NH_3-TPD, FT-IR, XPS, XRD, and SEM, it was shown that the chemocatalyst had Lewis/Brönsted acid centers, and its surface roughness could be well expanded to contact substrates. The corncob was initially hydrolyzed at 140 °C to obtain xylose-rich hydrolysate. Subsequently, SO_4^{2-}/SnO_2-FFS (3.6 wt.%) was used to catalyze the corn cob hydrolysate containing *D*-xylose (20.0 g/L) at a reaction temperature of 170 °C for 15 min. Additionally, $ZnCl_2$ (20.0 g/L) was added. Ultimately, furfural (93.8 mM, 70.5% yield) was produced in the deep eutectic solvent ChCl:maleic acid–water (DES_{MLA}–water = 10:90, v/v). A synergistic catalytic mechanism for transforming xylose-rich corncob-hydrolysate into furfural and byproducts were proposed using SO_4^{2-}/SnO_2-FFS as a chemocatalyst in DES_{MLA}–water containing $ZnCl_2$. Consequently, the efficient use of biochar SO_4^{2-}/SnO_2-FFS chemocatalysts for the sustainable synthesis of biobased furan compounds from biomass holds great promise in the future.

Keywords: biofuran; biochar catalyst; deep eutectic solvent; $ZnCl_2$; lignocellulose

1. Introduction

The continued use of non-renewable energy has raised people's awareness of natural resource depletion and global warming, shifting the focus of many countries to renewable energy [1,2]. It is known that lignocellulosic biomass (LB) is renewable and eco-friendly [3–5], which is considered a replacement for manufacture biofuels, materials, and high-value biobased compounds because it is plentiful, inexpensive, and promising [6–10]. As one of the important chemical intermediates, furfural (FAL) can be transformed from lignocellulose or its derivatives (e.g., xylan and *D*-xylose) via hydrothermal reaction [11]. In industrial production, FAL has been widely used. It has become a raw material for pharmaceuticals, industrial solvents, and numerous fuel additives [12]. By deeper processing, it can be transformed into various furan-based compounds (e.g., tetrahydrofuran, furoic acid, tetrahydrofuran, dihydropyran, acetyl furan, furfuryl alcohol, and furfurylamine) [13].

Generally, furfural (FAL) can be synthesized using both homogeneous and heterogeneous catalysts. Utilizing homogeneous catalysts such as hydrochloric acid, the highest

FAL yield from xylose could reach 37.5% [14]. Combining Lewis acid ($AlCl_3$) and Brønsted acid (HCl) as homogeneous catalysts for the catalytic conversion of untreated cellulose resulted in a remarkable FAL yield of 75% [15]. However, the use of homogeneous catalysis might accelerate equipment corrosion and lead to difficulties in recycling, posing challenges in terms of both economic efficiency and environmental sustainability [16]. Consequently, more eco-friendly heterogeneous catalysts have been employed for FAL synthesis [17,18]. Among these, sulfonated carbon derived from biomass has garnered attention due to its low cost, favorable thermal stability, and acidity. Using biochar-based SO_4^{2-}/SnO_2-CS as a catalyst, the corn stalk was transformed into FAL with a yield of 68% at a temperature of 170 °C for 0.5 h [19]. A total of 62% of FAL yield from *D*-xylose by carbonized and sulfonated teff straw was achieved in a toluene-water system at 170 °C for 0.5 h [20]. FAL was synthesized from corncob with sulfonated tin-loaded rice husk-activated carbon, and the yield of FAL was 40.9% [21]. Employing 4-BDS as a sulfonating agent, sucrose was carbonized into a solid acid catalyst, achieving a 61% FAL yield from a corn stalk in 100 min at a temperature of 200 °C [22]. This highlights the significant potential of using biomass as a catalyst carrier. Therefore, utilizing sulfonated carbon derived from biomass as a catalyst or carrier is essential for achieving high-value utilization of biomass.

Appropriate solvents also play a crucial role in enhancing FAL productivity [23]. Deep eutectic solvents (DESs) are defined as systems that combine eutectic Lewis (L) or Brönsted (B) acids with various ionic species [24]. Compared to traditional reaction systems, DESs offer several advantages including ease of preparation, non-toxicity, non-volatility, and reusability [25]. Utilizing choline chloride (ChCl)-lactic acid as a chemical catalyst and reaction medium, rice straw was transformed into FAL (12% yield) at 180 °C for 2 h [26]. Water–ChCl:oxalic acid was employed as a medium to convert corncob (CC) at 180 °C for 0.5 h, achieving a FAL yield of 46% [27]. ChCl-carboxylic acid containing H^+ and Cl^- might facilitate the dehydration of *D*-xylose to produce FAL [28]. Therefore, to further enhance the FAL yield, a synergistic catalysis approach using carboxylic acid-based DES and heterogeneous catalysts can be employed in the reaction system, providing a more environmentally friendly and efficient means of catalyzing biomass-derived *D*-xylose into FAL.

As one kind of bioresource, fish scales are rich in collagen, hydroxyapatite (HAP), and lipids [29]. According to the statistics, the fish processing industry produces about 30 million tons of fish waste each year, which contains 4% of fish scales. In recent years, fish scales have been used in the chemistry, medicine, and food industry [30]. As one main component of fish scales, HAP is a calcium phosphate biomaterial, which serves as an ideal catalyst carrier due to its ion exchange, adsorption, acid–base properties, and stability. Currently, there is limited information on the effective transformation of biomass or *D*-xylose to FAL using fish scale-based biochar solid acid catalysts. In order to realize the reuse of fresh fish scales (FFS) and improve the efficiency of FAL production by biomass catalysis, a novel biochar heterogeneous catalyst (SO_4^{2-}/SnO_2-FFS) was prepared using FFS as a carrier. The structure and morphology of SO_4^{2-}/SnO_2-FFS were characterized by NH_3-TPD, FT-IR, XPS, XRD, and SEM. In addition, an attempt was made to synthesize FAL from hydrolysate from CC in DES_{MLA}–water system using SO_4^{2-}/SnO_2-FFS as a heterogeneous catalyst. The process conditions for preparing FAL (such as SO_4^{2-}/SnO_2-FFS dosage, catalytic temperature, catalytic time, cycle time, chloride salt, etc.) were optimized. Finally, a sustainable strategy for the synthesis of FAL using fish scale-based heterogeneous catalysts in an environmentally friendly catalytic reaction system was established (Figure 1)

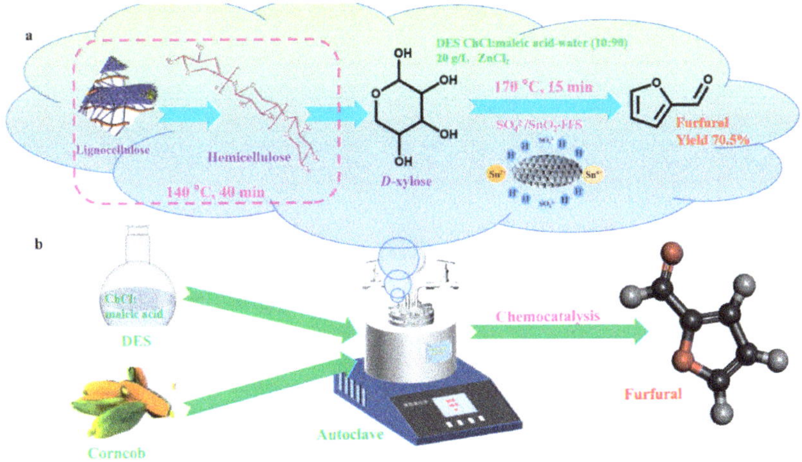

Figure 1. The process of catalytically converting the *D*-xylose solution obtained from the hydrolysis of corncob at 140 °C for 40 min into FAL using SO_4^{2-}/SnO_2-FFS in a DES-water system (**a**); a chemical strategy for synthesis of FAL from corncob with biochar-based heterogeneous chemocatalyst (**b**).

2. Results and Discussion

2.1. Characterization of Biochar-Based Catalyst SO_4^{2-}/SnO_2-FFS Using FFS as Biobased Carrier

FFS of carp is one of the underutilized fishery wastes [31]. Tin-based heterogeneous chemocatalysts have been proven to be able to utilize to dehydrate corncob-derived *D*-xylose to FAL. In this work, sulfonated tin-loaded heterogeneous chemocatalyst SO_4^{2-}/SnO_2-FFS was prepared by using carbonized FFS as biobased carrier. The surface area changes of SO_4^{2-}/SnO_2-FFS and FFS were tested by BET, and the results were displayed in Table 1. The surface change of the carrier FFS and catalyst SO_4^{2-}/SnO_2-FFS was compared. Specific surface area (SSA) of FFS was 12.4 m^2/g. SO_4^{2-}/SnO_2-FFS had bigger SSA (29.4 m^2/g). Large SSA might promote the contact between the catalytic active sites on solid acid catalyst and the substrate molecules, which would result in the enhancement of FAL productivity [32]. SO_4^{2-}/SnO_2-FFS appeared smaller pore size (4.7 nm) compared with the FFS pore size (14.5 nm), and a reduced pore volume of SO_4^{2-}/SnO_2-FFS (0.010 cm^3/g) was evident. The pore size of SO_4^{2-}/SnO_2-FFS was bigger than the molecular dynamics (MD) radius of FAL (0.68 nm) and *D*-xylose (0.86 nm) [33], suggesting that *D*-xylose might be beneficial to spread into the acid sites on the surface of SO_4^{2-}/SnO_2-FFS and was further dehydrated into FAL.

Table 1. Surface and pore difference of FFS and SO_4^{2-}/SnO_2-FFS.

Sample	BET Surface Area, m^2/g	Pore Volume, cm^3/g	Pore Size, nm
FFS	12.4	0.03	14.5
SO_4^{2-}/SnO_2-FFS	29.4	0.01	4.7

The SEM image depicted that the surface of FFS was comparatively smooth, while the surface of SO_4^{2-}/SnO_2-FFS was rough (Figure 2). It showed that SnO_2 and SO_4^{2-} were loaded on the carrier FFS, and enhanced the contact between SO_4^{2-}/SnO_2-FFS and *D*-xylose. As described in FT-IR (Figure 3), SO_4^{2-}/SnO_2-FFS had obvious surface compared to FFS. The peaks near 3421 cm^{-1} and 2922 cm^{-1} were attributed to the presence of hydroxyl groups and H^+ of -COOH stretching [34]. The peak about 1628 cm^{-1} corresponded to the stretching vibration of the amide in FFS [35]. The peak around 1404 cm^{-1} was associated with O-H stretching [36]. The peak near 1128 cm^{-1} corresponded to sulfonic acid groups

(SO_3H) [37]. The peaks around 741 cm^{-1} and 592 cm^{-1} were related to the inorganic phase, representing the bending mode of P-O groups, as well as hydroxyl stretching of the adsorption water [38]. The XRD spectra of FFS and SO_4^{2-}/SnO_2-FFS were displayed in Figure 4. The peak near 26.1° was associated with the crystal plane of amorphous carbon [39], and the peak around 31.9° was assigned to hydroxyl apatite [40]. The sharp characteristic peaks of crystalline SnO_2 were reflected at 25.9°, 33.8°, and 51.9° through the XRD analysis [41]. The observed XRD pattern was characteristic for HAP as reported in the literature [42]. XRD patterns of SO_4^{2-}/SnO_2-FFS also indicated that SnO_2 had been loaded. By using NH_3-TPD (temperature-programmed desorption of ammonia) (Figure 5), the acid properties of SO_4^{2-}/SnO_2-FFS were measured. Through the analysis with desorption of NH_3 under the various temperatures, the weak-, medium-, and strong-acid sites on the heterogeneous catalyst might be detected based on the temperature 100–200 °C, 200–400 °C, and 400–800 °C, respectively. From the observation of Figure 5, it could be deduced that SO_4^{2-}/SnO_2-FFS possessed a predominant type of acid site (weak acid site) only at 108 °C. As displayed in XPS image (Figure 6), SO_4^{2-}/SnO_2-FFS had been accompanied by Sn, and Sn had three valences (+4, +2, 0). The fraction of $Sn^{4+}3d_{5/2}$, $Sn^{2+}3d_{5/2}$, and $Sn^03d_{5/2}$ were 48.4%, 46.1%, and 5.5%, respectively.

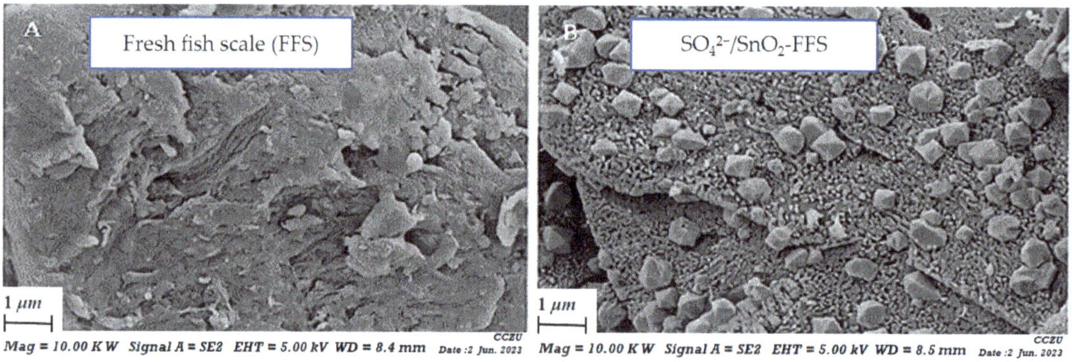

Figure 2. SEM image of FFS (**A**) and solid acid SO_4^{2-}/SnO_2-FFS (**B**).

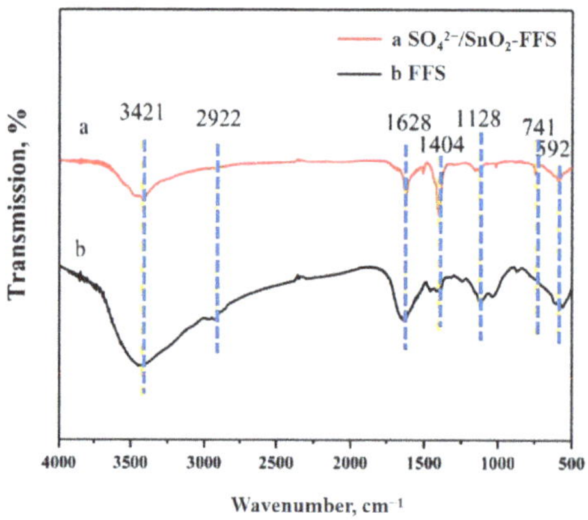

Figure 3. Fourier transform infrared spectroscopy (FT-IR) images of chemocatalyst SO_4^{2-}/SnO_2-FFS and FFS. FT-IR was conducted by using Nicolet IS50 (Thermo Scientific, Waltham, MA, USA) in the range of 4000–500 cm^{-1}.

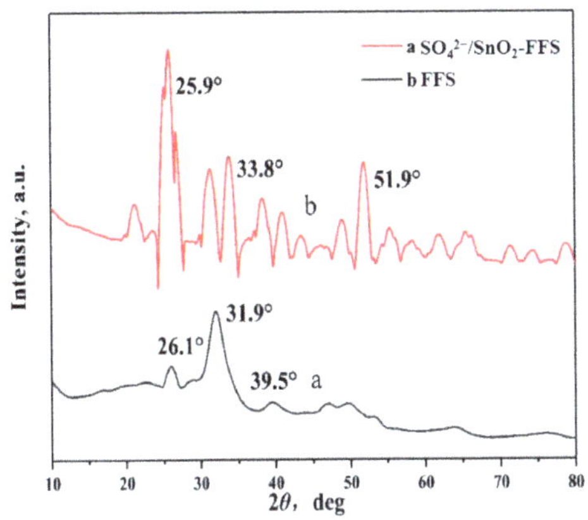

Figure 4. X-ray diffraction (XRD) images of chemocatalyst SO_4^{2-}/SnO_2-FFS and FFS (XRD was carried out by using a D/max-2500 (Japan) instrument (Rigaku Co., Akishima-shi, Japan)). The 2θ angle was scanned between 10° and 80°.

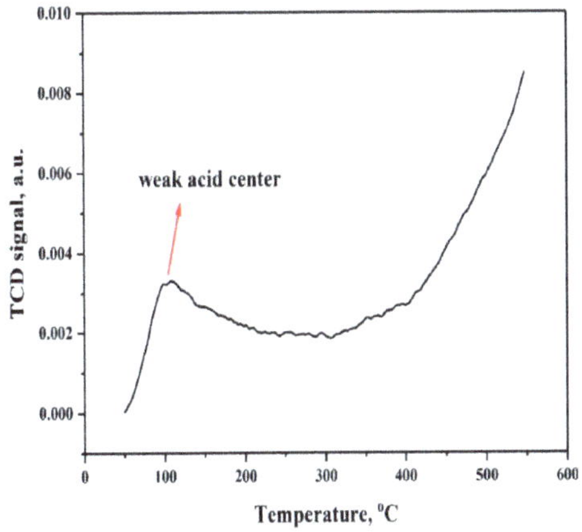

Figure 5. Temperature-programmed desorption of ammonia (NH_3-TPD) image of SO_4^{2-}/SnO_2-FFS. The temperature was scanned between 0–600 °C.

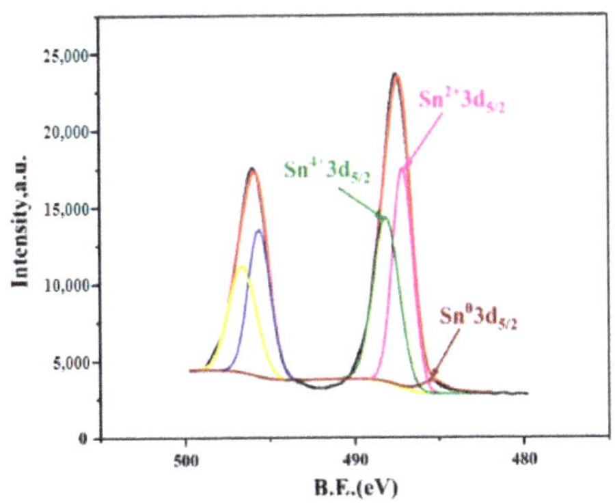

Figure 6. X-ray photoelectron spectroscopy (XPS) image of SO_4^{2-}/SnO_2-FFS. The B.E. (eV) was scanned between 480–500.

It has been reported that tin and sulfonic acid group have good catalytic activity for FAL production, forming lignocellulose and xylose-rich hydrolysate [43]. By comparing the surface structure, pore properties, and chemical groups, SO_4^{2-}/SnO_2-FFS preparation process could slightly alter the surface structure of FFS. The preparation of SO_4^{2-}/SnO_2-FFS involved the solvent immersion and sulfonation reaction by H_2SO_4, which would result in the dissolution of FFS and the disordered structure. Sn^{4+} as the Lewis acid center on the tin-based solid acid could promote the production of FAL [19].

2.2. Effects of SO_4^{2-}/SnO_2-FFS Load, Dehydration Temperature, and Reaction Duration on the FAL Generation

To promote the generation of FAL, three key parameters including SO_4^{2-}/SnO_2-FFS load (0–4.8 wt.%), dehydration temperature (160–190 °C), and reaction duration (10–50 min) were tested on the dehydration of xylose. As displayed in Figure 7a, when the loading of SO_4^{2-}/SnO_2-FFS increased from 0 to 3.6 wt.%, a significant enhancement in FAL yield was observed, and a maximum (48.8% yield) was obtained in the existence of 3.6 wt.% SO_4^{2-}/SnO_2-FFS at 160 °C in 15 min. With increasing the SO_4^{2-}/SnO_2-FFS content from 3.6 to 4.8 wt.%, the FAL yields did not change apparently. Significantly, the suitable load of SO_4^{2-}/SnO_2-FFS was 3.6 wt.%. Sulfonated montmorillonite (2.0 wt.%) transformed corncob to FAL (40% yield) under the temperature of 180 °C after 10 min [44]. Sulfonated argil (3.6 wt.%) transformed *D*-xylose (20 g/L) to FAL (44% yield) after 20 min under the temperature of 180 °C [45]. In this work, SO_4^{2-}/SnO_2-FFS (3.6 wt.%) catalyzed corncob hydrolysates containing *D*-xylose under the temperature of 160 °C after 15 min to generate FAL in the yield of 48.8% yield.

By comparing the yield of reaction at 160–190 °C for 10–50 min, the highest FAL yield (58.8%) was achieved at 170 °C for 15 min, and the reaction temperature was higher compared with the optimization measures in the previous step. When the dehydration temperature was above 170 °C, the dehydration activity was weakened, achieving a declined FAL yield. A higher performance temperature did not favor the formation of FAL [46], and a low FAL yield could be obtained [47]. Although a higher dehydration temperature might increase the dehydration activity for converting xylose into FAL, the unwanted side-reaction potentially would happen [48]. Some unnecessary substances might deposit in the catalytic centers on SO_4^{2-}/SnO_2-FFS, and significant influence the catalytic activity and the FAL yield was thusly decreased [19]. When the reaction duration was more than 15 min, the generation of FAL was weakened (Figure 7b). SO_4^{2-}/SnO_2-FFS had good catalytic activity to catalyze the synthesis of FAL from corncob-derived *D*-xylose within 15 min at 170 °C.

In order to analyze the impact of the SO_4^{2-}/SnO_2-FFS catalyst on FAL yield, xylose aqueous solution was employed as the substrate in a DES_{MLA}–water system, with the addition of $ZnCl_2$, and the reaction was conducted at 170 °C for 15 min. The results are presented in Table S1 (in Support Information), revealing that SnO_2-FFS and SO_4^{2-}/SnO_2 could give higher FAL yields compared to the control group without a catalyst. Furthermore, the FAL was achieved in the highest yield (70.5%) through the catalysis of SO_4^{2-}/SnO_2-FFS. This could be attributed to the synergistic effect of Lewis and Brønsted acidic sites in sulfonated SnO_2-FFS, resulting in more active acid sites [49,50]. In summary, the solid acid catalyst SO_4^{2-}/SnO_2-FFS employed in this study exhibited enhanced catalytic activity in biomass conversion to FAL.

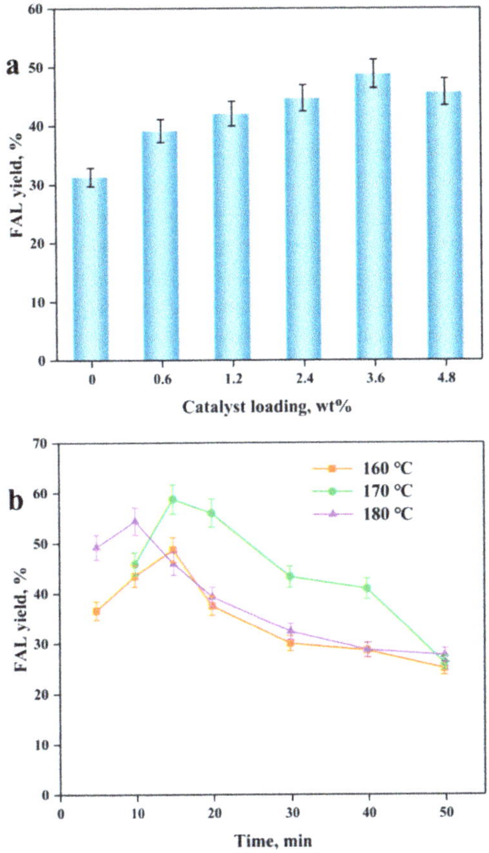

Figure 7. Effects of catalyst SO_4^{2-}/SnO_2-FFS loading (0–4.8 wt.%) on the FAL yield (xylose-rich hydrolysate containing 20 g/L xylose, 0–4.8 wt.% SO_4^{2-}/SnO_2-FFS, 160 °C, 15 min, 500 rpm) (**a**); effects of reaction temperature (160–180 °C) and duration (5–50 min) on the FAL yield (xylose-rich hydrolysate containing 20 g/L xylose, 3.6 wt.% SO_4^{2-}/SnO_2-FFS, 500 rpm) (**b**).

2.3. *Effects of DES–Water Systems on the FAL Formation*

DES acted as a reaction solvent in the catalytic transformation of lignocellulose and its derived sugars [51,52]. Five kinds of carboxylic acids, including maleic acid (MLA) (pKa = 1.9), malic acid (MA) (pKa = 3.46), lactic acid (LA) (pKa = 3.86), citric acid (CA) (pKa = 3.13), and tartaric acid (TA) (pKa = 3.04), were selected to synthesize DESs (DES_{MLA}, DES_{MA}, DES_{LA}, DES_{CA}, and DES_{TA}) with ChCl (hydrogen bond acceptor, HBA). Using 3.6 wt.% SO_4^{2-}/SnO_2-FFS as a catalyst, xylose-rich hydrolysates were catalyzed to generate FAL in a DES–water (20:80, v/v) system. The dissociation constant (pKa) values of carboxylic acids in DESs were close to those in water and also might be associated with the FAL yield [53]. When xylose was dehydrated into FAL in acidic DESs, the pKa value of carboxylic acid in DESs had a certain correlation with the yield of FAL. The results indicated that a lower pKa value would give a higher the yield of FAL (Figure 8a). The carboxylic acid with low pKa value in DES was beneficial to the FAL generation. As a hydrogen bond donor (HBD), MLA has the lowest pKa value (1.9), which facilitated to the FAL formation (FAL yield 63.0%). The dicarboxylic structure of MLA with stable positive anions might form by hydrogen bonds due to the action of protonation. The cis-anion structure could bind to ChCl to enhance the generation of a more stable liquid, which was necessary for transforming *D*-xylose efficiently into FAL. Additionally, with the increase in the -OH group in the carbon chain, the pKa value of carboxylic acid in DESs would be lower and a higher FAL yield would be obtained [54]. Using MLA (pKA = 1.9), CA (pKA = 3.13) and LA (pKA = 3.86) as HBDs, the prepared ChCl-based DESs had apparently influenced the FAL yields. In the water system, the yield of FAL was 58.8%, which was lower than the yield of FAL in DES_{MLA}–water (63.0% yield). The influence of the DES_{MLA}–H_2O volumetric ratio (0:1−4:6, v/v) was examined on the FAL generation through the conversion of *D*-xylose-rich corncob-hydrolysate into FAL with SO_4^{2-}/SnO_2-FFS (Figure 8b). In the DES_{MLA}–water system, the FAL yield exhibited a significant increase with the rising volumetric ratio of DES_{MLA}. When DES was absent from the reaction system, the FAL yield approached 60%. However, with an increase in DES_{MLA} volumetric ratio up to 10%, the yield of FAL exceeded 60%. The highest yield was achieved in DES_{MLA}–water (1:9, v/v). Furthermore, as the volumetric ratio of DES_{MLA} was further increased, the FAL yield displayed a noticeable decline. A high content of DES_{MLA} (> 10%) would increase the viscosity of the reaction medium, leading to the reduction in FAL yield. Accordingly, DES_{MLA}–water (1:9, v/v) was regarded as the most suitable reaction system.

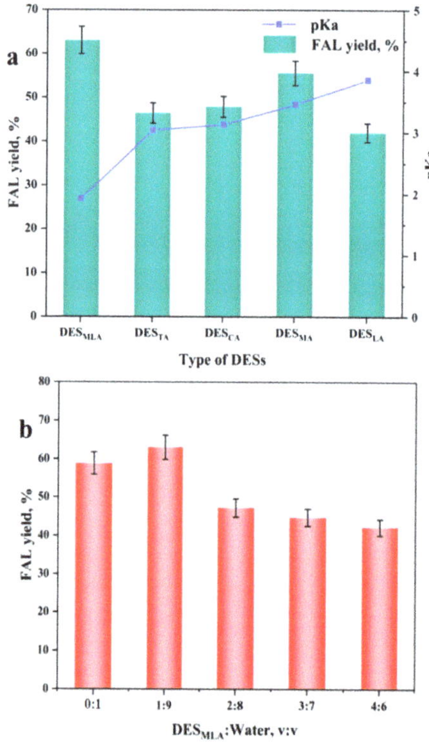

Figure 8. Effect of different DES type and the pKa value of different organic acid in DES on the FAL yield in DES–water (20:80, v/v) system (xylose-rich hydrolysate containing 20 g/L xylose, 3.6 wt.% SO_4^{2-}/SnO_2-FFS, 170 °C, 15 min, 500 rpm) (**a**); effect of DES_{MLA}–water system (0:1–7:3, v/v) on the FAL yield (xylose-rich hydrolysate containing 20 g/L xylose, 3.6 wt.% SO_4^{2-}/SnO_2-FFS, 170 °C, 15 min, 500 rpm) (**b**).

2.4. *Effect of Chloride Salts on the FAL Yield*

Chloride salts are frequently employed to stabilize the transition state and intermediate structures in catalytic processes, thereby reducing undesirable side reactions and significantly improving the yield of FAL [55]. Hence, different chloride salts (15 g/L) were individually supplemented into the catalytic system, which would result in the promotion of the dispersion and yield of FAL. Distinct from the control group, $SnCl_4$ and $LiCl_2$ could apparently hinder the production of FAL (Figure 9a). Most of the chloride salts ($CaCl_2$, $MgCl_2$, NH_4Cl, $AlCl_3$, KCl, $MnCl_2$, $BaCl_2$, and NaCl) did not significantly promote the catalytic activity. However, $NiCl_2$ and $ZnCl_2$ exhibited positive effects on the enhancement of catalytic reactions. Particularly, the addition of $ZnCl_2$ (15 g/L) facilitated the FAL formation and resulted in the highest FAL yield of 66.6%. Moreover, different loads of $ZnCl_2$ (0–30 g/L) were supplemented into DES_{MLA}–water mediums (Figure 9b), and the catalytic effect on the generation of FAL from *D*-xylose was tested. When the content of $ZnCl_2$ was 20 g/L, the highest FAL yield reached 70.5% for 15 min in DES_{MLA}–water (170 °C). In the absence of $ZnCl_2$, the FAL yield was only 63%. The addition of $ZnCl_2$ enhanced the selectivity of FAL production from xylan-rich biomass, reducing the occurrence of side reactions and the formation of by-products [56]. As a result, this led to a further increase in the overall FAL yield.

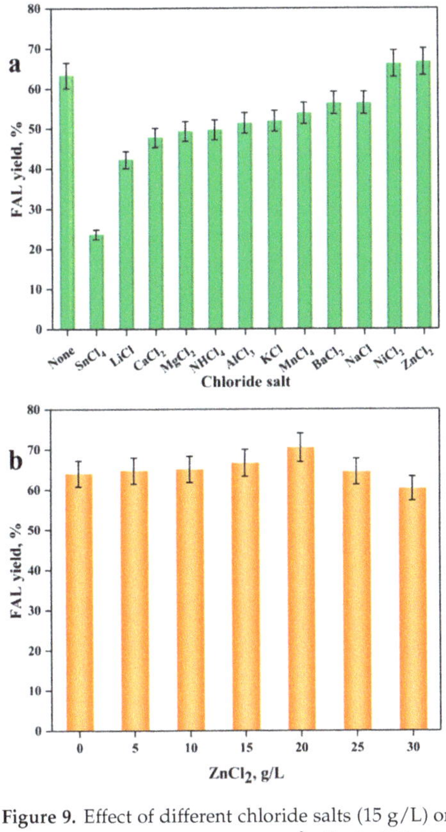

Figure 9. Effect of different chloride salts (15 g/L) on the FAL yield (xylose-rich hydrolysate containing 20 g/L xylose, 3.6 wt.% SO_4^{2-}/SnO_2-FFS, DES_{MLA}–water system (1:9, *v*/*v*), 170 °C, 15 min, 500 rpm) (**a**); effect of $ZnCl_2$ dosage (0–30 g/L) on the FAL yield (xylose-rich hydrolysate containing 20 g/L xylose, 3.6 wt.% SO_4^{2-}/SnO_2-FFS, DES_{MLA}–water system (1:9, *v*/*v*), 170 °C, 15 min, 500 rpm) (**b**).

To sum up, the yield of FAL was up to 70.5% from *D*-xylose-rich corncob-hydrolysate (20 g/L) by 3.6 wt.% SO_4^{2-}/SnO_2-FFS and 20 g/L $ZnCl_2$ in DES_{MLA}–water (1:9, *v*/*v*) system. Various solid acids prepared by different carriers were used to catalyze the synthesis of FAL from biomass or xylose. Using Sn-adamellite or Sn-sepiolite as a catalyst, biomass treated by an alkalic solution was catalyzed into FAL under acidic conditions, and the yields were below 60% [57]. Sn-zeolite, SO_4^{2-}/SnO_2-CS, and Sn-GP were also employed to transform lignocellulose to FAL in the yield of 52.3%, 68.2%, and 47.3%, respectively [16,19,29]. Coconut shell-activated carbon-based biochar solid acid transformed sugarcane bagasse to FAL with a yield of 49% [11]. Accordingly, SO_4^{2-}/SnO_2-FFS had high catalytic ability for the catalysis of biomass to yield FAL (70.5% yield). Notably, this was the first report that fish scale-based solid acids could be efficiently utilized to catalyze biomass into FAL with high yield in DES_{MLA}–water.

2.5. Recyclability of SO_4^{2-}/SnO_2-FFS and DES_{MLA}–Water

The main benefit of using solid acid catalysts is that they can be easy to recycle and reuse [58]. In order to test the stability of SO_4^{2-}/SnO_2-FFS used in *D*-xylose dehydration to prepare FAL in DES_{MLA}–water, a five-cycle experiment was carried out. Before each use, the recovered catalyst was regenerated by sulfonation. As displayed in Figure 10, FAL could reach 70.5% by using fresh SO_4^{2-}/SnO_2-FFS. Then, the yield of FAL dropped to 66.6% at 170 °C for 15 min in the DES_{MLA}–water system at the first reuse of the catalyst, and the FAL yield was weakened gradually to 54.2% after five cycles. The catalytic ability of SO_4^{2-}/SnO_2-FFS decreased slightly but it still could achieve a yield of FAL up to 54%, which had a favorable thermostability. Furthermore, the reduction in catalytic ability might be related to the loss of the catalytic component SnO_2. Sn-zeolite was reused seven times, and the FAL yields were declined from 52% (1st batch) to 37% (seventh batch) [16]. After three cycles of Sn-MMT, the yield of FAL was weakened by 12% [59]. These results displayed that SO_4^{2-}/SnO_2-FFS had good reusability and high thermostability. The good reusability of SO_4^{2-}/SnO_2-FFS could effectively reduce operating costs. Furthermore, DES_{MLA} also could be reused. After being extracted and purified, DES_{MLA}–water was recycled for the next batch. It could be reused five times without a remarkable impact on the productivity of FAL. In conclusion, well recyclability and stability of DES_{MLA}–water showed high potential for industrial production of FAL.

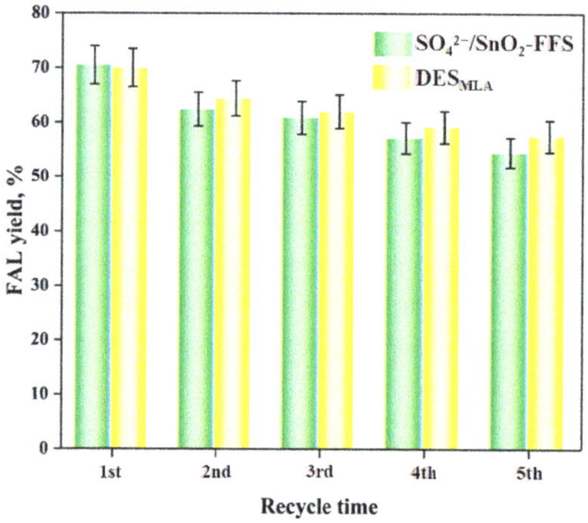

Figure 10. Recycled of SO_4^{2-}/SnO_2-FFS and DES_{MLA} catalyzed xylose-rich hydrolysate to FAL (xylose-rich hydrolysate containing 20 g/L xylose, 3.6 wt.% SO_4^{2-}/SnO_2-FFS, DES_{MLA}–water system (1:9, *v*/*v*), 20 g/L $ZnCl_2$, 170 °C, 15 min, 500 rpm).

2.6. Mass Flow from Corncob to FAL

In Figure 11, the mass flow from CC to FAL in DES_{MLA}–water system was summarized. The 100 g of CC consisted of 41.5 g of glucan, 31.5 g of xylan, 22.5 g of lignin and 4.5 g of others. 1 L of *D*-xylose hydrolysate (containing 20 g *D*-xylose) was obtained after acidolysis of 100 g CC. After carbonization, sulfonation, and loading on FFS, SO_4^{2-}/SnO_2-FFS was finally obtained. The FAL preparation was generally carried out in a 10 L stainless steel reactor which contained 20 g of *D*-xylose, 36 g of catalyst, 0.1 L DES_{MLA}, 0.9 L water and 20 g $ZnCl_2$, then the reaction was well-distributed and mixed at 170 °C for 15 min by stirring at 500 rpm. In the DES_{MLA}–water system, 20 g of *D*-xylose was converted into 9.0 g of FAL. The yield of FAL reached up to 70.5%. This result of mass balance confirmed that CC could be utilized to produce FAL effectively via SO_4^{2-}/SnO_2-FFS. In the fish processing industry, 1 million tons of fish scale are produced each year [60]. Fish scale is an economic, abundant, and sustainable bioresource. In this study, the carbonized fish scale was utilized as a carrier to synthesize biochar-based heterogeneous SO_4^{2-}/SnO_2-FFS catalyst for producing FAL via the dehydration of *D*-xylose in a sustainable reaction system (DES_{MLA}–water). The waste FFSs were prepared into biochar-based solid acid and catalyzed to produce FAL, which realized the efficient utilization of FFSs. A core-shell sulfonated tin-loaded diatomite catalyst could be utilized to prepare FAL from xylose with a yield of 66% [61]. SO_4^{2-}/SnO_2-FFS, which was produced by utilizing carbonized FFS as a carrier, transformed xylose-rich hydrolysate to FAL with a yield of 70.5% at 170 °C for 15 min in DES_{MLA}–water. Apparently, the abundant and renewable fish scale could be utilized to prepare solid acid for the catalytic synthesis of FAL from biomass hydrolysates efficiently.

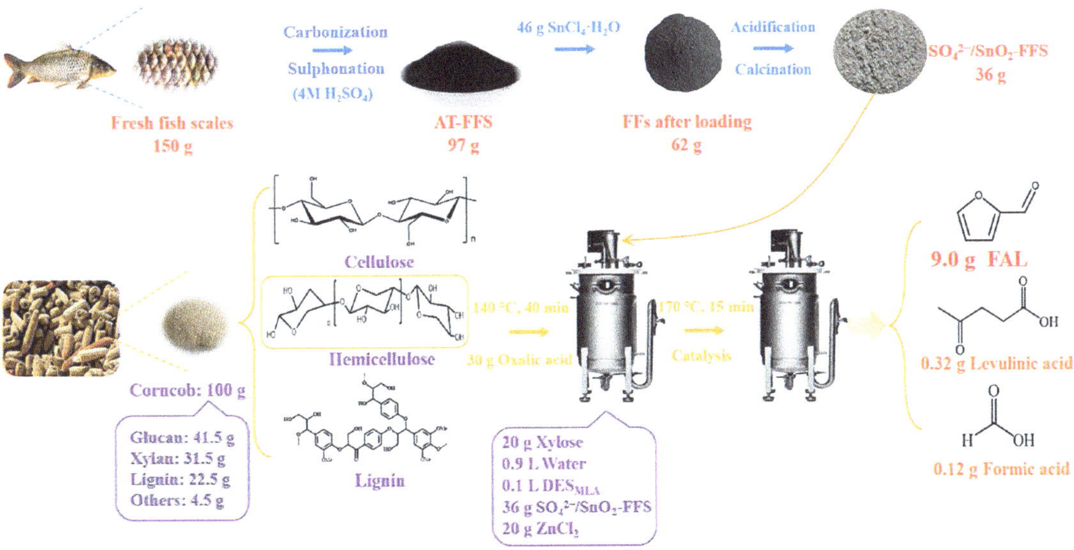

Figure 11. Mass flow for corncob (CC) to FAL.

2.7. *Proposed Mechanism for Synergistic Catalysis of D-Xylose-Rich Hydrolysate into FAL in DES_{MLA}–Water*

In this study, the biochar catalyst SO_4^{2-}/SnO_2-FFS is a biomass-based carbon material with a pore size of 4.5 nm (Table 1), high surface area, and contained uniform load content (Figure 1). In the SO_4^{2-}/SnO_2-FFS catalyst, the induction effect of S=O might cause the electron cloud on the Sn-O to deviate wildly, which would strengthen the L acid site [62]. The hydroxyl group (-OH) of HAP in FFSs might be replaced by fluoride, chloride, and carbonate ions to form fluoroapatite or chloroapatite. The Ca^{2+} can be replaced with various metal ions through an ion exchange reaction to generate apatite corresponding to metal ions. Sn ions had been successfully loaded into SO_4^{2-}/SnO_2-FFS, which might have a positive effect on the catalytic efficiency of FAL [63]. Furthermore, the L acid site on SO_4^{2-}/SnO_2-FFS had a strong attraction to the electrons of water molecules, which could dissociate the L acid site and then form the B acid site. In the dehydration reaction,

DES_{MLA} could be utilized as both catalyst and solvent with acidic and ionic properties to transform biomass-derived sugars to FAL.

In an aqueous system, it was speculated that SO_4^{2-}/SnO_2-FFS and DES_{MLA} had a synergistic catalytic action in the production of FAL (Figure 12). According to the HPLC analysis results, the reaction solution contained *D*-xylose, glucose, cellobiose, formic acid, levulinic acid, and FAL (data not shown). Sn ions and acid ions were uniformly supported on the catalyst carrier. The proposed catalytic mechanism was as follows: the loop-opening reaction of *D*-xylose from *D*-xylose hydrolysate of CC produced chain structure, and the L acid site of SO_4^{2-}/SnO_2-FFS cooperated with acyclic *D*-xylose to generate more reactive isomer through hydrogen transfer reaction. FAL was further synthesized by loop-closing and dehydration by B acid. The pore size of the FFS carbon base was suitable for contact between *D*-xylose and the active center, which might promote the transformation of *D*-xylose. Moreover, the HAP component in FFS was prone to ion exchange reactions with tin ions, and tin ions had a certain attraction to sulfate ions and chloride ions, which improved the acidity. ChCl might improve the proton availability and activity in the system under acidic conditions. Additionally, Cl^- from ChCl could be beneficial to the isomerization of the *D*-xylose molecule and strengthen its attraction on the SO_4^{2-} site, which might increase the reaction rate [64]. It was worth noting that chloride salt might accelerate the transformation of *D*-xylose from lignocellulose to FAL [65]. Hence, DES_{MLA} played a synergistic role in FAL production when SO_4^{2-}/SnO_2-FFS was used as a chemocatalyst. In addition, under this system, glucose from hydrolysate might be isomerized and dehydrated to form HMF, which could be also partially decomposed into formic acid and levulinic acid.

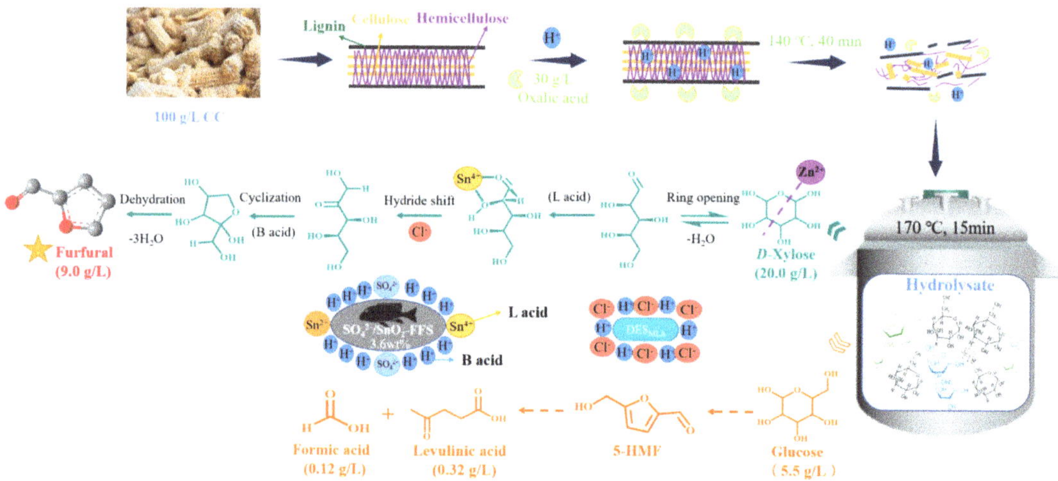

Figure 12. Proposed catalytic mechanism for catalyzing *D*-Xylose hydrolysate of CC into FAL and byproduct at 170 °C for 15 min with 20 g/L $ZnCl_2$ in DES_{MLA}–water (1:9, *v*/*v*) system.

Clean production of biobased chemicals from renewable lignocellulose is essential in a sustainable biorefinery [66]. Lignocellulose, with its wide range of sources, low cost, and sustainable production, can be derived from various biomass materials (e.g., straw, bagasse, wheat husk, etc.) for the production of FAL [67]. In China, about 900 million tons of crop waste are produced each year, 11.1% of which is CC [65]. In addition, FFS as a waste bioresource has good biocompatibility and degradability, and excellent mechanical properties due to their highly ordered layered microstructure and composition similar to human hard tissue, which has been utilized in tissue engineering, biological filling, sewage processing, and flexible electronics [30]. In this study, FFS was first used as a carrier to prepare biochar chemocatalyst (SO_4^{2-}/SnO_2-FFS) for FAL production. The yield of

FAL (70.5%) was obtained by catalyzing the CC-hydrolysate containing xylose to FAL in DES_{MLA}–water system using SO_4^{2-}/SnO_2-FFS as catalyst, verifying that available, cheap, and renewable FFS could be used as a bio-based carrier for the synthesis of highly catalytic activity of heterogeneous biochar catalysts to transform biomass-derived *D*-xylose into FAL in a high yield. Compared with the conditions or yields of FAL production from biomass or xylose catalyzed by other solid acids, the yield of FAL by SO_4^{2-}/SnO_2-FFS was higher, the reaction time was shorter, and the reaction solution could be reused, which would not only save energy, but also be more friendly to the environment. Apparently, the prepared biochar catalyst catalyzed biomass for FAL production has full application prospects.

3. Materials and Methods

3.1. Reagents and Materials

Fresh fish scale of carp was collected from a fishery in Weifang (Weifang, China). Choline chloride (ChCl), malic acid (MA), maleic acid (MLA), lactic acid (LA), Citric acid (CA), tartaric acid (TA), $MgCl_2$, NaCl, KCl, $FeCl_3$, $AlCl_3$, $CrCl_3$, $CuCl_2$, $CoCl_2$, NH_4Cl, $ZnCl_2$, furfural (FAL), $SnCl_4{\cdot}5H_2O$, ammonia, ethanol, sulfuric acid (H_2SO_4), and other chemicals were bought from other commercial sources (Shanghai, China).

3.2. Synthesis of DESs

DES was prepared according to the procedure as previously reported [68]. The seven DESs used in this study were DES_{LA}, DES_{MA} DES_{CA}, DES_{MLA}, and DES_{TA}. These DESs were prepared by mixing ChCl and a corresponding organic acid at 80 °C. The ratio of ChCl to organic acid was 1:1 (mol/mol).

3.3. Preparation of Xylose-Rich Hydrolysate from Corncob and SO_4^{2-}/SnO_2-FFS Catalyst

Xylose-rich hydrolysate from corncob was produced by hydrothermal reaction. Corncob was hydrolyzed with DES_{MLA} at 140 °C for 40 min. Then, xylose-rich hydrolysate was obtained by filtration. By concentration, the xylose in hydrolysate reached 20 g/L.

FFS was thoroughly washed with tap water, followed by boiling in deionized water. Subsequently, the boiled mixture was filtered to remove gelatin, proteins, and other impurities. The obtained wet fish scale was then dried in a 60 °C oven for 16 h. The desiccated solid was finely ground using a milling machine to achieve particle sizes between 60 and 80 mesh. Finally, the powdered fish scale was placed in a muffle furnace and subjected to a temperature of 300 °C for 1 h, resulting in the formation of biochar. The obtained powders were immersed in 4 M H_2SO_4 at 60 °C for 4 h for sulfonation. The sulfonated solid powders were obtained by filtration and rinsed continuously with deionized water until neutral. The acid-treated FFS (AT-FFS) was blended with $SnCl_4{\cdot}5H_2O$ and anhydrous ethanol, and then ammonia (25.0 wt.%) was slowly dripped into this mixture to regulate the pH to 6.0. This generated colloidal liquor was dried in an oven (70 °C) for 12 h and then oven dried at 90 °C for another 12 h. The collected powders were steeped in 500 mM H_2SO_4 for 180 min. The acidic solid powders were separated by filtration. The powders ware dried for 12 h in an oven (80 °C), and then calcined in a muffle furnace (500 °C) for 4 h. Finally, the formed SO_4^{2-}/SnO_2-FFS was collected for further use.

3.4. Transformation of Xylose-Rich Hydrolysate into FAL by SO_4^{2-}/SnO_2-FFS in DES–Water System

In 40 mL DES-water system, xylose-rich hydrolysate (20 g/L) was mixed with SO_4^{2-}/SnO_2-FFS in a 100 mL stainless-steel autoclave (TGYF-A-0.1L, Zhengzhou Huate Instrument Equipment Co., Ltd., Zhengzhou, China) by stirring (500 rpm). After the xylose-rich corncob-hydrolysate was catalyzed with SO_4^{2-}/SnO_2-FFS (0–4.8 wt.%) at 160–180 °C for 10–50 min in DES–water (0:1–4:6, *v*/*v*) containing chlorite salts (0–30.0 g/L), the autoclave was quenched in an ice-water bath to room temperature. FAL yields were calculated as below equation:

$$\text{FAL yield } (\%) = \frac{\text{FAL produced (g)}}{D\text{-Xylose (g)}} \times \frac{150}{96} \times 100$$

3.5. Reuse of SO_4^{2-}/SnO_2-FFS and DES

To evaluate the activity and stability of solid acid, SO_4^{2-}/SnO_2-FFS was recovered and repeatedly reused five times to catalyze *D*-xylose into FAL. After each use, SO_4^{2-}/SnO_2-FFS was extracted by suction flask, washed thoroughly with deionized water, and dried in an oven (60 °C). Then it was calcined in the muffle furnace to remove product residues and further sulfonation. The recovered SO_4^{2-}/SnO_2-FFS was used in the next batch of reaction to measure its effect. Each batch was performed for 15 min in an autoclave (170 °C) containing 40 mL of DES−water (10:90, *v*/*v*), 20.0 g/L of *D*-xylose, and 20.0 g/L of $ZnCl_2$. Subsequently, the reaction media were extracted with ethyl acetate three times, and the reaction liquid was reused five times.

3.6. Analytical Methods

Through FTIR, XRD, and BET and XPS [69], the diversity between SO_4^{2-}/SnO_2-FFS and FFS was discovered. FAL was measured with HPLC as reported in reference [45]. NH_3-TPD was carried out by using a ChemiSorb 2720 chemisorption system (Micromeritics, Norcross, GA, USA) to measure the acid strength of SO_4^{2-}/SnO_2-FFS. SO_4^{2-}/SnO_2-FFS (0.10 g) was treated for 120 min with helium (300 °C). After the NH_3 adsorption for 60 min at 50 °C, SO_4^{2-}/SnO_2-FFS was pretreated with helium to remove excessive NH_3 until a stable baseline was obtained. SO_4^{2-}/SnO_2-FFS was heated to 850 °C for NH_3 desorption at a flow rate of 30–50 mL per minute and a heating rate of 10 °C per minute.

4. Conclusions

In this work, the abundant biobased FFS was used as a carrier to prepare the biochar SO_4^{2-}/SnO_2-FFS catalyst. After the acid hydrolysis of corncob, xylose-rich corncob-hydrolysate was used as a substrate to produce FAL through the catalysis with biochar SO_4^{2-}/SnO_2-FFS catalyst. In DES_{MLA}–water (DES_{MLA}, 10 vol%), the FAL yield was obtained at 70.5% by SO_4^{2-}/SnO_2-FFS (3.6 wt.%) with $ZnCl_2$ (20.0 g/L) for 15 min under the temperature of 170 °C. The potential catalytic mechanism for the production of FAL from corncob catalyzed by SO_4^{2-}/SnO_2-FFS in DES_{MLA}–water was proposed. An efficient method for the synthesis of biofuran from lignocellulose was successfully developed by using a biochar solid acid catalyst in a green reaction system.

Supplementary Materials: The following supporting information can be downloaded at: https://www.mdpi.com/article/10.3390/catal13091277/s1. Table S1. The FAL production from xylose catalyzed by FFS-based solid acids in DES_{MLA}–water (1/9, *v*/*v*) system.

Author Contributions: Methodology, software, validation, formal analysis, investigation, resources, data curation, writing—original draft preparation, L.Y., Y.L. and Y.W.; visualization, supervision, writing—review and editing, Y.H.; Methodology, formal analysis, investigation, resources, data curation, C.M. All authors have read and agreed to the published version of the manuscript.

Funding: This research was kindly funded by Open Project Program of the State Key Laboratory of Bioreactor Engineering.

Data Availability Statement: Not applicable.

Acknowledgments: The authors thank the Analysis and Testing Center (Changzhou University) for the analysis of solid acid with FT-IR, SEM, XRD, BET analysis, NH_3-TPD, and XPS.

Conflicts of Interest: The authors declare no conflict of interest.

Abbreviations

DES	deep eutectic solvent
LB	lignocellulosic biomass
FFS	fresh fish scale
L	Lewis
B	Brönsted
ChCl	choline chloride
CC	corncob
HAP	hydroxyapatite
MA	malic acid
MLA	maleic acid
LA	lactic acid
CA	citric acid
TA	tartaric acid
HBD	hydrogen bond donor
XPS	X-ray photoelectron spectroscopy
HPLC	high performance liquid chromatography
FAL	furfural
FOL	furfuryl alcohol
XRD	X-ray diffraction
BET	Brunner–Emmet–Teller measurements
TG	thermogravimetric analysis
SEM	scanning electron microscopy
FT-IR	Fourier transform infrared spectroscopy

References

1. Chen, X.J.; Guo, T.; Mo, X.; Zhang, L.D.; Wang, R.F.; Xue, Y.N.; Fan, X.L.; Sun, S.L. Reduced nutrient release and greenhouse gas emissions of lignin-based coated urea by synergy of carbon black and polysiloxane. *Int. J. Biol. Macromol.* **2023**, *231*, 123334. [CrossRef]
2. He, Y.C.; Liu, F.; Di, J.H.; Ding, Y.; Zhu, Z.Z.; Wu, Y.Q.; Chen, L.; Wang, C.; Xue, Y.F.; Chong, G.G. Effective enzymatic saccharification of dilute NaOH extraction of chestnut shell pretreated by acidified aqueous ethylene glycol media. *Ind. Crops Prod.* **2016**, *81*, 129–138. [CrossRef]
3. Hu, L.; Wu, Z.; Jiang, Y.T.; Wang, X.Y.; He, A.Y.; Song, J.; Xu, J.M.; Zhou, S.Y.; Zhao, Y.J.; Xu, J.X. Recent advances in catalytic and autocatalytic production of biomass-derived 5-hydroxymethylfurfural. *Renew. Sustain. Energy Rev.* **2020**, *134*, 110317. [CrossRef]
4. Chen, X.J.; Guo, T.; Yang, H.C.; Zhang, L.D.; Xue, Y.N.; Wang, R.F.; Fan, X.L.; Sun, S.L. Environmentally friendly preparation of lignin/paraffin/epoxy resin composite-coated urea and evaluation for nitrogen efficiency in lettuce. *Int. J. Biol. Macromol.* **2022**, *221*, 1130–1141. [CrossRef]
5. Jia, W.; Zhang, J.; Yu, X.; Feng, Y.; Wang, Q.; Sun, Y.; Tang, X.; Zeng, X.; Lin, L. Insight into the Mars-van Krevelen mechanism for production 2,5-diformylfuran over FeNx@C catalyst. *Biomass Bioenergy* **2022**, *156*, 106320. [CrossRef]
6. He, J.; Huang, C.; Lai, C.; Wang, Z.; Yuan, L.; Ragauskas, A.; Yan, Y.; Yong, Q. Revealing the mechanism of lignin re-polymerization inhibitor in acidic pretreatment and its impact on enzymatic hydrolysis. *Ind. Crops Prod.* **2022**, *179*, 114631. [CrossRef]
7. Hu, L.; Jiang, Y.; Wu, Z.; Wang, X.Y.; He, A.Y.; Xu, J.X.; Xu, J.M. State-of-the-art advances and perspectives in the separation of biomass-derived 5-hydroxymethylfurfural. *J. Clean. Prod.* **2020**, *276*, 124219. [CrossRef]
8. Ma, S.; Chen, B.; Zeng, A.; Li, Z.; Tang, X.; Sun, Y.; Lin, L.; Zeng, X. Chemical structure change of lignin extracted from bamboo biomass by maleic acid. *Int. J. Biol. Macromol.* **2022**, *221*, 986–993. [CrossRef] [PubMed]
9. Amini, E.; Valls, C.; Roncero, M.B. Ionic liquid-assisted bioconversion of lignocellulosic biomass for the development of value-added products. *J. Clean. Prod.* **2021**, *326*, 129275. [CrossRef]
10. Zhou, Z.; Ju, X.; Zhou, M.; Xu, X.; Fu, J.; Li, L. An enhanced ionic liquid-tolerant immobilized cellulase system via hydrogel microsphere for improving in situ saccharification of biomass. *Bioresour. Technol.* **2019**, *294*, 122146. [CrossRef]
11. Li, Q.; Ma, C.L.; Zhang, P.Q.; Li, Y.Y.; Zhu, X.; He, Y.C. Effective conversion of sugarcane bagasse to furfural by coconut shell activated carbon-based solid acid for enhancing whole-cell biosynthesis of furfurylamine. *Ind. Crops Prod.* **2021**, *160*, 113169. [CrossRef]
12. Ramírez, E.; Bringué, R.; Fité, C.; Iborra, M.; Tejero, J.; Cunill, F. Role of ion-exchange resins as catalyst in the reaction-network of transformation of biomass into biofuels. *J. Chem. Technol. Biotechnol.* **2017**, *92*, 2775–2786. [CrossRef]
13. Xia, Z.-H.; Zong, M.-H.; Li, N. Catalytic synthesis of 2,5-bis(hydroxymethyl)furan from 5-hydroxymethylfurfual by recombinant Saccharomyces cerevisiae. *Enzym. Microb. Technol.* **2020**, *134*, 109491. [CrossRef] [PubMed]

14. Yemiş, O.; Mazza, G. Acid-catalyzed conversion of xylose, xylan and straw into furfural by microwave-assisted reaction. *Bioresour. Technol.* **2011**, *102*, 7371–7378. [CrossRef]
15. Mittal, A.; Pilath, H.M.; Johnson, D.K. Direct conversion of biomass carbohydrates to platform chemicals: 5-hydroxymethylfurfural (HMF) and furfural. *Energy Fuels* **2020**, *34*, 3284–3293. [CrossRef]
16. Li, Y.Y.; Li, Q.; Zhang, P.Q.; Ma, C.L.; Xu, J.H.; He, Y.C. Catalytic conversion of corncob to furfuryl alcohol in tandem reaction with tin-loaded sulfonated zeolite and NADPH-dependent reductase biocatalyst. *Bioresour. Technol.* **2021**, *320*, 124267. [CrossRef] [PubMed]
17. Chatterjee, A.; Hu, X.; Lam, F.L.Y. Modified coal fly ash waste as an efficient heterogeneous catalyst for dehydration of xylose to furfural in biphasic medium. *Fuel* **2019**, *239*, 726–736. [CrossRef]
18. Chen, X.; Li, Z.; Zhang, L.; Wang, H.; Qiu, C.; Fan, X.; Sun, S. Preparation of a novel lignin-based film with high solid content and its physicochemical characteristics. *Ind. Crops Prod.* **2021**, *164*, 113396. [CrossRef]
19. Ji, L.; Tang, Z.; Yang, D.; Ma, C.; He, Y.C. Improved one-pot synthesis of furfural from corn stalk with heterogeneous catalysis using corn stalk as biobased carrier in deep eutectic solvent–water system. *Bioresour. Technol.* **2021**, *340*, 125691. [CrossRef]
20. Yang, Q.; Tang, Z.; Xiong, J.; He, Y. Sustainable chemoenzymatic cascade transformation of corncob to furfuryl alcohol with rice husk-based heterogeneous catalyst UST-Sn-RH. *Catalysts* **2022**, *13*, 37. [CrossRef]
21. Dulie, N.W.; Woldeyes, B.; Demsash, H.D. Synthesis of lignin-carbohydrate complex-based catalyst from *Eragrostis tef* straw and its catalytic performance in xylose dehydration to furfural. *Int. J. Biol. Macromol.* **2021**, *171*, 10–16. [CrossRef] [PubMed]
22. Zhang, T.; Li, W.; Xu, Z.; Liu, Q.; Ma, Q.; Jameel, H.; Chang, H.-M.; Ma, L. Catalytic conversion of xylose and corn stalk into furfural over carbon solid acid catalyst in γ-valerolactone. *Bioresour. Technol.* **2016**, *209*, 108–114. [CrossRef] [PubMed]
23. Xu, D.; Tang, W.; Tang, Z.; He, Y. An Efficient Strategy for Chemoenzymatic Conversion of Corn Stover to Furfuryl Alcohol in Deep Eutectic Solvent ChCl: PEG10000−Water Medium. *Catalysts* **2023**, *13*, 467. [CrossRef]
24. Smith, E.L.; Abbott, A.P.; Ryder, K.S. Deep eutectic solvents (DESs) and their applications. *Chem. Rev.* **2014**, *114*, 11060–11082. [CrossRef]
25. Zuo, M.; Wang, X.; Jia, W.; Zhu, Y.; Zeng, X.; Lin, L. Efficient 5-hydroxymethylfurfural synthesis from carbohydrates and food wastes in aqueous-natural deep eutectic solvent (A-NADES) with robust Al_2O_3 or $Al(OH)_3$. *Fuel* **2022**, *369*, 125062. [CrossRef]
26. Li, A.L.; Hou, X.D.; Lin, K.P.; Zhang, X.; Fu, M.H. Rice straw pretreatment using deep eutectic solvents with different constituents molar ratios: Biomass fractionation, polysaccharides enzymatic digestion and solvent reuse. *J. Biosci. Bioeng.* **2018**, *126*, 346–354. [CrossRef]
27. Bu, C.Y.; Yan, Y.X.; Zou, L.H.; Ouyang, S.P.; Zheng, Z.J.; Ouyang, J. Comprehensive utilization of corncob for furfuryl alcohol production by chemo-enzymatic sequential catalysis in a biphasic system. *Bioresour. Technol.* **2021**, *319*, 124156. [CrossRef]
28. Yong, K.J.; Wu, T.Y.; Lee, C.B.T.L.; Lee, Z.J.; Liu, Q.; Jahim, J.M.; Zhou, Q.; Zhang, L. Furfural production from biomass residues: Current technologies, challenges and future prospects. *Biomass Bioenergy* **2022**, *161*, 106458. [CrossRef]
29. Mukkanti, V.B.; Tembhurkar, A.R. Defluoridation of water using adsorbent derived from the *Labeo rohita* (rohu) fish scales waste: Optimization, isotherms, kinetics, and thermodynamic study. *Sustain. Chem. Pharm.* **2021**, *23*, 100520. [CrossRef]
30. Qin, D.; Bi, S.; You, X.; Wang, M.; Cong, X.; Yuan, C.; Yu, M.; Cheng, X.; Chen, X.-G. Development and application of fish scale wastes as versatile natural biomaterials. *Chem. Eng. J.* **2022**, *428*, 131102. [CrossRef]
31. Yao, Q.F.; Zhou, D.S.; Yang, J.H.; Huang, W.T. Directly reusing waste fish scales for facile, large-scale and green extraction of fluorescent carbon nanoparticles and their application in sensing of ferric ions. *Sustain. Chem. Pharm.* **2020**, *17*, 100305. [CrossRef]
32. Tang, Z.; Li, Q.; Di, J.; Ma, C.; He, Y.-C. An efficient chemoenzymatic cascade strategy for transforming biomass into furfurylamine with lobster shell-based chemocatalyst and mutated ω-transaminase biocatalyst in methyl isobutyl ketone-water. *Bioresour. Technol.* **2023**, *369*, 128424. [CrossRef] [PubMed]
33. Romo, J.E.; Bollar, N.V.; Zimmermann, C.J.; Wettstein, S.G. Conversion of sugars and biomass to furans using heterogeneous catalysts in biphasic solvent systems. *ChemCatChem* **2018**, *10*, 4805–4816. [CrossRef] [PubMed]
34. Bhaumik, R.; Mondal, N.K.; Chattoraj, S. An optimization study for defluoridation from synthetic fluoride solution using scale of Indian major carp Catla (*Catla catla*): An unconventional biosorbent. *J. Fluor. Chem.* **2017**, *195*, 57–69. [CrossRef]
35. Ramadoss, P.; Arul, K.T.; Ramya, J.R.; Begam, M.R.; Chandra, V.S.; Manikandan, E. Enhanced mechanical strength and sustained drug release of gelatin/keratin scaffolds. *Mater. Lett.* **2017**, *186*, 109–112. [CrossRef]
36. George, A.M.; Tembhurkar, A.R. Analysis of equilibrium, kinetic, and thermodynamic parameters for biosorption of fluoride from water onto coconut (*Cocos nucifera* Linn.) root developed adsorbent. *Chin. J. Chem. Eng.* **2019**, *27*, 92–99. [CrossRef]
37. Kim, H.; Yang, S.; Kim, D.H. One-pot conversion of alginic acid into furfural using Amberlyst-15 as a solid acid catalyst in γ-butyrolactone/water co-solvent system. *Environ. Res.* **2020**, *187*, 109667. [CrossRef]
38. Xia, D.; Liu, Y.; Cheng, X.; Gu, P.; Chen, Q.; Zhang, Z. Temperature-tuned fish-scale biochar with two-dimensional homogeneous porous structure: A promising uranium extractant. *Appl. Surf. Sci.* **2022**, *591*, 153136. [CrossRef]
39. Zhang, W.; Sun, M.; Yin, J.; Lu, K.; Schwingenschlögl, U.; Qiu, X.; Alshareef, H.N. Accordion-like carbon with high nitrogen doping for fast and stable K ion storage. *Adv. Energy Mater.* **2021**, *11*, 2101928. [CrossRef]
40. Ferrairo, B.M.; Mosquim, V.; de Azevedo-Silva, L.J.; Pires, L.A.; Padovini, D.S.S.; Magdalena, A.G.; Fortulan, C.A.; Lisboa-Filho, P.N.; Rubo, J.H.; Borges, A.F.S. Production of bovine hydroxyapatite nanoparticles as a promising biomaterial via mechanochemical and sonochemical methods. *Mater. Chem. Phys.* **2023**, *295*, 127046. [CrossRef]

41. Teng, X.; Si, Z.; Li, S.; Yang, Y.; Wang, Z.; Li, G.; Zhao, J.; Cai, D.; Qin, P. Tin-loaded sulfonated rape pollen for efficient catalytic production of furfural from corn stover. *Ind. Crops Prod.* **2020**, *151*, 112481. [CrossRef]
42. Piccirillo, C.; Pullar, R.C.; Tobaldi, D.M.; Castro, P.M.L.; Pintado, M.M.E. Hydroxyapatite and chloroapatite derived from sardine by-products. *Ceram. Int.* **2014**, *40*, 13231–13240. [CrossRef]
43. Jia, Q.; Teng, X.; Yu, S.; Si, Z.; Li, G.; Zhou, M.; Cai, D.; Qin, P.; Chen, B. Production of furfural from xylose and hemicelluloses using tin-loaded sulfonated diatomite as solid acid catalyst in biphasic system. *Bioresour. Technol. Rep.* **2019**, *6*, 145–151. [CrossRef]
44. Huang, Y.; Liao, X.; Deng, Y.; He, Y. Co-catalysis of corncob with dilute formic acid plus solid acid SO_4^{2-}/SnO_2-montmorillonite under the microwave for enhancing the biosynthesis of furfuralcohol. *Catal. Commun.* **2019**, *120*, 38–41. [CrossRef]
45. Xue, X.X.; Ma, C.L.; Di, J.H.; Huo, X.Y.; He, Y.C. One-pot chemo-enzymatic conversion of *D*-xylose to furfuralcohol by sequential dehydration with oxalic acid plus tin-based solid acid and bioreduction with whole-cells. *Bioresour. Technol.* **2018**, *268*, 292–299. [CrossRef]
46. He, Y.C.; Jiang, C.X.; Chong, G.G.; Di, J.H.; Wu, Y.F.; Wang, B.Q.; Xue, X.X.; Ma, C.L. Chemical-enzymatic conversion of corncob-derived xylose to furfuralcohol by the tandem catalysis with SO_4^{2-}/SnO_2-kaoline and *E. coli* CCZU-T15 cells in toluene-water media. *Bioresour. Technol.* **2017**, *245*, 841–849. [CrossRef]
47. Bhaumik, P.; Dhepe, P.L. Exceptionally high yields of furfural from assorted raw biomass over solid acids. *RSC Adv.* **2014**, *4*, 26215–26221. [CrossRef]
48. Hua, D.; Ding, H.; Liu, Y.; Li, J.; Han, B. Dehydration of xylose to furfural over imidazolium-based ionic liquid with phase separation. *Catalysts* **2021**, *11*, 1552. [CrossRef]
49. Fihri, A.; Len, C.; Varma, R.S.; Solhy, A. Hydroxyapatite: A review of syntheses, structure and applications in heterogeneous catalysis. *Coord. Chem. Rev.* **2017**, *347*, 48–76. [CrossRef]
50. Weingarten, R.; Tompsett, G.A.; Conner, W.C.; Huber, G.W. Design of solid acid catalysts for aqueous-phase dehydration of carbohydrates: The role of Lewis and Brønsted acid sites. *J. Catal.* **2011**, *279*, 174–182. [CrossRef]
51. Tang, X.; Zuo, M.; Li, Z.; Liu, H.; Xiong, C.; Zeng, X.; Sun, Y.; Hu, L.; Liu, S.; Lei, T.; et al. Green processing of lignocellulosic biomass and its derivatives in deep eutectic solvents. *ChemSusChem* **2017**, *10*, 2696–2706. [CrossRef]
52. Zhang, S.; Ma, C.; Li, Q.; Li, Q.; He, Y.C. Efficient chemoenzymatic valorization of biobased D-fructose into 2,5-bis(hydroxymethyl)furan with deep eutectic solvent Lactic acid:Betaine and *Pseudomonas putida* S12 whole cells. *Bioresour. Technol.* **2022**, *344*, 126299. [CrossRef] [PubMed]
53. Pan, L.; Li, Q.; Tao, Y.; Ma, C.; Chai, H.; Ai, Y.; He, Y.-C. An efficient chemoenzymatic strategy for valorisation of corncob to furfuryl alcohol in CA: Betaine-water. *Ind. Crops Prod.* **2022**, *186*, 115203. [CrossRef]
54. Shen, J.; Gao, R.; He, Y.-C.; Ma, C. Efficient synthesis of furfural from waste biomasses by sulfonated crab shell-based solid acid in a sustainable approach. *Ind. Crops Prod.* **2023**, *202*, 116989. [CrossRef]
55. Sadula, S.; Oesterling, O.; Nardone, A.; Dinkelacker, B.; Saha, B. One-pot integrated processing of biopolymers to furfurals in molten salt hydrate: Understanding synergy in acidity. *Green Chem.* **2017**, *19*, 3888–3898. [CrossRef]
56. Hu, B.; Xie, W.L.; Wu, Y.T.; Liu, J.; Ma, S.W.; Wang, T.P.; Zheng, S.; Lu, Q. Mechanism study on the formation of furfural during zinc chloride-catalyzed pyrolysis of xylose. *Fuel* **2021**, *295*, 120656. [CrossRef]
57. Yang, D.; Ma, C.; Peng, B.; Xu, J.; He, Y.C. Synthesis of furoic acid from biomass via tandem pretreatment and biocatalysis. *Ind. Crops Prod.* **2020**, *153*, 112580. [CrossRef]
58. Faruque, M.O.; Razzak, S.A.; Hossain, M.M. Application of heterogeneous catalysts for biodiesel production from microalgal oil—A review. *Catalysts* **2020**, *10*, 1025. [CrossRef]
59. Li, H.; Ren, J.; Zhong, L.; Sun, R.; Liang, L. Production of furfural from xylose, water-insoluble hemicelluloses and water-soluble fraction of corncob via a tin-loaded montmorillonite solid acid catalyst. *Bioresour. Technol.* **2015**, *176*, 242–248. [CrossRef]
60. Muhammad, N.; Gao, Y.; Iqbal, F.; Ahmad, P.; Ge, R.; Nishan, U.; Rahim, A.; Gonfa, G.; Ullah, Z. Extraction of biocompatible hydroxyapatite from fish scales using novel approach of ionic liquid pretreatment. *Sep. Purif. Technol.* **2016**, *161*, 129–135. [CrossRef]
61. Cai, D.; Chen, H.; Zhang, C.; Teng, X.; Li, X.; Si, Z.; Li, G.; Yang, S.; Wang, G.; Qin, P. Carbonized core-shell diatomite for efficient catalytic furfural production from corn cob. *J. Clean. Prod.* **2021**, *283*, 125410. [CrossRef]
62. Zhu, C.; Wang, K.; Luo, J.; Tian, B.; Sun, J.; Liu, X.; Zhu, W.; Zou, Z. Solid superacid SO_4^{2-}-$S_2O_8^{2-}/SnO_2$-Nd_2O_3-catalyzed esterification of α-aromatic amino acids. *Mol. Catal.* **2023**, *535*, 112833. [CrossRef]
63. Ni, J.; Di, J.; Ma, C.; He, Y.C. Valorisation of corncob into furfuryl alcohol and furoic acid via chemoenzymatic cascade catalysis. *Bioresour. Bioprocess.* **2021**, *8*, 113. [CrossRef]
64. Enslow, K.R.; Bell, A.T. The role of metal halides in enhancing the dehydration of xylose to furfural. *ChemCatChem* **2015**, *7*, 479–489. [CrossRef]
65. Zha, J.; Fan, B.; He, J.; He, Y.-C.; Ma, C. Valorization of biomass to furfural by chestnut shell-based solid acid in methyl isobutyl ketone–water–sodium chloride system. *Appl. Biochem. Biotechnol.* **2022**, *194*, 2021–2035. [CrossRef]
66. Mujtaba, M.; Fernandes Fraceto, L.; Fazeli, M.; Mukherjee, S.; Savassa, S.M.; Araujo de Medeiros, G.; do Espírito Santo Pereira, A.; Mancini, S.D.; Lipponen, J.; Vilaplana, F. Lignocellulosic biomass from agricultural waste to the circular economy: A review with focus on biofuels, biocomposites and bioplastics. *J. Clean. Prod.* **2023**, *402*, 136815. [CrossRef]
67. Arevalo-Gallegos, A.; Ahmad, Z.; Asgher, M.; Parra-Saldivar, R.; Iqbal, H.M. Lignocellulose: A sustainable material to produce value-added products with a zero waste approach—A review. *Int. J. Biol. Macromol.* **2017**, *99*, 308–318. [CrossRef]

68. Singh, M.B.; Kumar, V.S.; Chaudhary, M.; Singh, P. A mini review on synthesis, properties and applications of deep eutectic solvents. *J. Indian Chem. Soc.* **2021**, *98*, 100210. [CrossRef]
69. Ni, J.; Li, Q.; Gong, L.; Liao, X.L.; Zhang, Z.J.; Ma, C.; He, Y. Highly efficient chemoenzymatic cascade catalysis of biomass into furfurylamine by a heterogeneous shrimp shell-based chemocatalyst and an ω-transaminase biocatalyst in deep eutectic solvent–water. *ACS Sustain. Chem. Eng.* **2021**, *9*, 13084–13095. [CrossRef]

catalysts

MDPI

Article

Bacterial Cellulose and Biodegradable Superbase for Heterogeneous Transesterification to Alkyl Esters

Cristina Ionela Gogoaşă [1,2], Cristian Eugen Răducanu [1,*], Laura Elisabeta Petraş [1,3], Doinița Roxana Cioroiu Tîrpan [4], Gabriel Vasilievici [5], Andreea Luiza Mîrţ [5], Tănase Dobre [1,6] and Oana Cristina Pârvulescu [1]

1 Chemical and Biochemical Engineering Department, National University of Science and Technology POLITEHNICA Bucharest, 1-7 Gheorghe Polizu St., 011061 Bucharest, Romania; cristina_ionela_92@yahoo.com (C.I.G.); laurapetras@yahoo.ro (L.E.P.); tghdobre@gmail.com (T.D.); oana.parvulescu@yahoo.com (O.C.P.)
2 Laropharm, 145 A Alexandriei St., Bragadiru, 077025 Ilfov, Romania
3 Petromidia, 215 Navodari Blvd., 900527 Constanta, Romania
4 Department of Chemistry and Chemical Engineering, Ovidius University of Constanta, 124 Mamaia Blvd., 900527 Constanta, Romania; tirpanroxana@gmail.com
5 National Institute for Research Development for Chemistry and Petrochemistry-ICECHIM, 202 Spl. Independentei, S6, 060021 Bucharest, Romania; gvasilievici@icechim.ro (G.V.); luiza.mirt@icechim.ro (A.L.M.)
6 Technical Sciences Academy of Romania, 26 Dacia Blvd., 030167 Bucharest, Romania
* Correspondence: cristianrdcn1@yahoo.com

Abstract: Heterogeneous catalysts, basic, acidic or bifunctional, can catalyze transesterification reactions where the raw material has a significant content of FFA fatty acids, such as used cooking oils or other lipid-based residues, which do not have the purity required for homogeneous catalysis, in which case the purity of the triglycerides above 99.5% is the first condition for the initiation of the reaction, to avoid saponification. In this work, a green supported catalyst was developed, using bacterial cellulose as catalytic support and biodegradable superbase as a chemical compound, for transesterification reaction to obtain alkyl esters, yielding over 99% of its content at 70 °C temperature and 7.5% catalyst loading (1.5/20 *w*/*w* catalyst:oil). A Plackett—Burman design was used for screening experiments to explore the main effect in terms of catalytic activity and performance of the triglyceride conversion reaction.

Keywords: bacterial cellulose; heterogeneous catalyst; superbases; transesterification; Plackett–Burman design; methyl esters

1. Introduction

Biofuels should be non-disputable alternatives to fossil fuels, as biofuel technologies already scaled up, or in pilot phase (involving production of biodiesel, bioethanol or biogas) are waiting to be integrated in the fuel matrix of any economy, lowering the pressure on its energy system.

Considering the multiple disadvantages of the production, refining, and use of fossil fuels, starting with their depletion, or the environmental problems due to CO_2 emissions, one can instead resort to technologies for the production of biofuels or synthetic fuels, which can be produced as much as necessary, benefiting from the knowledge of the last 15–20 years of new techniques and technologies [1], gaining in the end its own energy dependence, fuels or otherwise.

Regarding the insertion of biofuels into the fuel matrix of the economy, it should be noted that, for several years, solutions have also started to come from the large companies producing petro-fuels, and two directions are worth mentioning; the first would be the one where they comply with the biofuel content regulations at the gas station pump, ranging from 6 to 10% ethanol (E6/E10) or B6/B10, while the second is the solution they proposed in regarding awareness of climate change issues, namely hydrotreating or hydroprocessing [2–5], the option that some oil companies are considering to address, or even commercially produce, so-called green diesel.

Hydroprocessing of vegetable oils involves hydrodeoxygenation of triglycerides and hydrorefining of derived products in the presence of catalysts and H_2 environment. This type of process leads to obtaining mixtures and liquid hydrocarbons (HC), usually *n*-paraffins C15–C18, with a boiling point below that specific to fossil diesel, which are also called green diesel or bio-dehydrogenated diesel. In most cases, the hydrodeoxygenation process takes place on Ni/Co/Mo-Al_2O_3-type catalysts or other proprietary patented catalysts [2] to obtain long chains of n-paraffins, while the hydro-isomerization step is carried out by using specific, specially designed catalysts to obtain specific cold flow properties by converting *n*-paraffins to *iso*-paraffins. This hydro-isomerization is the most critical stage because obtaining the flow properties for the produced green diesel must not compromise the cetane number or other properties required for its use in aviation, for example.

This technology is already applied commercially [5] and, among the existing solutions, we also find the possibility of retrofitting in the case of some industrial platforms with installations for fossil fuels. The implementation of these operations does not require additional installations and can be easily integrated into the infrastructure of some refineries, with several companies expressing their interest in this green diesel production technology.

Biofuels are, on the other hand, chemical compounds of synthesis. Homogeneously catalyzed commercial biodiesel is an excellent choice, therefore, to complement fossil fuel (or even replace it in some cases, e.g., agriculture operation involving diesel engine machinery, interurban merchandise transport, or in areas of temperate continental clime). This process converts, mostly, fresh soy, sun flower, or rapeseed oil (considered though conflictual raw material [6]) to the final main product, alkyl esters. Improvements have been made in the last years regarding safety or environmental consideration [7]. Still, alkyl esters producing technologies can also sustain changes, in terms of raw material nonconflictual destination, reduced stage number of the process, which contributes further to minimizing several process costs, and even a better control of end-process waste waters. Such changes may involve using lipid wastes of all kinds, from waste cooking oils, animal factory waste, or municipal waste; basically, every waste containing lipid can be transformed into a transesterification raw material process, all of them conditioned by its availability as raw material (feedstock process) for production capacity ranging between 20,000 t/year for local, small communities and up to 100–1,000,000 t/year, similar to those of fossil fuel.

Although mature as a technology, the commercial production of biodiesel could still undergo another injection of new techniques, such as that of heterogeneous catalysis, which is insufficiently widespread but can be helpful in increasing biodiesel production, based on long-term and precise strategies on a certain (broad) economy market segment—agriculture, transport, and residential heating—and depending on how much this biofuel can replace or supplement fossil fuel. The transition to the second (or third) generation of biofuel will only complete this circle of green solutions offered for this in climate change issues [8].

A different technique can also be approached in terms of transesterification catalysis, particularly in heterogeneous base catalysts [9], able to transform such second-generation-type raw materials and able to yield 98% conversion at alcohol:oil (25:1), catalyst loading (1.5% *w*/*v*), and temperature 65 °C [9]; ~98.5% biodiesel using methanol: oil molar ratio of 9:1 in 1.5 h reaction time at 65 °C reaction temperature of catalyst loading of 10 wt.% (to oil) [10]; or 98% fatty acids methyl esters content for 2.5 wt.% catalyst loading and 1:9 oil:methanol molar ratio at 65 °C reaction temperature [11]. Other advantages of this approach are the ease of separation of the final product, the ability to build a heterogeneous catalyst adapted to a certain type of waste raw material, and perhaps cost reductions, obtained from the first two [10,11]. If the heterogeneous catalyst is built from a catalytic support and combined with a compound (basic in our case), the key lies in choosing a specific type of catalytic support from many others (there are a multitude of options), compatible with the basic or acid catalyst, it being rather difficult to find an exact fit for these two components.

Among this multitude of suitable materials capable of being transformed into a catalytic support, bacterial cellulose (BC) has the versatility required for this type of approach, if considered as a catalytic support.

As can be found in many studies, the versatility of BC is special. The characteristics of the cellulosic matrix or the absence of hemicellulose [12] provide multiple options in functionalization strategies. The use of BC falls into many fields, with just as many applications. Applications in the biomedical field can be developed around the characteristics and properties of BC [13,14] and can go from the construction and use as a material biocompatible with human tissue, respectively, in different forms of treatments [15] up to the drug carrier, helping to transport it [16].

Due to the inclusive improvement of BC production methods, whether it is about its production in a static medium or especially if a mobile medium is used [17], the applications for which BC can be used, based on its mechanical properties as well as special ones that can help it incorporate classic materials, include the textile industry or the leather industry [17]. Different membranes can be obtained by functionalizing BC membranes or are capable of permeation or simple filtration [18–20].

The functionalization strategy of BC membranes can include at least two different approaches, the first being the use of this membrane after the post-production treatment process and after total dehydration and the second involves membrane pre-drying treatments, aiming to replace water molecules with functional molecules, capable of easily binding to compounds with the required properties [21]. In this context, a heterogeneous green catalyst was developed in this work, within the concept of green chemistry, in terms of environmentally friendly, nontoxic, and highly biodegradable nonconflictual materials, by using the newly produced solid material as a matrix to combine it with a superbase also having green chemistry characteristics and used to heterogeneously catalyze transesterification reactions, yielding over 99% methyl ester content, while a Plackett–Burman design was used for screening experiments to explore the main effect in terms of catalytic activity and triglyceride conversion reaction performance.

2. Results and Discussion

2.1. Scanning Electron Microscope (SEM) Analysis of Heterogeneous Catalyst and Catalytic Support

The surface morphology study for the bacterial cellulose membranes was performed using scanning electron spectroscopy (SEM) for each type of material stu. Figure 1 shows images for a membrane (simple, nontreated BC) surface with morphological characteristics often found in similar studies: a cellulose fibrillar network, yet unaltered (Figure 1c), with whole fibers intercrossed, in short, a typical BC surface morphology.

Figure 2 shows images of BC_{Gu} membrane with slight structural changes ((Figure 2b,d)—very slight modification if compared with the images for plain *BC* from Figure 1c with images from Figure 2a,c) and nanocrystals starting to aggregate on the membrane surface.

Visible morphological changes of catalyst $BC_{Gu\text{-}4K24\,h}$ can be observed in SEM images in Figure 3a compared to both the surface non-altered by alkalinity of BC shown in Figure 1c and the slightly modified BC surface morphology shown in Figure 2.

The shape of the 3D nanocrystals can be observed, different from that noticed in Figure 2c, whose formation may be due to the reaction product between the two strong bases. It should also be noted that the heterogeneous catalyst shown in Figure 3 (SEM image of its surface) was kept for 24 h after the impregnation reaction in a 4K $BC_{Gu\text{-}K}$ solution because, compared to the surface morphology of the BC_{Gu} catalyst, significant changes in the dispersion of basic aggregates on its surface are easy to spot, introducing us to a first variable to follow when developing the strategy of combining the two, catalyst and catalytic support.

A similar approach can be discussed after comparing Figures 4 and 5 with Figure 6, in terms of adsorption capacity vs. adsorption time strategy, which is probably time-dependent in this impregnation reaction. The catalyst loading is more visible on the BC surface shown in the SEM image of Figure 5 with a functionalization time of 24 h than on the BC surface in the SEM image of Figure 4, where, although the functionalization time is 96 h, the BC was treated in 8 K solution, while the BC in the SEM image of Figure 6 shows the highest presence of superbase nanocrystals dispersed on the BC surface.

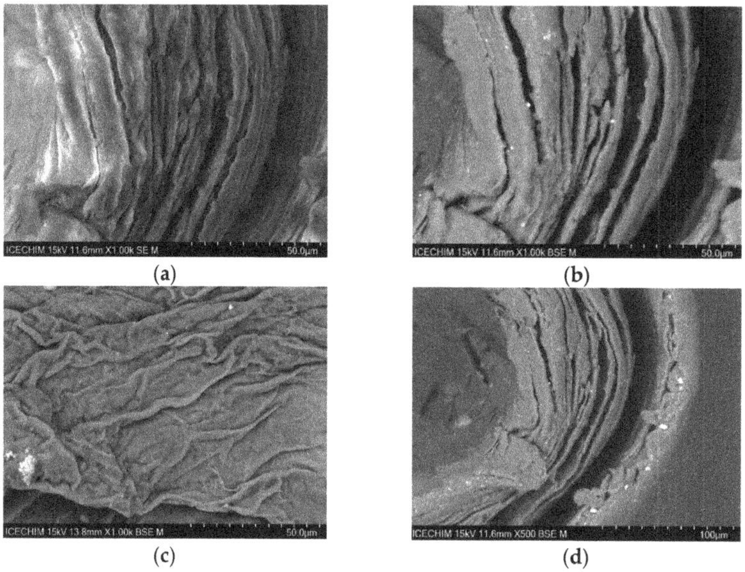

Figure 1. SEM image of catalytic support (*BC*) before functionalization, common *Acetobacter* sp. strain specific surface: (**a**,**b**,**d**) pore network (channels); (**c**) membrane surface with visible cellulosic network fibrils.

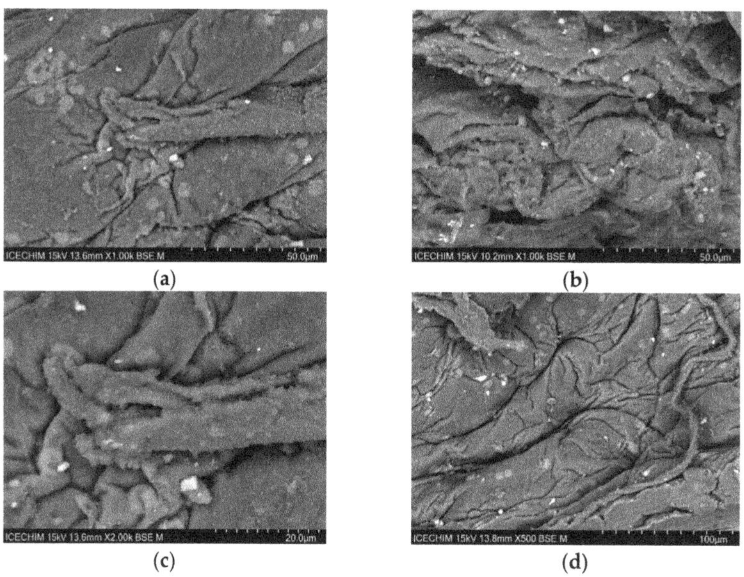

Figure 2. SEM catalytic support (BC_{Gu}) after functionalization in guanidine alcoholic solution: (**a**,**c**,**d**) loaded/dispersed guanidine nanocrystals; (**b**) membrane surface altered by alkalinity.

Figure 3. SEM image of catalyst ($BC_{Gu\text{-}4K24\,h}$): (**a**,**b**) microcrystals of alkaline compound dispersed; (**c**) altered surface, newly formed pore channel; (**d**) intercalated catalyst and fibrillar cellulose.

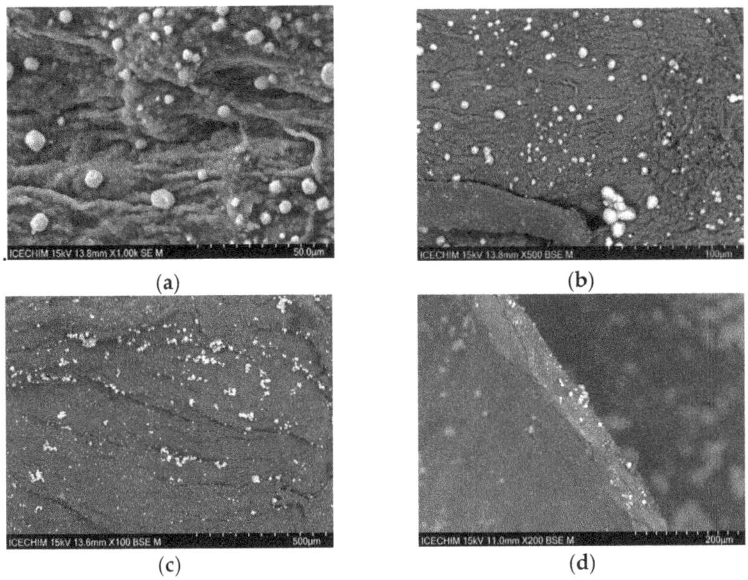

Figure 4. SEM image of catalyst ($BC_{Gu\text{-}4K96\,h}$): nanoparticle aggregation due to higher impregnation time (**b**,**c**); (**a**,**d**) improved distribution on the surface due to a possible better adsorption.

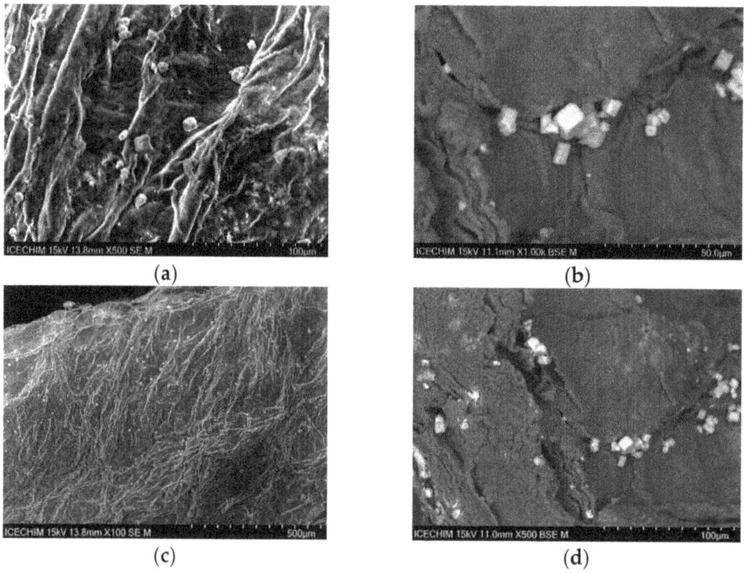

Figure 5. SEM image of catalyst ($BC_{Gu\text{-}8K24\,h}$): (**a**,**c**) fibrillar network (intercrossed) and nanoparticle-wrapped; (**b**,**d**) pores acting as sites for basic nanoparticle aggregates.

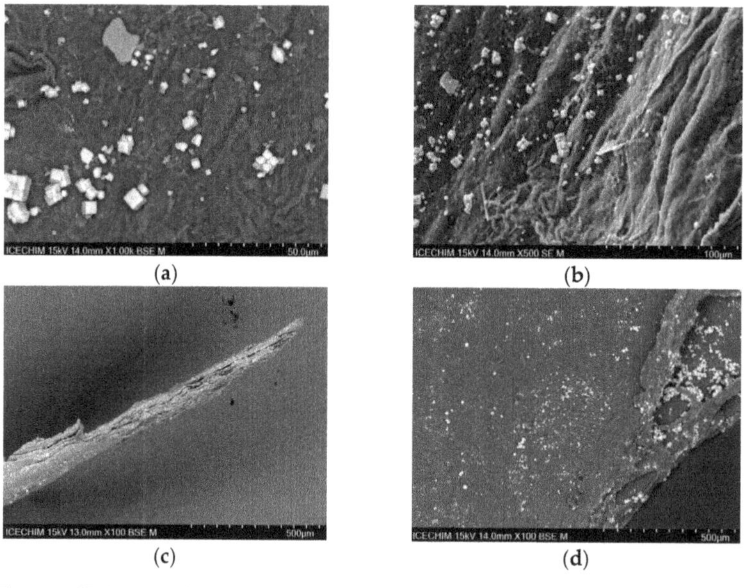

Figure 6. SEM image of catalyst ($BC_{Gu\text{-}8K96\,h}$): (**a**,**b**) highly visible 3D shape of basic nanocrystals (**b**,**c**); (**c**,**d**) enhanced distribution on the surface due to higher nanoparticle adsorption.

2.2. Fourier Transform Infrared Spectroscopy for Heterogeneous Catalyst and Catalytic Support

FT-IR analysis was conducted to evaluate potential interactions and functional groups present both in the nonfunctionalized BC materials and the treated BC (basic heterogeneous catalyst). Figure 7 shows the characteristic peaks of a simple BC membrane, similar to those found in the literature data, common to the membranes produced by different strains [22–24], in a range of 4000–400 cm^{-1}, and OH groups at the wavelength of 3342 cm^{-1} can be observed due to its stretching vibrations and C–H bonds at 2899 cm^{-1} wavelength band due to stretching vibration of aliphatic C–H bonds.

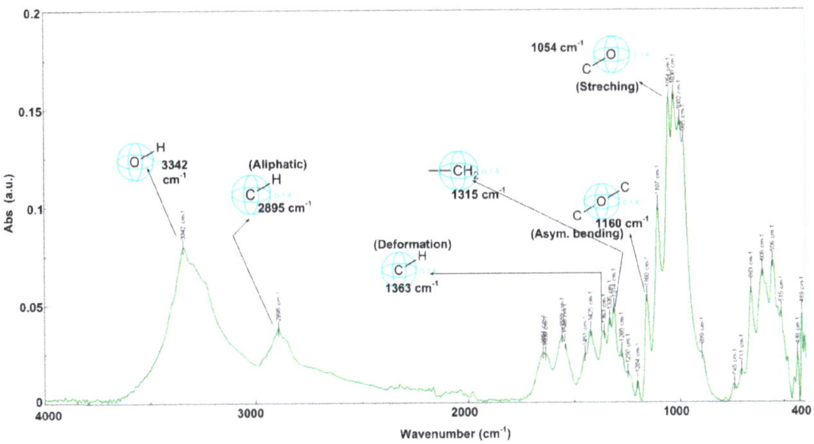

Figure 7. FT-IR spectra of simple BC: particular potential functional groups specific to a simple BC membrane from *Acetobacter* sp. strain.

Characteristic peaks can also be observed at wavelength 1054 cm^{-1}, which can be attributed to C–O bond stretching; at 1365.35 cm^{-1} wavelengths and 1432 cm^{-1}, which can be attributed to C–H deformation; and at 1432 cm^{-1} wavelength for potential C–H symmetric bending, [22–24] while peaks at 1159 cm^{-1} can be attributed to asymmetric bending of the C–O–C bonds [22–24].

Figures 8 and 9 show new peaks following the functionalization of BC of potential N–H bonds, characteristic at wavelength bands of 3230–3285 cm^{-1} from the stretching vibration of N–H groups, linked to guanidinium cations [23], as well as the characteristic peaks for potential C–N bonds at 1368–1370 cm^{-1} from the hydrogen bond vibration [23], as expected to be influenced by guanidine treatment.

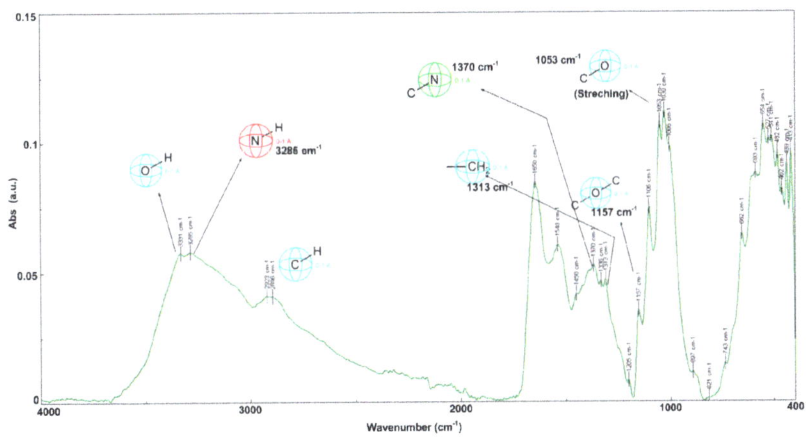

Figure 8. FT-IR spectra of BC from *Acetobacter* sp. strain: functional groups specific to functionalization with guanidine/potassium hydroxide ($BC_{Gu\text{-}4K}$).

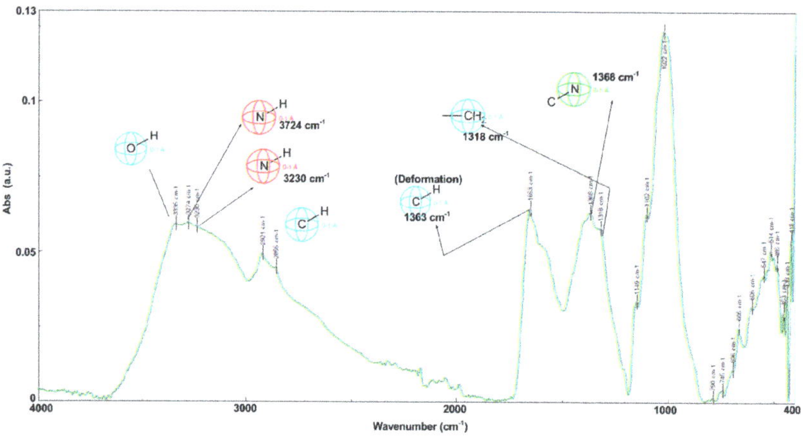

Figure 9. FT-IR spectra of BC from *Acetobacter* sp. strain: functional groups specific to functionalization with guanidine/potassium hydroxide ($BC_{Gu\text{-}8K}$).

Figure 10 summarizes the evaluation of potential chemical bonds and functional groups, before and after the preparation of the catalyst, allowing us to observe how the chemical bonds potentially formed on the bare catalytic support were affected by the superbase impregnation by weakening the hydrogen bonds, allowing the development of new ones but with the intended characteristics in terms of strong basicity due to both the amino- and imino-functional group.

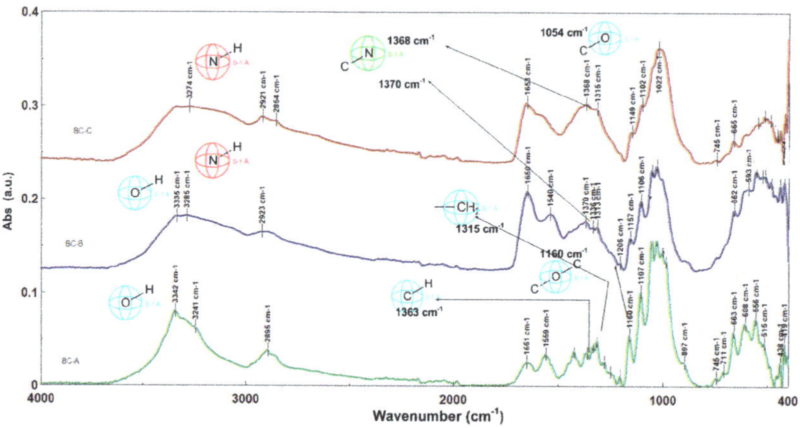

Figure 10. FT-IR summarizing spectra of BC from *Acetobacter* sp. strain: functional groups specific for BC, ($BC_{Gu\text{-}4K}$), and ($BC_{Gu\text{-}48K}$).

2.3. Predicted Responses and Process Factor Optimization

The effects of process factors, i.e., KOH concentration in impregnation solution (*A*), mean surface area of catalyst particle (*B*), agitation speed (*C*), initial alkoxide temperature (*D*), methanol-to-oil molar ratio (*E*), initial oil temperature (*F*), and impregnation time (*G*), on FAME phase yield (Y_{pred}) and soap content in the FAME phase (S_{pred}) were quantified using statistical models described by Equations (1) and (2), where regression coefficients were determined based on experimental data summarized in Table 1 (exp. 1–8). According to Equation (1), higher levels of *F* and lower levels of the other factors lead to higher values of $Y_{pred} = Y$. Equation (2) indicates higher values of $S_{pred} = S$ at higher levels of A, C, and D, as well as at lower levels of B, E, F, and G. Equations (1) and (2) can be used to predict the process responses for values of process factors within ranges considered in the experimental study.

$$Y_{pred} = Y = 103.7 - 163.7A - 0.830B - 0.003C - 0.021D - 0.060E + 0.113F - 0.057G \quad (1)$$

$$S_{pred} = S = 3027 + 7311A - 81.61B + 1.076C + 12.19D - 132.7E - 7.884F - 6.587G \quad (2)$$

Table 1. Levels of process factors and responses.

Exp.	*A* (g/mL)	*B* (cm²)	*C* (rpm)	*D* (°C)	*E* (mol/mol)	*F* (°C)	*G* (h)	*Y* (%)	*S* (ppm)
1	0.053	6.250	300	70	12	70	96	89.08	1303
2	0.053	6.250	0	25	6	25	96	86.32	1583
3	0.053	0.640	300	70	6	25	24	93.12	3386
4	0.053	0.640	0	25	12	70	24	99.81	1364
5	0.027	6.250	300	25	12	25	24	93.32	1394
6	0.027	6.250	0	70	6	70	24	98.81	2061
7	0.027	0.640	300	25	6	70	96	99.31	1819
8	0.027	0.640	0	70	12	25	96	93.92	1603
9	0.027	0.640	0	25	12	70	96	99.91	721.9
10	0.027	0.640	0	25	12	70	96	99.88	704.8
11	0.027	0.640	0	25	12	70	96	99.79	691.4

(*A*) KOH concentration in impregnation solution; (*B*) mean surface area of catalyst particle; (*C*) agitation speed; (*D*) initial alkoxide temperature; (*E*) methanol-to-oil molar ratio; (*F*) initial oil temperature; (*G*) impregnation time; (*Y*) FAME phase yield determined using Equation (4); (*S*) soap content determined using Equation (5).

A desirability function approach was used to determine the optimum factor levels to maximize the FAME phase yield ($Y_{pred} = Y$). A desirability function, $d(Y)$, is defined by Equation (3), where L_Y and U_Y (Figure 11) are the lower and upper limits of response Y [25,26]. A completely desirable value of response corresponds to $d(Y) = 1$, whereas $d(Y) = 0$ represents an undesirable value.

$$d(Y) = \begin{cases} 0 & \text{if } Y < L_Y \\ \frac{Y - L_Y}{U_Y - L_Y} & \text{if } L_Y \leq Y \leq U_Y \\ 1 & \text{if } Y > U_Y \end{cases} \quad (3)$$

Profiles for predicted values of Y and for desirability function, which are shown in Figure 11, indicate that the optimum values of process factors for maximizing the response Y are the low levels of A, B, C, and D and high levels of E, F, and D specified in Table 1 (exp. 1–8), as follows: A_{low} = 0.027 g/mL, B_{low} = 0.640 cm², C_{low} = 0 rpm, D_{low} = 25 °C, E_{high} = 12 mol/mol, F_{high} = 70 °C, and G_{high} = 96 h. Under these optimum conditions, the process response is Y_{opt} = 99.96% and desirability function is $d(Y_{opt}) = 1$.

In order to validate the statistical model described by Equation (1), three experimental runs (exp. 9–11 in Table 1) were performed at optimum values of process factors. The difference between the mean value of these replicates ($Y_{mn,opt}$ = 99.86%) and that predicted by Equation (1), i.e., Y_{opt} = 99.96%, is statistically nonsignificant (the modulus of t-statistic (−2.75) is lower than its critical value (2.92 for one tail and 4.30 for two tails at a significance level of 0.05)), which demonstrates the model validity. The values of soap content for exp. 9–11, calculated using Equation (2), are also included in Table 1. The difference between the mean value of these replicates and that predicted by Equation (2), i.e., 706.0 ppm and 699.7 ppm, is statistically nonsignificant (t-statistic of 0.71 is lower than its critical value), which proves the validity of the statistical model described by Equation (2).

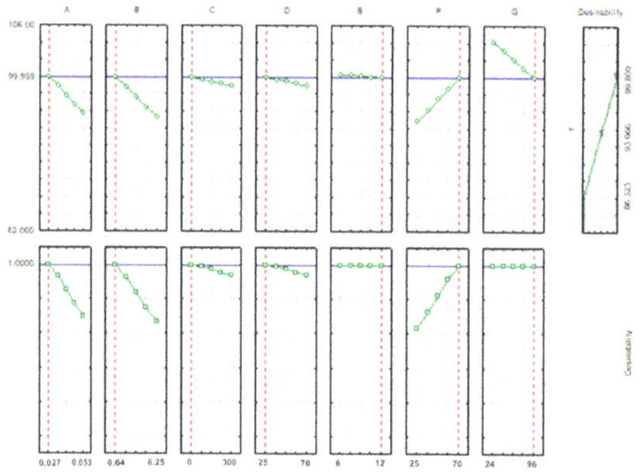

Figure 11. Profiles for predicted values of biodiesel yield (Y_{pred} = Y) and related desirability; (*A*) KOH concentration in impregnation solution; (*B*) mean surface area of catalyst particle; (*C*) agitation speed; (*D*) initial alkoxide temperature; (*E*) methanol-to-oil molar ratio; (*F*) initial oil temperature; (*G*) impregnation time.

2.4. *Gas Chromatography–Mass Spectrometry (GC-MS) Analysis*

GC-MS analysis for alkyl ester identification was carried out for exp. 1 and exp. 8 from present experimental research, and FAME total proportion and chemical composition can be seen in Figures 12 and 13. Figure 12 shows that GC-MS analysis of alkyl esters from exp. 1 led to the identification of 15 different components, with 9,12-octadecadienoic acid (*Z*,*Z*)-methyl ester as the major component (51.07%), 9-octadecenoic acid methyl ester, (*E*)- as the second major component (38.45%), and hexadecanoic acid methyl ester as the third major component (9.65%). The remaining 12 components were found having under 1% methyl hexadec-9-enoate (0.29%), methyl tetradecanoate (0.22%), linolenic acid methyl ester (0.1), linoleic acid ethyl ester (0.06%), 4-penten-2-ol (0.04%), heptadecanoic acid methyl ester (0.04%), *cis*-10-heptadecenoic acid methyl ester (0.04%), pentadecanoic acid methyl ester (0.03), and octanoic acid methyl ester (0.01%).

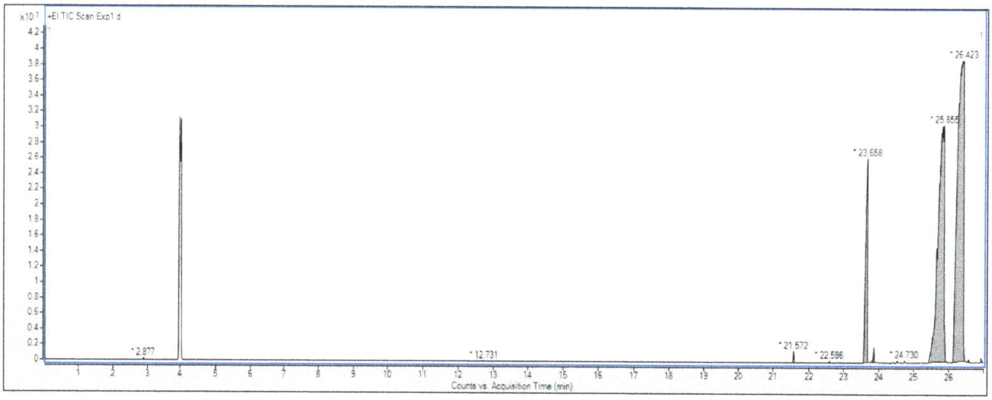

Figure 12. GC-MS analysis for the composition of the alkyl esters in exp. 1.

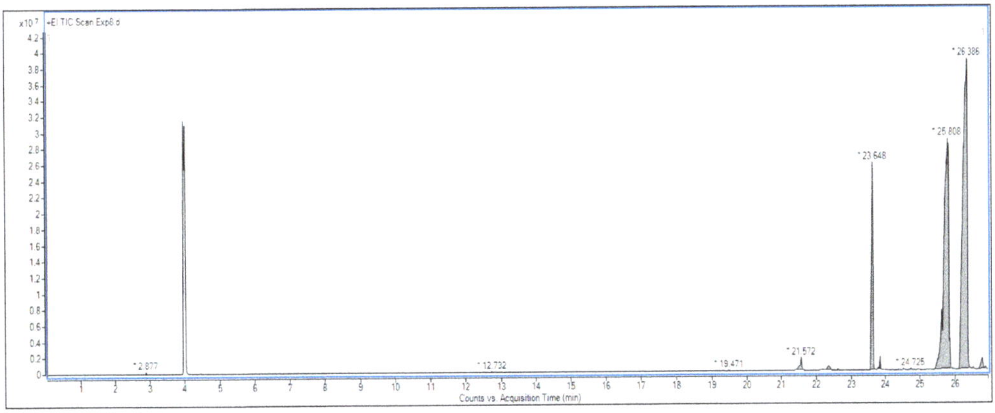

Figure 13. GC-MS analysis for the composition of the alkyl esters in exp. 8.

Figure 13 shows that GC-MS analysis of alkyl esters from exp. 8 led again to the identification of 15 different components, with 9,12-octadecadienoic acid (*Z*,*Z*)- methyl ester as the major component (49.44%), 9-octadecenoic acid methyl ester, (*E*)- as the second major component (37.23%), and hexadecanoic acid methyl ester as the third major component (10.95%).

The remaining 12 components were found having under 1% and being slightly different in composition (components and percentage): docosanoic acid methyl ester (0.81%), methyl tetradecanoate (0.67%), palmitoleic acid (0.33%), 9-hexadecenoic acid methyl ester, (Z) (0.30%), 4-penten-2-ol (0.05%), 7-hexadecenoic acid methyl ester, (Z) (0.04%), heptadecanoic acid methyl ester (0.04%), *cis*-10-heptadecenoic acid methyl ester (0.04%), octanoic acid methyl ester (0.03%), methyl 13-methyltetradecanoate (0.03 b%), undecanoic acid, 10-methyl-, methyl ester (0.02%), and 2,4-decadienal (0.02%).

Although they are present in small proportions, rather considerable traces, the presence of docosanoic acid methyl ester, palmitoleic acid (0.33%), 9-hexadecenoic acid methyl ester, (Z), 7-hexadecenoic acid methyl ester, (Z), methyl 13-methyltetradecanoate (0.03 b%), undecanoic acid, 10-methyl-, methyl ester, and 2,4-decadienal provides important data regarding alkyl ester composition, considering these compounds were missing in exp. 1 alkyl ester chemical composition.

3. Materials and Methods

3.1. Materials

Potassium hydroxide pellets (VWR BDH Chemicals, Prague, Czech Republic), methanol 99.9% (Avantor, Gliwice, Poland), guanidine hydrochloride (VWR Chemicals, Radnor, PA, USA), 2-propanol 100% (VWR Chemicals, Paris, France), bromphenol blue (Sigma–Aldrich, Darmstadt, Germany), and waste cooking oil were the materials used in this experimental research.

BC membranes (99% water content) were obtained in static culture from *Acetobacter* sp. strain isolated from traditionally fermented vinegar in the Mass Transfer Laboratory of the Chemical and Biochemical Engineering Department of National University of Sciences and Technologies POLITEHNICA Bucharest [16,27].

3.2. Bacterial Cellulose Production

BC production was performed using a method from previous work [16,27].

Depending on the application, the production time of a BC membrane can be in the range of 7–30 days or even longer. At the daily temperature of 25–30 °C, or at 18–24 °C at night (a temperature considered in the optimal range for the development of BC culture) without thermostating, the period of 7–10 days was sufficient to obtain a BC membrane of

110 mm diameter and of 8 mm thickness [21], which, following post-production treatments, after the drying operation, reached a membrane of 80 mm diameter and of 2.5 mm thickness for a moisture content of 99.4% (experimental). In the present work, the period allocated to the process of obtaining BC was intentionally extended to more than 25 days, resulting in the production of the BC membrane shown in Figure 14b,c.

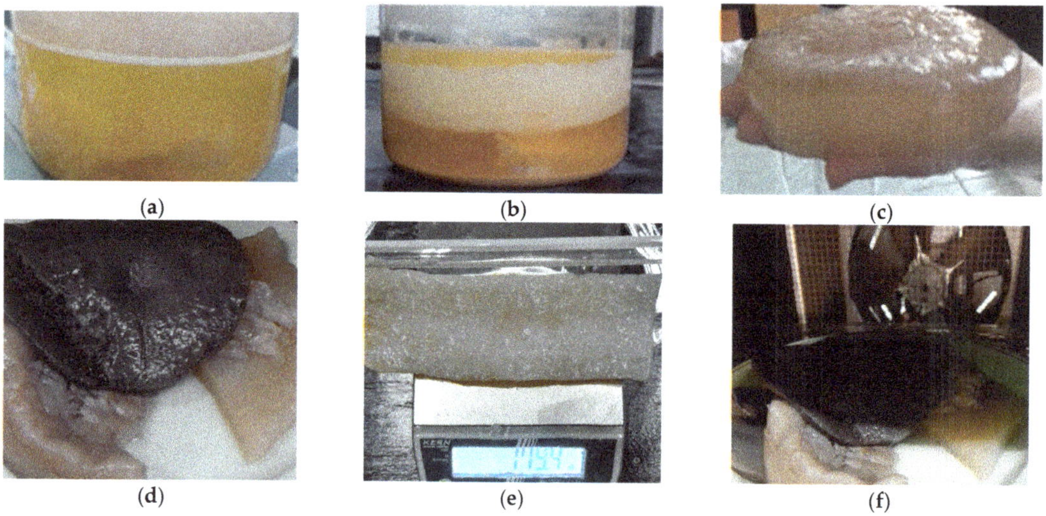

Figure 14. Bacterial cellulose production: (**a**) culture medium, 6 days; (**b**) after 29 days; (**c**,**d**,**f**) after weak acid treatment, followed by neutralization; (**e**) under weak acid treatment, ready for functionalization.

After production, the BC can be preserved for a long period of time if kept in a weak acid solution, while its further use involves a fairly standard neutralization method before it is functionalized. For this work, this post-treatment was approached similarly as part of the catalyst construction strategy. After neutralization BC was dried to remove even traces of moisture at 70 °C in a Biovita DEH450 (Figure 14f) dehydrator (voltage of 220–240 V 50 Hz and power of 420–500 W) and then deposited in desiccator. At this stage, this membrane of BC was considered the solid material used for the subsequent functionalization to catalytic support.

Following our strategy developed for the construction of the catalyst, the particles (pieces) of BC of specific dimensions were prepared (Figure 15), after which the impregnation solutions were prepared. These solutions were made in two steps; in the first step, a methanol solution of guanidine hydrochloride and potassium hydroxide of Gu-KOH 1:1 molar ratio was prepared at reflux by heating the reaction mixture under stirring for 6 h at 60 °C and keeping it for 24 h to complete the reaction. In the second stage, the alcoholic guanidine solution (Gu_a) was divided and used for BC functionalization.

Figure 15. Dried BC: (**a**,**b**) from dehydrator, large dimensions; (**c**,**d**) large pieces according to experimental plan; (**e**,**f**) small particles according to experimental plan.

3.3. Catalyst Preparation

Four Gu_a solutions were later used to prepare the final basic compound for combining with the catalytic support by impregnation: two mixtures for each particle type by adding KOH 1:18.75 w/v to alcoholic guanidine solution (denoted 8K) kept at reflux for 6 h at 60 °C, under stirring, and left for 24 h to complete the reaction and another two different mixtures by adding 1:37.5 w/v KOH in alcoholic guanidine solution (denoted 4K), reacted in a similar way (Table 2), with all four reactions performed in respect with experimental research design. At this point, the catalyst was considered prepared and ready to be tested.

Table 2. *BC* and *BC* catalyst types obtained after functionalization.

Exp.No.	Catalytic Supp./Catalyst	Comments	Observations
1	BC	BC, plain, simple	Evaluation
2	BC_{Gu}	BC in alcoholic guanidine, without KOH	Evaluation
3	$BC_{Gu\text{-}4K24h}$	BC/superbase (Gu-KOH, 4K, impregnation time 24 h) *	Heterogeneous catalyst
4	$BC_{Gu\text{-}4K96h}$	BC/superbase (Gu-KOH, 4K, impregnation time 96 h)	Heterogeneous catalyst
5	$BC_{Gu\text{-}8K24h}$	BC/superbase (Gu-KOH, 8K, impregnation time 24 h) **	Heterogeneous catalyst
6	$BC_{Gu\text{-}4K96h}$	BC/superbase (Gu-KOH, 8K, impregnation time 96 h)	Heterogeneous catalyst

* KOH 1:18.75 w/v in alcoholic guanidine solution, ** KOH 1:37.5 w/v in alcoholic guanidine solution.

3.4. Catalyst Characterization

Specific analysis was performed for both the characterization of the catalytic support and the heterogeneous catalyst.

3.4.1. Scanning Electron Microscope (SEM) Analysis

The morphology of the guanidine-based catalyst loading in section and on the surface was evaluated using a TM4000Plus Tabletop Scanning Electron Microscope (Hitachi, Tokyo, Japan) at 15 kV in standard vacuum conditions (according to the TM4000Plus Software vacuum settings). The SEM images were obtained using both backscattered electron (BSE) and/or secondary electron (SE) detectors at multiple magnifications.

3.4.2. Fourier Transform Infrared Spectroscopy

Fourier transform infrared spectroscopy (ATR-FTIR) was performed in the range of 4000–400 cm^{-1} with a Vertex 80 spectrometer (Bruker Optik GMBH, Billerica, MA, USA). An attenuated total reflectance (ATR) accessory was used.

3.5. Catalyst Testing and Analysis

3.5.1. Catalyst Testing in Transesterification Reactions to Alkyl Esters

Testing of the heterogeneous catalyst involved a set of transesterification reactions to alkyl esters under conditions that were specifically selected to be similar to commercial biodiesel production.

3.5.2. Analysis of Main Product in Transesterification Reactions to Alkyl Esters

Alkyl (methyl) ester characterization was performed by gas chromatography–mass spectrometry (GC-MS) analysis using the Agilent 7890 A GC-MS/MS TRIPLE QUAD system. A capillary column Agilent DB-WAX of 30 m length, 0.25 mm internal diameter, and 0.25 μm film thickness was used, having helium at a flow rate of 1 mL/min as the carrier gas. The starting oven temperature was 40 °C; then, a gradual increas up to 210 °C was performed, with an increase of 3 °C/min between 40 and 80 °C, which was kept for 2 min, and then of 15 °C/min up to 210 °C, where it was held again for 2 min. The temperature for GC injector was 250 °C, while the MS detector was set at 150 °C. The temperature for the transfer line was maintained at 280 °C. Electron ionization (EI) mode was used for MS detection, at 70 eV, with a mass scan range of m/z 50–450. The identification of peaks within the analyzed samples was performed according to NIST MS database.

3.5.3. Soap Content in FAME Test Reaction Product

Soap formation during the transesterification reaction to alkyl esters is inevitable [28]. Regardless of whether the methanol is in excess and the reaction is shifted to the desired triesters and regardless of the percentage of FFA reduction or how effectively any traces of water were removed, if you have methanol and NaOH reacting (homogeneous catalyst), you will also have soap. Studies showed, however, that, using heterogeneous catalyst, a lower soap content value can be attained, this being another advantage beside an easier main reaction product separation, thus allowing the alkyl esters to enter the biodiesel standard in this respect. In this context, soap content was determined while testing the catalyst in transesterification reaction in this work.

3.5.4. Independent and Dependent Process Variables

KOH concentration in impregnation solution (*A*), mean surface area of catalyst particle (*B*), agitation speed (*C*), initial alkoxide temperature (*D*), methanol-to-oil molar ratio (*E*), initial oil temperature (*F*), and impregnation time (*G*) were selected as process independent variables (factors). Process dependent variables (responses) were FAME phase yield (*Y*) and final soap content in the FAME phase (*S*), calculated using Equations (4) and (5), where m_{FAME}, m_{oil}, and m_S are the masses of FAME phase, oil, and soap.

$$Y = 100\frac{m_{FAME}}{m_{oil}} \quad (4)$$

$$S = 10^6\frac{m_S}{m_{FAME}} \quad (5)$$

3.5.5. Experimental Design, Statistical Analysis, and Optimization

Experimental design, statistical analysis, and process factor optimization were performed using STATISTICA 10 software (Stat Soft Inc., Tulsa, OK, USA). According to a

Plackett–Burman design, eight experimental runs (1–8 in Table 1) were performed at two levels of process factors.

4. Conclusions

BC used as a catalytic support was able, after impregnation with a guanidine/potassium hydroxide superbase, to heterogeneously catalyze the transesterification of untreated used cooking oil, yielding over 99% methyl esters and, considering the percentage of 96.5% standard biodiesel for its content, it can be said (from this approach alone) that biodiesel is what has been achieved under process conditions similar to those of commercial biodiesel. Further studies and experimental research are needed on the response of BC to different treatments, applied especially after its production and before similar functionalization considering the multitude of variables that need to be addressed in terms of impregnation conditions, targeted chemisorption, or process factors' optimal number of variables.

Author Contributions: Conceptualization, C.I.G., C.E.R., L.E.P., T.D. and O.C.P.; methodology, C.I.G., C.E.R., L.E.P., D.R.C.T., T.D. and O.C.P.; validation, T.D. and O.C.P.; formal analysis, O.C.P.; investigation, C.I.G., C.E.R., L.E.P., D.R.C.T., G.V. and A.L.M.; data curation, D.R.C.T., G.V. and A.L.M.; writing—original draft preparation, C.I.G., C.E.R., L.E.P., D.R.C.T., T.D. and O.C.P.; writing, C.E.R. and O.C.P.; visualization, D.R.C.T., G.V. and A.L.M.; supervision, T.D. and O.C.P. All authors have read and agreed to the published version of the manuscript.

Funding: This work has been funded by the European Social Fund from the Sectoral Operational Programme Human Capital 2014–2020, through the Financial Agreement with the title "Training of PhD students and postdoctoral researchers in order to acquire applied research skills—SMART", Contract no. 13530/16.06.2022—SMIS code: 153734.

Data Availability Statement: Data is contained within the article.

Acknowledgments: This research was funded by the Ministry of Research, Innovation, and Digitization through Program 1—Development of the national research and development system, Subprogram 1.2—Institutional Performance—Projects to finance excellence in RDI, Grant no. 15PFE/2021.

Conflicts of Interest: The authors declare no conflict of interest.

References

1. Knothe, G.; Van Gerpen, J. Chapter 2—The History of Vegetable Oil-Based Diesel Fuells. In *The Biodiesel Handbook*, 1st ed.; Knothe, G., Van Gerpen, J., Krahl, J., Eds.; AOCS Press Publishing: Champaigne, IL, USA, 2005; pp. 12–24. [CrossRef]
2. Liu, Y.; Boyás, R.S.; Murata, K.; Minowa, T.; Sakanishi, K. Hydrotreatment of vegetable oil to produce bio-hydrogenated diesel and liquefied petroleum gas fuel over catalysts containing sulfided Ni–Mo and solid acids. *Energy Fuel* **2011**, *25*, 4675–4685. [CrossRef]
3. Murata, K.; Liu, Y.; Inaba, M.; Takahara, I. Production of synthetic diesel by hydrotreatment of Jatropha oils using Pt–Re/H-ZSM-5 catalyst. *Energy Fuel* **2010**, *24*, 2404–2409. [CrossRef]
4. Veriansyah, B.; Han, J.Y.; Kim, S.K.; Hong, S.A.; Kim, Y.J.; Lim, J.S.; Shu, Y.W.; Oh, S.G.; Kim, J. Production of renewable diesel by hydroprocessing of soybean oil: Effect of catalysts. *Fuel* **2011**, *94*, 578–585. [CrossRef]
5. Hydrotreating Catalysts and Technologies for All Crude Oil Fractions. Available online: http://www.topsoe.com/processes/hydrotreating (accessed on 30 September 2023).
6. Demirbas, A.; Bafail, A.; Ahmad, W.; Sheikh, M. Biodiesel production from non-edible plant oils. *Energy Explor. Exploit.* **2016**, *34*, 290–318. [CrossRef]
7. Castro, E.; Strætkvern, K.O.; Romero-García, J.M.; Martín, C. Pretreatment and Bioconversion for Valorization of Residues of Non-Edible Oilseeds. *Agronomy* **2023**, *13*, 2196. [CrossRef]
8. Neto, J.M.; Komesu, A.; Da Silva Martins, L.H.; Gonçalves, V.O.O.; De Oliveira, J.A.R.; Rai, M.; Ingle, A.P. Chapter 10—Third generation biofuels: An overview. In *Sustainable Bioenerg*; Rai, M., Ingle, A.P., Eds.; Elsevier: Amsterdam, The Netherlands, 2019; pp. 283–298. [CrossRef]
9. Bhatia, K.S.; Gurav, R.; Choi, T.R.; Kim, H.J.; Yang, S.Y.; Hong, H.S.; Park, J.Y.; Park, Y.L.; Han, Y.H.; Choi, Y.K.; et al. Conversion of waste cooking oil into biodiesel using heterogenous catalyst derived from cork biochar. *Bioresour. Technol.* **2020**, *302*, 122872. [CrossRef]
10. Singh, H.; Ali, A. Esterification as well as transesterification of waste oil using potassium imbued tungstophosphoric acid supported graphene oxide as heterogeneous catalyst: Optimization and kinetic modeling. *Renew. Energy* **2023**, *207*, 422–435. [CrossRef]

11. Chinglenthoiba, C.; Das, A.; Vandana, S. Enhanced biodiesel production from waste cooking palm oil, with NaOH-loaded Calcined fish bones as the catalyst. *Environ. Sci. Pollut. Res.* **2020**, *27*, 15925–15930. [CrossRef]
12. Poddar, M.K.; Pritam Kumar Dikshit, P.K. Recent development in bacterial cellulose production and synthesis of cellulose based conductive polymer nanocomposites. *Nano Select* **2021**, *2*, 1605–1628. [CrossRef]
13. Sharma, P.; Mittal, M.; Yadav, A.; Aggarwal, N.K. Bacterial cellulose: Nano-biomaterial for biodegradable face masks—A greener approach towards environment. *Environ. Nanotechnol. Monit. Manag.* **2023**, *19*, 100759. [CrossRef] [PubMed]
14. Bueno, F.; Fultz, L.; Husseneder, C.; Keenan, M.; Sathivel, S. Biodegradability of bacterial cellulose polymer below the soil and its effects on soil bacteria diversity. *Polym. Degrad. Stab.* **2023**, *217*, 110535. [CrossRef]
15. Fernando, G.; Torres, F.G.; Solene Commeaux, S.; Omar, P.; Troncoso, O.P. Biocompatibility of Bacterial Cellulose Based Biomaterials. *J. Funct. Biomater.* **2012**, *3*, 864–878. [CrossRef]
16. Păvăloiu, R.D.; Stroescu, M.; Pârvulescu, O.C.; Dobre, T. Composite hydrogels of bacterial cellulose-carboxymethyl cellulose for drug release. *Rev. Chim.* **2014**, *65*, 948–951.
17. Parvulescu, O.P.; Isopencu, G.; Busuioc, C.; Raducanu, C.; Mocanu, A.; Deleanu, I.; Stoica-Guzun, A. Chapter 17—Antimicrobial bacterial cellulose composites as textile materials. In *Antimicrobial Textiles from Natural Resources*; Mondal, I.H., Ed.; Woodhead Publishing: Sawston, UK, 2021; pp. 513–556. [CrossRef]
18. Dai, Q.; Liu, J.; Zheng, J.; Fu, B. Immobilization of UiO-66-NH2 into Bacterial Cellulose Aerogels for Efficient Particulate Matter Filtration. *Sustainability* **2023**, *15*, 13382. [CrossRef]
19. Tahir, D.; Karim, M.R.A.; Hu, H.; Naseem, S.; Rehan, M.; Ahmad, M.; Zhang, M. Sources, Chemical Functionalization, and Commercial Applications of Nanocellulose and Nanocellulose Based Composites: A Review. *Polymers* **2022**, *14*, 4468. [CrossRef] [PubMed]
20. Wahab, R.M.A.E.; Ghanem, A.F.; Sheir, D.H.; Selim, M.M.; Zaher, F.A.; Badawy, A.A. Assessment of a green zeolite/bacterial cellulose nanocomposite membrane as a catalyst to produce biodiesel from waste cooking oil. *Energy Sources Part A Recovery Util. Environ. Eff.* **2023**, *45*, 10350–10365. [CrossRef]
21. Raducanu, C.E. New Solutions Regarding Catalytic Transesterification Integration with Separation in Biodiesel Technology. Ph.D. Thesis, Chemical and Biochemical Engineering Department, Faculty of Chemical Engineering and Biotechnologies, National University of Sciences and Technologies POLITEHNICA Bucharest, Bucharest, Romania, 2020.
22. Drozd, M.; Dudzic, D.; Pietraszko, A. Crystal structure, differential scanning calorimetric, infrared spectroscopy and theoretical studies of C(NH2)2(NH)CH2@CHCOOH noncentrosymmetric crystal. *Spectrochim. Acta Part A Mol. Biomol. Spectrosc.* **2013**, *105*, 135–148. [CrossRef]
23. Haynes, W.M.; Lide, D.R.; Bruno, T.J. Chapter 8—Analytical Chemistry. In *Handbook of chemistry and Physics: A Ready-Reference Book of Chemical and Physical Data*, 97th ed.; CRC Press: Boca Raton, FL, USA, 2016; Volume 1, pp. 43–48.
24. Nelli Atykyan, N.; Victor Revin, V.; Shutova, V. Raman and FT-IR Spectroscopy investigation the cellulose structural differences from bacteria Gluconacetobacter sucrofermentans during the different regimes of cultivation on a molasses media. *AMB Express* **2020**, *10*, 84. [CrossRef]
25. Dima, A.D.; Pârvulescu, O.C.; Mateescu, C.; Dobre, T. Optimization of substrate composition in anaerobic co-digestion of agricultural waste using central composite design. *Biomass Bioenerg.* **2020**, *138*, 105602. [CrossRef]
26. Drăghici-Popa, A.-M.; Boscornea, A.C.; Brezoiu, A.-M.; Tomas, S.T.; Pârvulescu, O.C.; Stan, R. Effects of extraction process factors on the composition and antioxidant activity of blackthorn (*Prunus spinosa* L.) fruit extracts. *Antioxidants* **2023**, *12*, 1897. [CrossRef]
27. Dobre, T.; Stoica, A.; Pârvulescu, O.C.; Stroescu, M.; Iavorschi, G. Factors influence on bacterial cellulose growth in static reactors. *Rev. Chim.* **2008**, *59*, 591–594. [CrossRef]
28. Chanakaewsomboon, I.; Tongurai, C.; Photaworn, S.; Kungsanant, S.; Nikhom, R. Investigation of saponification mechanisms in biodiesel production: Microscopic visualization of the effects of FFA, water and the amount of alkaline catalyst. *J. Environ. Chem. Eng.* **2020**, *8*, 103538. [CrossRef]

MDPI AG
Grosspeteranlage 5
4052 Basel
Switzerland
Tel.: +41 61 683 77 34

Catalysts Editorial Office
E-mail: catalysts@mdpi.com
www.mdpi.com/journal/catalysts

www.ingramcontent.com/pod-product-compliance
Lightning Source LLC
Chambersburg PA
CBHW061507210926
51850CB00008B/1841